SECOND EDITION

Recombinant DNA and Biotechnology
A Guide for Students

SECOND EDITION

Recombinant DNA and Biotechnology

A Guide for Students

Helen Kreuzer, Ph.D.
Carolina Biological Supply Company
Burlington, North Carolina

Adrianne Massey, Ph.D.
A. Massey & Associates
Chapel Hill, North Carolina

ASM PRESS
Washington, D.C.

Address editorial correspondence to ASM Press, 1752 N Street, NW, Washington, DC 20036 USA

Send orders to ASM Press, P.O. Box 605, Herndon, VA 20172 USA
Phone: 800-546-2416; 703-661-1593
Fax: 703-661-1501
Email: books@asmusa.org
Online: www.asmpress.org

Copyright © 2001   ASM Press
American Society for Microbiology
1752 N St. NW
Washington, DC 20036-2804

ISBN 1-55581-176-0

10   9   8   7   6   5   4   3   2   1

Cover figure: Genetic scientists with plasmids in the background (VU/©K. G. Murti)

To Lindsay, Marilyn, and Sherri

About the Authors

Helen Kreuzer describes herself as cross-trained in both science and education. She has degrees from the University of Alabama in chemistry, microbiology, and education and received her Ph.D. from Duke University in molecular genetics. During the 3 years that she coordinated the North Carolina Biotechnology Center's teacher education program, she developed many of the teaching activities presented in this book. She has also worked in academic research and served on the faculty of a 4-year liberal arts college, where she taught genetics and molecular biology to undergraduates. Dr. Kreuzer joined the staff of the Carolina Biological Supply Company in 1995, where she is developing new molecular biology teaching materials, including wet labs, models, and videos. She has taught biotechnology workshops to high school teachers and college faculty at sites across the country.

Adrianne Massey has been involved with scientific research, education, workforce training, and public policy for over 15 years. Before starting her own company in 1997, she was vice president for education and training at the North Carolina Biotechnology Center, where she directed the center's award-winning public education and workforce training programs. She received her Ph.D. in zoology from North Carolina State University and has taught biology and evolutionary ecology to undergraduate and graduate students. In addition to publishing articles on her research in evolutionary ecology, she has been an invited contributor to a number of government reports on biotechnology. She has been involved in international biosafety treaty negotiations, served as the science advisor for the PBS series *BREAK-THROUGH: Television's Journal of Science and Medicine,* and developed interactive exhibits for science and technology museums. Dr. Massey is also a frequent speaker, both nationally and internationally, on the science, applications, and policy issues of biotechnology.

Contents

Preface

We live in very exciting times. It seems that every day, newspapers announce a new and exciting finding about life at the molecular level. Many of these findings elicit great hope because they open avenues for curing diseases, feeding the burgeoning world population, improving our criminal justice system, and preventing environmental problems.

The pace of discovery keeps increasing because scientific understanding begets new technologies that generate more scientific understanding.

Scientific breakthroughs of the last 30 years have furnished biologists with a bag full of new research tools—molecular techniques that provide answers to old questions and create a whole set of new questions. These techniques are commonly grouped under the inclusive heading "biotechnology." Scientists are using the tools of biotechnology to tackle problems in all branches of biology, from cellular and molecular biology to ecology and evolution.

The pace of biological discovery is influenced not only by breakthroughs in biological understanding and biologically based technologies but also by advances in other scientific and technical disciplines. The information technology revolution of the last decade has played a significant role in accelerating the speed of discovery in biology. The availability of scientific information over the Internet has revolutionized how scientists ask questions, manipulate data, and share information. In the past, scientists learned about one another's findings through individual journal articles and seminars. There was no easy way to compare findings from many laboratories at once, and no one had instant access to another laboratory's data. Now, anyone with Internet access can instantly compare vastly complex collections of information, such as sequence information from laboratories around the world, and search thousands of scientific journals for information on topics of their choice. Computer databases, data manipulation programs, and online availability, often referred to collectively

as "bioinformatics," have changed the world of biological research.

Biotechnology provides us with more than useful research tools. It touches all of us in many ways, for we are using the diverse technologies that constitute biotechnology to produce an even broader array of products. Thanks to monoclonal antibody technology and DNA probes, we have more than 500 new tests for diagnosing plant, animal, and human diseases and for assessing food safety. We are using bioprocessing technology to exploit the biochemical machinery of microbes to manufacture new pharmaceuticals, degrade environmental pollutants, and synthesize enzymes useful in the textile, food processing, chemical, and paper industries. Tissue culture technology is allowing us to create skin and cartilage in the laboratory. Genetic engineering is helping us combat cancer and making possible the development of more nutritious crops that are also resistant to insects, diseases, and environmental stresses.

Even though biotechnology includes many different technologies, in this text, as in the first edition, we focus primarily on those technologies that are based on our knowledge of DNA and the DNA-RNA-protein interconnection. To provide conceptual grounding for understanding the technologies and their applications, we also include information on some of the science fundamental to DNA-based technologies.

The first part of this book, *Laying the Foundation*, provides basic information on the topics covered by the activities in the remainder of the book. In chapters 1 and 2, we introduce the entire scope of biotechnology before narrowing our focus to DNA-based technologies for the remainder of the book. Chapter 3 provides a historical and conceptual context for the detailed discussions of molecular biology and DNA-based technologies given in chapters 4 and 5. We hope you will refer to *Laying the Foundation* much as you would a biology textbook.

Parts II and III, *Classroom Activities* and *Societal Issues,* resemble a laboratory manual or workbook. Part II provides wet and dry laboratory activities for learning both the basic science of molecular genetics and the hands-on techniques of DNA- and protein-based technologies. Part III includes both background information and activities for exploring the relationship among science, technology, and society.

We have divided Part II into five sections.

- The first section, *DNA Structure and Function,* focuses primarily on basic molecular biology, specifically, the structure and function of DNA. The integral relationship between structure and function in the natural world is a basic unifying principle in biology.
- In the second group of activities, *Manipulation and Analysis of DNA,* we shift the focus from basic science to some of the hands-on techniques commonly used in DNA-based technologies.
- The activities in the third group, *Transfer of Genetic Information,* deal with both the basic science of the methods microbes use to transfer genetic information and the laboratory techniques that exploit these natural means of gene transfer.
- The fourth section, *Bioinformatics and Evolutionary Analysis of Proteins,* focuses on proteins and the use of amino acid sequence information to study evolution. Genotypic and phenotypic variation are fundamental concepts in biology. Evolution is the single most important unifying principle in biology and must be understood at all levels, from molecular to ecosystems, to develop a true appreciation and understanding of biological systems.
- The fifth section, *Molecular Biology and Genetics,* makes the connection between molecular biology and classical genetics. It explores the biochemistry underlying coat color in Labrador retrievers and then shows how similar genetic and biochemical variation gives rise to several traits seen in humans.

Science and technology don't exist in a social vacuum. They exert a powerful influence on society, and society, in turn, alters how science is conducted and how technologies are developed. Because biotechnology has generated wide public debate about a number of important issues, we would be remiss if we described only the scientific foundations and technological applications of biotechnology and excluded its attendant societal issues.

All too often, discussions about controversial social issues are merely emotional exchanges of opinions that may have nothing to do with facts. These unproductive debates contribute to rather than alleviate confusion. Objectively analyzing the complex relationship among science, technology, and society is essential for guiding the wise and fair development and use of biotechnology.

In *Societal Issues,* we offer background information and tools for rationally analyzing and discussing a number of difficult issues. We begin by asking readers to consider the complex relationship and interdependence of scientific understanding, technological development, and societal structure. In chapters 35 and 36, we introduce the science of risk assessment and risk-benefit analysis, and we provide suggestions for guiding debate on the risks and benefits of biotechnology. Chapters 37 to 39 move to more difficult questions about the bioethical ramifications of biotechnology. To facilitate productive discussion, we provide a model for guiding thoughtful consideration of bioethical dilemmas. Finally, in chapter 40, we describe career options in biotechnology, highlighting different types of jobs and giving their educational requirements.

The final segment of the book includes templates to use in carrying out the activities (see Appendix A) and a glossary.

In this book, we want not only to provide information and activities but also to share our understanding and appreciation of biology with you. Our understanding comes from years spent in an academic environment. During that time, we were fortunate enough to be able to immerse ourselves in biology. It takes years of single-minded pursuit of knowledge for knowledge to be transformed into understanding. With that understanding comes a very deep respect and reverence for the workings of the natural world. If we convey just a fraction of that reverence to you, then we will have succeeded.

Helen Kreuzer
and Adrianne Massey

Laboratory Biosafety

Handling Microorganisms in the Laboratory

Escherichia coli is a normal inhabitant of the digestive tract. Of the many strains of *E. coli,* some inhabit the human gut, and others reside in animals. A few strains of *E. coli* cause significant disease in humans and have received attention from the media in recent years.

The laboratory strains of *E. coli* discussed in this book, MM294, cI, cII, CR63, and B$_E$, have been used in the laboratory for years and do not normally cause disease. MM294 is reported to be ineffective at colonizing the human digestive tract and so is especially harmless in that respect. However, all of these strains could cause infection if introduced into an open wound or into the eye. It is therefore very important to use aseptic technique (see page xiii) when handling the organisms. Students should never eat, drink, smoke, or apply cosmetics in the laboratory. Many instructors require that students wear protective goggles while working in the laboratory.

When transferring cultures of *E. coli,* keep pipette tips away from the face to avoid inhaling any aerosol that might be created. If a student contaminates his or her hands with a culture, wash immediately. Avoid contaminating any cuts with bacterial culture, and keep all bacteria away from the eyes. If a student believes he or she may have contaminated an area of broken skin, wash that area immediately. If a student gets bacteria in his or her eyes, use the eyewash fountain to rinse the eyes, and call a physician's office for further advice.

Agrobacterium tumefaciens does not cause disease in humans. However, the same precautions should be taken with it as with *E. coli.*

Bacteriophage T4 is harmless to humans. It cannot infect human cells. It should be handled in the same manner as *E. coli,* mostly to avoid inadvertent contamination of laboratory *E. coli* cultures.

Disinfect all plates and cultures before disposing of them.

Disinfecting

Keep disinfectant solutions available in the laboratory in squeeze bottles. These solutions can be 2% Lysol, 70% ethanol, rubbing alcohol, or other special disinfectants.

Before carrying out any experiments involving bacteria or phage, wipe down the laboratory bench with disinfectant solution. At the end of the laboratory period, wipe down the benchtop with disinfectant again. Clean up any spills involving organisms immediately, and disinfect the area thoroughly.

After the laboratory period, disinfect all materials that have come in contact with bacteria or phage (micropipette tips, pipettes, agar plates, culture tubes, flasks, etc.) either with pressurized steam or by soaking them in concentrated disinfectant. To use steam, place all biological waste in an autoclave bag, and sterilize it in an autoclave or pressure cooker. To use disinfectant, soak all contaminated materials for at least 15 min in either 10% Lysol or 15 to 20% chlorine bleach. Drain the liquid. Place trash in a plastic bag, and put the bag in the trash. Thoroughly rinse any containers to be reused. Clean the containers as usual, and then sterilize as needed.

It is not necessary to disinfect materials that have come in contact only with DNA and restriction enzymes (for example, from the gel electrophoresis laboratories).

Regulations for Recombinant DNA Work

The National Institutes of Health (NIH) oversees and regulates all research involving transfer of DNA between species. Rules for conducting recombinant DNA research are published as the NIH *Guidelines for Research Involving Recombinant DNA Molecules.* Certain kinds of recombinant DNA research are designated as exempt from these guidelines. Provision III-D-3 states that "The following molecules are exempt from these guidelines . . . those that consist entirely of DNA from a prokaryotic host, including its indigenous plasmids or viruses when propagated only in the host (or a closely related strain of the same species) or when transferred to another host by well-established physiological means." *Under this guideline, all the experiments and DNA molecules used in this book are exempt and may be conducted in a high school setting.*

If students wish to pursue further research involving recombinant DNA, instructors should make sure that the work involves only exempt molecules and procedures or should make arrangements for the students to work in an NIH-approved laboratory.

Selected Reading

Horn, T. M. 1992. *Working with DNA and Bacteria in Precollege Science Classrooms.* National Association of Biology Teachers, Reston, Va. An excellent resource.

Aseptic Technique

General Information

In experiments with microorganisms, it is essential to avoid contamination of the experiment by other microbes. Contamination could cause the procedure to fail. More important, contaminating organisms cannot be guaranteed to be harmless. Avoiding contamination is most important when cultures are being inoculated. If a contaminating microbe should find its way into the growth medium at the beginning of the growth period, the contaminant could grow along with the intended experimental organism.

To prevent microbiological contamination, a system of laboratory practice called "sterile technique" or "aseptic technique" is used. Proper aseptic technique in the laboratory minimizes the risk of contamination.

There are a few principles to remember when learning aseptic technique:

*   An object or solution is sterile only if it contains *no* living thing.
*   In general, objects and solutions are sterile only if they have been treated (autoclaved, irradiated, etc.) to kill contaminating microorganisms.
*   Any sterile surface or object that comes in contact with a nonsterile thing is no longer considered sterile.
*   Air is not sterile (unless it was sterilized inside a closed container).

Acting on those principles, we first sterilize all containers and solutions to be used when culturing microorganisms. Thereafter, material is transferred between sterile containers with sterile tools (sterile pipettes, micropipettes with sterile tips, sterile inoculating loops, etc.) in such a way as to minimize exposure to outside air and avoid contact with nonsterile surfaces and items.

The keys to good aseptic technique are as follows.

*   Keep lids off sterile containers for the shortest time possible.
*   Pass the mouths of open containers through a flame. Flaming warms the air at the opening, creating positive pressure and preventing contaminants from falling into the tube. Even plastic items can be flamed briefly.
*   Hold open containers at an angle whenever possible to prevent contaminants from falling in.

Specific Techniques

Use of glass or plastic pipettes

Glass pipettes are put into containers or wrapped and then autoclaved. Plastic pipettes are purchased presterilized in individual wrappers. To use a pipette, remove it from its wrapper or container by the end opposite the tip. Do not touch the lower two-thirds of the pipette. Do not allow the pipette to touch any laboratory surface. Draw the lower length of the pipette through a Bunsen burner flame. Insert only the untouched lower portion of the pipette into a sterile container.

Using test or culture tubes

Sterilize test tubes with lids or caps on. When you open a sterile tube, touch only the outside of the cap, and do not set the cap on any laboratory surface. Instead, hold the cap with one or two fingers while you complete the operation, and then replace it on the tube. This technique usually requires some practice, especially if you are simultaneously opening tubes and operating a sterile pipette. If you are working with a laboratory partner, one person can operate the test tube and the other can operate the pipette.

After you remove the cap from the test tube, pass the mouth of the tube through a flame. If possible, hold the open tube at an angle. Put only sterile objects into the tube. Complete the operation as quickly as you reasonably can, and then flame the mouth of the tube again. Replace the lid.

Inoculating loops and needles

Inoculating loops and needles are the primary tools for transferring microbial cultures. Loops and needles are sterilized by flaming. Put the business end of the tool directly into a Bunsen or alcohol burner flame, and hold it there until the end glows bright red. Withdraw the tool from the flame. The tool is now hot and sterile. Count to 5 or 10 to let it cool, and then transfer the organisms.

If you are moving organisms from an agar plate, touch an isolated colony with the transfer loop. Be sure your inoculating loop is cool before you do this. Replace the plate lid. Open and flame the culture tube, and inoculate the medium in it by stirring the end of the transfer tool in the medium. If you are removing cells from a liquid culture, insert the loop into the culture. The loop may hiss. If so, wait until the hissing stops, move the loop a little in the culture, then withdraw it.

Even if you cannot see any liquid in the loop, there will be enough cells there to inoculate a plate or a new liquid culture.

Transferring large volumes

If you don't have to be careful about the volume you transfer, a pure culture or sterile solution can be transferred to a sterile container or new sterile medium by pouring. Remove the cap or lid from the solution to be transferred. Thoroughly flame the mouth of the container, holding it at an angle as you do so. Remove the lid from the target container. Hold the container at an angle and flame it, if possible (you may not be able to if you are holding many items in your hands). Quickly and neatly pour the contents from the first container into the second. Flame the mouth of the second container. Replace the lid.

If you must transfer an exact volume of liquid, use a sterile pipette or a sterile graduated cylinder. When using a sterile graduated cylinder, complete the transfer as quickly as you reasonably can to minimize the time the sterile liquid is exposed to the air.

PART I

Laying the Foundation

Laying the Foundation contains background information for conceptual grounding in both the science fundamental to DNA-based technologies and the technologies themselves. We introduce the broad scope of biotechnology before narrowing our focus to the historical development and scientific underpinnings of the biotechnologies that are DNA based.

An Overview of Biotechnology

Introduction

When one reads or hears the word "biotechnology," the word "revolution" is often close behind. This combination of words is appropriate in many ways, for advancements in biotechnology will revolutionize major aspects of our lives and our relationship with the natural world.

In the field of human health, biotechnology will bring new ways to diagnose, treat, and prevent diseases. In agriculture, every aspect, from the seed placed in the ground to the food that appears on our tables, will be affected. Biotechnology is often touted as an environmental savior of sorts, for it will offer new, cleaner, renewable energy sources, methods for detecting and cleaning up environmental contamination, and products and processes that are more environmentally benign than some we have used before.

Even though all of us are certain that biotechnology is important, most of us are not sure we know exactly what biotechnology is. Such confusion is understandable, for "biotechnology" is an ambiguous term. To make matters even worse, it is used differently by different people. So what is biotechnology anyway?

Biotechnology Definitions

Defining "biotechnology" is actually very easy. Break it into its root words, "bio" and "technology," and you have the following definition:

Biotechnology: the use of living organisms to solve problems or make useful products.

After reading that definition, you may question the appropriateness of the phrase "biotechnology revolution." We domesticated plants and animals 10,000 years ago (Figure 1.1). For thousands of years we have used microbes such as yeasts and bacteria to make useful food products like bread, wine, cheese, and yogurt. Virtually all antibiotics come from microbes, as do the vitamins added to breakfast cereals and the en-

zymes used in manufacturing processes as diverse as making high-fructose corn syrup and stone-washed jeans. In agriculture, we have used microbes since the 19th century to control insect pests and have inoculated the soil with nitrogen-fixing bacteria to improve crop yield. Microbes have been used extensively in sewage treatment for decades. Certain vaccines are based on the use of live, but weakened, viruses or bacteria.

So why are people talking about a biotechnology revolution? Why go to the trouble to coin a new phrase for activities we have engaged in for ages?

The answer is this. During the 1960s and 1970s, our knowledge of cellular and molecular biology reached the point where we could begin to manipulate organisms at those levels. Manipulating organisms to our advantage is not new. What is new is *how* we are manipulating them. Before, when we manipulated organisms, we had little or no understanding of the mechanisms underlying the manipulations. Our manipulations were hit-or-miss ventures. Now, we understand our manipulations at the most basic level, the molecular level. As a result, we can predict what effect our manipulations will have, and we can direct the change we want with great specificity.

So, "biotechnology" in the new sense of the word can be defined as follows:

"New" biotechnology: the use of cells and biological molecules to solve problems or make useful products.

Biological molecules are the large macromolecules unique to living organisms (see Table 1.1). The biological molecules most often utilized in biotechnology today are **nucleic acids,** such as **DNA** and **RNA,** and **proteins**. As our understanding of the roles other biological molecules play in cellular structure and activities increases, we are finding, and no doubt will continue to find, new ways to use these molecules to our advantage.

Figure 1.1 Painting depicting Egyptian agriculturists 6,000 years ago. (Photograph courtesy of the Bettmann Archive.)

For example, we have long thought of **carbohydrates** as comparatively simple biological molecules, important in energy storage and cell structure, but lacking the complexity to carry out specialized functions that require molecular recognition and coordination. We have recently discovered that we underestimated the importance of carbohydrates in complex molecular interactions. For instance, carbohydrates stimulate immune responses by increasing the numbers and activi-

Table 1.1 The four classes of biological molecules

Biological molecule (polymer)	Repeating subunit (monomer)
Proteins	Amino acid
Carbohydrates	Simple sugar
Nucleic acid	Nucleotide
Lipids[a]	

[a]Lipids are a heterogeneous group of compounds that are grouped together because they are soluble in organic solvents and insoluble in water. Lipids exhibit a great deal of variety in chemical structure, size, and complexity, so they are not as tidy a class to describe as classes of other biological molecules are. Listing a single repeating subunit is not possible. The most useful fact about the biochemistry of lipids is that they are primarily hydrocarbons, that is, chains of carbon molecules bound to hydrogens.

ties of white blood cells. They also act as road signs for the cells and chemicals involved in the inflammatory response. They are important for other physiological functions requiring cell-to-cell or molecule-to-molecule recognition, including antigen-antibody interactions. Both viruses and bacterial pathogens recognize and bind to carbohydrates on their host cell surfaces. Therefore, the ability to manipulate carbohydrate linkages may also lead to new types of biological products.

Biotechnology: A Collection of Technologies

Some of the confusion surrounding the word "biotechnology" could be eliminated by simply changing the singular noun to its plural form, "biotechnologies," because biotechnology is not a singular entity. Instead, biotechnology is a collection of technologies, all of which utilize cells and biological molecules (Figure 1.2).

Developing technologies that use cells and biological molecules in place of whole multicellular organisms allow us to capitalize on a critical aspect of life at the cellular and molecular level: the extraordinary specificity of the interactions. Because of this specificity, the tools and techniques of biotechnology are quite precise and are tailored to operate in known, predictable ways.

Any one of the biotechnologies can produce a wide variety of products, and any one product may have diverse applications. For example, xanthan gum, currently produced by bacteria through bioprocessing technology, can be used to thicken salad dressings or to clean residual oil from oil wells. A genetically engineered bacterium could synthesize a protein-digesting enzyme that both dissolves blood clots and unstops drains.

The Technologies and Their Uses

What are some of these technologies that use cells and biological molecules, and how are we using them?

Monoclonal antibody technology

Monoclonal antibody (MCAb) technology uses cells of the immune system that make proteins called **antibodies**. Your immune system is composed of a number of cell types that work together to locate and destroy substances that invade your body. One type of immune system cell, the **B lymphocyte,** responds to invaders by producing antibodies that bind to the foreign substance with extraordinary specificity. We are

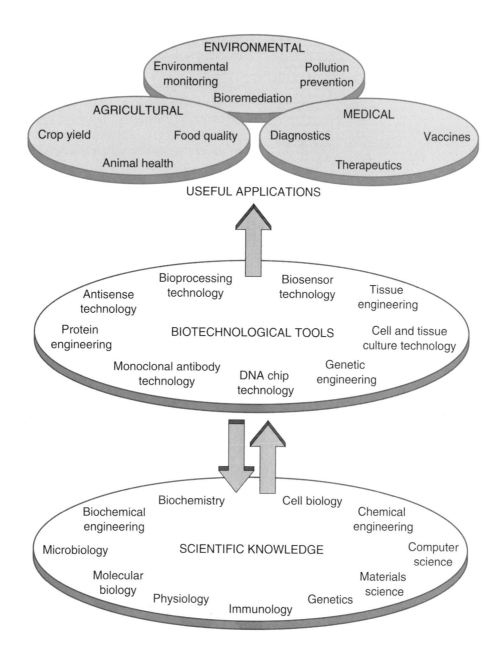

Figure 1.2 Synthesis of scientific and technical knowledge from many academic disciplines has produced the set of enabling technologies we call biotechnology. Any one technology will be applied to a number of industries to produce an even broader array of products.

now harnessing the ability of B lymphocytes to make these very specific antibodies.

Because of their specificity, MCAb are powerful tools for detection, quantification, and localization. Measurements based on MCAb are fast, accurate, and extremely sensitive because of this specificity.

Diagnostic and Therapeutic Uses of MCAb
The substances that MCAb detect, quantify, and localize are remarkably varied and are limited only by the

substance's ability to trigger the production of antibodies. Home pregnancy kits use an MCAb that binds to a hormone produced by the placenta. MCAb are currently being used to diagnose a number of infectious diseases such as strep throat and gonorrhea. Because cancer cells differ biochemically from normal cells, we can make MCAb that detect cancers by binding selectively to tumor cells (Figure 1.3). In addition to diagnosing diseases in humans, MCAb are being used to detect plant and animal diseases, food contaminants, and environmental pollutants.

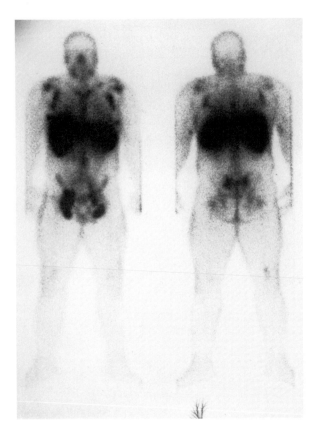

Figure 1.3 Radiolabeled monoclonal antibodies confirm that this patient's cutaneous T-cell lymph cancer involves the lymph nodes and skin. The antibodies collect in the cancerous lymph nodes of the armpits, neck, and groin, and a strong outline of the patient's body verifies skin involvement. The liver and spleen are darkened because these organs normally collect antibodies. (Photograph courtesy of Dr. Jorge Carrasquillo, National Cancer Institute.)

Because of the specificity of MCAb, one molecule with therapeutic potential can be separated from a mixture of thousands of other molecules. Ultimately, we hope to use MCAb to treat many types of cancer and other diseases. By tagging a radioisotope or toxin to the tumor antibody, we can deliver these tumor-killing agents directly to our target and bypass healthy cells. We are also using MCAb to treat autoimmune diseases, to prevent complications following heart bypass surgery, and to protect children at high risk from viral respiratory infections.

Bioprocessing technology

Bioprocessing technology uses living cells or components of their biochemical machinery doing what they normally do: synthesizing products, breaking down substances, and releasing energy. The living cells most frequently used are one-celled microorganisms, such as bacteria and yeasts, or mammalian cells; the cellular components most often used are proteins called **enzymes**.

Enzymes are essential for life. They catalyze all cellular biochemical reactions, most of which would occur much too slowly to support life. Through enzyme-catalyzed reactions, organisms break down large organic molecules to obtain energy and to generate a supply of chemical building blocks for making new molecules. We are using the cells and enzymes of other organisms to commercially manufacture chemical compounds, to generate energy, and to break down chemical pollutants for us (Table 1.2).

Fermentation and Mammalian Cell Culture

We commercially manufacture a wide variety of biotechnology products through large-scale **fermentation** and mammalian **cell culture,** two types of bioprocessing technologies that rely on cellular enzymes (Figure 1.4). Because of the scale of the production systems, these technologies represent triumphs of both chemical engineering and molecular biology.

Figure 1.4 Large-scale mammalian cell culture and microbial fermentation, carried out in bioreactors such as this, are biologically based manufacturing processes that utilize the biochemical machinery of cells to manufacture useful products. (Photograph courtesy of Covance, Inc., Research Triangle Park, N.C.)

Table 1.2 Examples of the range of bioprocessing products and typical organisms used to manufacture them

Product	Typical organism
Bulk organics	
Acetone/butanol	*Clostridium acetobutylicum*
Biomass	
Starter cultures and yeast for food and agriculture	Lactic acid bacteria or baker's yeast
Single-cell protein	*Candida utilis*
Organic acids	
Citric acid	*Aspergillus niger*
Lactic acid	*Lactobacillus delbrueckii*
Amino acids	
Glutamic acid	*Corynebacterium glutamicum*
Lysine	*Brevibacterium flavum*
Microbial transformations	
Sorbitol to sorbose (vitamin C)	*Acetobacter suboxydans*
Steroids	*Rhizopus arrhizus*
Antibiotics	
Penicillins	*Penicillium chrysogenum*
Cephalosporins	*Cephalosporium acremonium*
Tetracyclines	*Streptomyces aureofaciens*
Extracellular polysaccharides	
Xanthan gum	*Xanthomonas campestris*
Dextran	*Leuconostoc mesenteroides*
Enzymes	
α-amylase	*Bacillus amyloliquefaciens*
Pectinase	*Aspergillus niger*
Vitamins	
B_{12}	*Propionibacterium shermanii*
Riboflavin	*Eremothecium ashbyii*
Pigments	
Beta-carotene	*Blakeslea trispora*
Vaccines	
Diphtheria	*Corynebacterium diphtheriae*
Poliomyelitis	Monkey kidney or human cells
Rubella	Hamster kidney cells
Hepatitis B	Recombinant yeast
Therapeutic proteins	
Insulin	Recombinant *Escherichia coli*
Growth hormone	Recombinant *Escherichia coli*
Erythropoietin	Recombinant mammalian cells
Factor VIII-C	Recombinant mammalian cells
Monoclonal antibodies	Hybridoma cells

The oldest and most familiar bioprocessing technology is **microbial fermentation.** Originally, the microbial fermentation products we used were derived from the series of enzyme-catalyzed reactions that microbes use to break down glucose. In the process of metabolizing glucose to acquire energy, microbes synthesize by-products we can use: carbon dioxide for leavening bread, ethanol for brewing wine and beer, lactic acid for making yogurt, and acetic acid (vinegar) for pickling foods (Figure 1.5A and B).

Now we have extended our use of the rich biochemical machinery of microbes beyond the metabolic pathway for glucose breakdown. We use microbial fermentation to synthesize an extraordinary array of products, including antibiotics, amino acids, hormones, vitamins,

industrial solvents, pesticides, food processing aids, pigments, enzymes, enzyme inhibitors, and pharmaceuticals.

As our experience with microbial production of recombinant therapeutic proteins has grown, we have learned that some of these proteins must be produced by mammalian cells to be therapeutically effective. As a result, our reliance on mammalian cell production systems has grown and is expected to increase even more in the future.

Biodegradation

Microbes and the enzymes they use to break down organic molecules are helping us clean up certain environmental problems: oil spills, toxic waste sites,

A

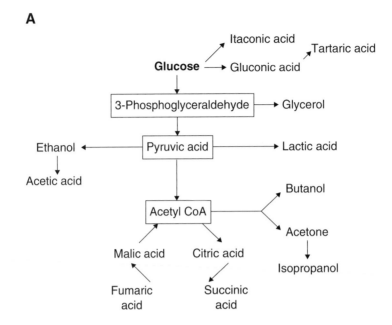

B

Organic chemical	Microbial source(s)	Industrial uses
Ethanol	*Saccharomyces*	Industrial solvent, fuel, beverages
Acetic acid	*Acetobacter*	Industrial solvent, rubber, plastics, food acidulant (vinegar)
Citric acid	*Aspergillus*	Food, pharmaceuticals, cosmetics, detergents
Gluconic acid	*Aspergillus*	Pharmaceuticals, food, detergent
Glycerol	*Saccharomyces*	Solvent, sweetener, printing, cosmetics, soaps, antifreezes
Isopropanol	*Clostridium*	Industrial solvent, cosmetic preparations, antifreeze, inks
Acetone	*Clostridium*	Industrial solvent, intermediate for many organic chemicals
Lactic acid	*Lactobacillus, Streptococcus*	Food acidulant, fruit juice, soft drinks, dyeing, leather treatment, pharmaceuticals, plastic
Butanol	*Clostridium*	Industrial solvent, intermediate for many organic chemicals
Fumaric acid	*Rhizopus*	Intermediate for synthetic resins, dyeing, acidulant, antioxidant
Succinic acid	*Rhizopus*	Manufacture of lacquers, dyes, and esters for perfumes
Malic acid	*Aspergillus*	Acidulant
Tartaric acid	*Acetobacter*	Acidulant, tanning, commercial esters for lacquers, printing
Itaconic acid	*Aspergillus*	Textiles, paper manufacture, paint

Figure 1.5 (A) Useful metabolic products of glucose breakdown provided by various microorganisms. (B) Chemicals currently produced by microbial fermentation of glucose and their industrial applications.

and leakage from underground storage tanks. The use of microbial populations to clean up pollution is known as **bioremediation**. Probably the best-known example of bioremediation is the use of oil-eating bacteria to clean up oil spills such as the *Exxon Valdez* spill in Alaska's Prince William Sound in 1989 and spills in Iraq after the 1991 Gulf War (Figure 1.6).

We are also exploring the possibility of using plants to treat contamination caused by sources of pollution such as contaminated wastewater from certain indus-

trial manufacturing facilities. This practice, which is called **phytoremediation**, uses certain plants to remove, destroy, or sequester contaminants. Some plants secrete enzymes that break down the contaminants. Other plants act as sponges for pollutants such as organic solvents, petroleum derivatives, and toxic metals such as lead, zinc, cadmium, and mercury. Some cultivars of these plants can tolerate high levels of radioactive elements.

In the future we may be able to use sewage and agricultural refuse as renewable energy sources by ex-

Figure 1.6 *Exxon Valdez* oil spill, Alaska. Rocks on the beach before (right) and after (left) bioremediation. (Photograph courtesy of the U.S. Environmental Protection Agency.)

Figure 1.7 Stages of plant cell culture from callus to plantlet.

ploiting microbes that degrade these organic compounds and, in the process, release energy.

Cell culture technology

Cell culture technology is the growing of cells in appropriate nutrients in laboratory containers or in large **bioreactors** in manufacturing facilities.

Plant Cell Culture

Plant cell culture is an essential aspect of plant biotechnology. The centrality of cell culture to plant biotechnology stems from a property unique to plant cells, their totipotency, or the potential to generate an entire multicellular plant from a single differentiated cell (Figure 1.7).

We genetically engineer plants at the level of the single cell. When a leaf cell is genetically engineered to contain a useful trait such as resistance to insect pests, that cell must develop into a whole plant if it is to be useful to farmers. This regeneration is accomplished through cell and tissue culture.

Animal Cell Culture

Plant cell culture is not the only type of cell culture being applied to agriculture. Using insect cell culture to grow viruses that infect insects may enable us to broaden the application of viruses as **biological control** agents. Mammalian cell culture is also being used in livestock breeding. Large numbers of bovine zygotes from genetically superior bulls and cows can be produced and cultured before being implanted into surrogate cows.

The medical community uses animal cell culture to study such topics as the safety and efficacy of pharmaceutical compounds, the molecular mechanism of viral infection and replication, the toxicity of compounds, and basic cell biochemistry.

Embryonic Stem Cell Culture

Recently, scientists have successfully cultured a unique type of human cell, **embryonic stem cells**. This feat deserves particular attention because of the potential medical benefits it may provide and the ethical issues it may raise.

Your body has a number of tissue-specific cells that are not specialized to perform a specific function. When these permanently immature cells, which are called stem cells, divide, they give rise to one stem cell that remains immature and one cell that becomes specialized. Unspecialized cells in the liver can differentiate into one of a number of specialized liver cells, such as cells that produce bile or epithelial cells that line the bile duct. Stem cells found in the bone marrow can produce all blood cell types as well as muscle, cartilage, and bone cells. But liver cells will not differentiate into white blood cells, nor will bone marrow cells become specialized for bile production.

Embryonic stem cells can give rise to virtually any type of cell. This complete developmental plasticity sets them apart from other stem cells and opens the possibility of using them therapeutically. For example, if we develop the capability to control the differentiation of human embryonic stem cells, we may be able to produce replacement cells to treat diabetes,

Parkinson's disease, and heart disease, among others. We discuss the laboratory methods for producing embryonic stem cells and the ethical issues this process may raise in chapter 2.

Tissue engineering technology

Tissue engineering, a marriage of cell biology and materials science, allows us to create semisynthetic tissues in the laboratory. These tissues consist of biodegradable scaffolding material plus living cells.

The most basic forms of tissue engineering use natural biological materials, such as collagen, for scaffolding. For example, one of the first products developed with this technology, two-layer skin, is made by infiltrating a collagen gel with fibroblasts, allowing them to grow, multiply, and become the dermis, then adding a layer of keratinocytes to serve as the epidermis (Figure 1.8). In other tissue engineering methods, the scaffolding, made of a synthetic polymer, is shaped and then placed in the body where new tissue is needed. Adjacent cells, stimulated by the appropriate growth factors, invade the scaffolding, which is eventually degraded and absorbed.

Figure 1.8 A surgeon prepares to apply a sheet of artificially produced human skin to a patient. (Photograph courtesy of VU/©SIU.)

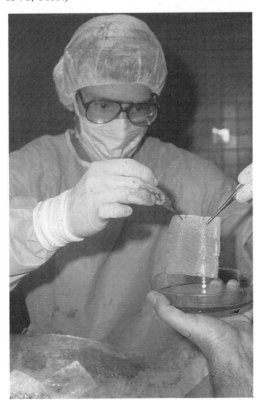

The more sophisticated tissue engineering methods use implants composed of synthetic polymer scaffolding spiked with cells grown in the laboratory. The scaffolding creates the space for tissue formation, transports the cells into the organisms, and guides the structural development of the replacement tissue.

Simple tissues, such as skin and cartilage, were the first to be engineered successfully. Ultimately, the goal is to create complex organs, consisting of a number of tissue types, to serve as replacements for diseased or injured organs.

Biosensor technology

Biosensor technology represents the joining of molecular biology and microelectronics. A **biosensor** is a detecting device composed of a biological substance linked to a transducer (Figure 1.9). The biological substance might be a microbe, a single cell from a multicellular animal, or a cellular component such as an enzyme or an antibody. Biosensors allow us to measure substances that occur at extremely low concentrations.

How do biosensors work? Biosensors generate digital electronic signals by exploiting the specificity of biological molecules. When the substance we want to measure collides with the biological detector, the transducer produces a tiny electrical current. This electrical signal is proportional to the concentration of the substance.

Figure 1.9 Schematic drawing of a simple biosensor. A biosensor is a detecting device that utilizes an immobilized biological sensing element and a transducer to produce an electrical signal proportional to the concentration of the substance to be detected.

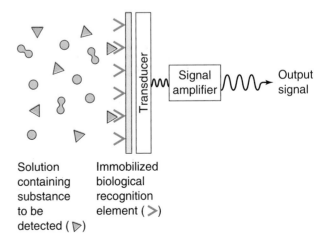

Biosensors are being developed for uses as varied as measuring the nutritional value, freshness, and safety of food; providing point-of-care analyses of blood gases, electrolyte concentrations, and blood-clotting capability in emergency rooms and intensive care units; monitoring industrial processes in real time with immediate feedback for process control; locating and measuring pollutants; and detecting minute quantities of substances in blood. By coupling a glucose biosensor to an insulin infusion pump, the correct blood concentration of glucose could be maintained at all times in diabetics.

Genetic engineering technology

Genetic engineering technology is often referred to as recombinant DNA (rDNA) technology. Recombinant DNA is made by joining or recombining genetic material from two different sources. In nature, genetic material is constantly recombining. Each of the following is just one of many ways nature joins genetic material from two sources:

- when **crossing over** occurs between **homologous** maternal and paternal chromosomes during gamete formation
- when egg and sperm fuse during fertilization
- when bacteria exchange genetic material through **conjugation, transformation,** and **transduction**

In each of these examples of natural recombination, when genetic material from two different sources is combined, the result is increased genetic variation. The genetic variation that exists in nature has provided the raw material for evolutionary change driven by natural selection or artificial selection imposed by humans.

Using Existing Genetic Variation in Selective Breeding

As soon as humans domesticated organisms, we began to use selective breeding to alter their genetic makeup to suit our needs. Certain individuals in a population had traits, and therefore genes, we valued, and we chose these individuals to serve as parents for the next generation. By selecting certain genetic variants from a population and excluding others, we intentionally directed the recombining of genetic material. As a result, we radically changed the genetic makeup of the organisms we domesticated. If you question the extent to which human-imposed ("artificial") selection has altered the genetic makeup of organisms, see the photograph showing today's corn next to its presumed ancestor, teosinte (Figure 1.10).

Figure 1.10 Teosinte, the presumed ancestor of corn, next to a modern variety of field corn.

So existing genetic variation has been a valuable natural resource that humans have exploited for centuries. The tools and knowledge we need to make selective breeding more predictable and more precise have been continually evolving. Genetic engineering is the next step in that continuum.

Generating Novel Genetic Variation with Genetic Engineering

The term "recombinant DNA technology," or genetic engineering, refers to the precise molecular techniques that join specific segments of DNA molecules from different sources. We recombine DNA by using **restriction enzymes** (also called restriction endonucleases) designed to cut and join DNA in predictable ways. To ferry the DNA into the target organism, we usually use bacteria and viruses that transport DNA in nature, or we use their DNA molecules.

Therefore, in addition to directing the recombining of genetic material through the intentional joining of eggs and sperm (pollen in plants) in selective breeding, we can now recombine genetic material with greater precision by working at the molecular level.

Selective Breeding versus Genetic Engineering

Many scientists view genetic engineering as simply an extension of selective breeding, because both techniques join genetic material from different sources to create organisms that possess useful new traits. However, even though genetic engineering and selective breeding bear a fundamental resemblance to one another, they also differ in important ways (Table 1.3 and Figure 1.11).

In genetic engineering, we move single genes whose functions we know from one organism to another, while in selective breeding, sets of genes of unknown function are transferred. By increasing the precision and certainty of our genetic manipulations, the risk of

producing organisms with unexpected traits decreases. The trial-and-error approach of selective breeding is circumvented.

In selective breeding, we have crossbred organisms in the same species, different species, and, sometimes, different genera. In genetic engineering, no taxonomic barriers exist. Our genetic improvement of organisms is no longer restricted to the small pool of genetic variation within which we can manipulate breedings. A desirable gene from any organism may theoretically be placed in a second organism no matter how distantly related the two are. The potential to move genes between organisms gives us great flexibility, because now we have greater access to nature's astounding genetic diversity.

Case Study: *Bacillus thuringiensis*

An example of the flexibility genetic engineering gives us is the work being conducted on a variety of products that contain a gene from the bacterium ***Bacillus thuringiensis,*** or Bt for short. Bt is a naturally occurring organism that produces a protein that kills certain insects that ingest it but does not harm other organisms, such as fish, birds, or mammals. The degree of selectivity of the Bt toxin is even more remarkable than being limited to insects. Each Bt strain is toxic only to a distinct group of insects. Certain Bt

Table 1.3 Differences in selective breeding and genetic engineering

Parameter	Selective breeding	Genetic engineering
Level	Whole organism	Cell or molecule
Precision	Sets of genes	Single gene
Certainty	Genetic change poorly characterized	Gene well characterized
Taxonomic limitation	Usable only within and between species, sometimes between genera	None

Figure 1.11 Schematic representation of the movement of a gene for a desired trait (gold circle) to a crop plant from its wild relative. Two methods of gene transfer are depicted: selective breeding and genetic engineering. Note that in selective breeding, sets of genes of unknown function are transferred from the wild relative to the crop plant. In genetic engineering, a single gene of known function is transferred.

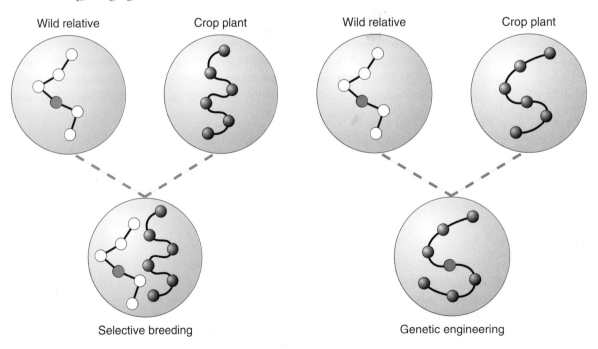

strains are toxic to butterflies and moths, while others are toxic to beetles and still others to mosquitoes. Each of these strains produces a slightly different version of the toxic protein.

Bt has been used for years as a biological control agent. Now the genes that code for the selectively toxic proteins in various Bt strains have been identified, and scientists are using the genes in a variety of ways.

- Genetic engineers have modified Bt by increasing the number of copies of the endotoxin gene. The modified Bt can produce 10 times as much toxin.
- They have moved the Bt endotoxin gene into a different bacterium that forms a protective capsule around the protein when the bacterium is killed. Within the capsule, the endotoxin is protected against ultraviolet light and therefore persists in the field much longer than Bt does.
- They have taken the Bt gene and moved it into bacteria that exist in symbiotic relationships with crop plants. For example, bacteria that naturally live inside cornstalks have been given the Bt gene to help protect corn from the European corn borer. In a separate venture, a bacterium that lives on corn roots received the Bt gene and protected the plants against corn root worm. In both cases, genetic engineers moved the gene into microbes whose lifestyles make them perfect candidates for pest control but that do not make pesticidal compounds.
- Finally, genetic engineers have moved the gene into a wide variety of plants: tomato, cotton, potato, and corn (Figure 1.12).

Protein engineering technology

Protein engineering technology will often be used in conjunction with genetic engineering to improve existing proteins and to create proteins not found in nature. In theory, we should eventually be able to create any protein from scratch. As of now, research efforts are aimed primarily at modifying existing proteins. Why would we want to improve existing proteins? The answer to this question varies with the protein.

Enzyme Engineering

Enzymes are beautifully designed for the role they play in nature as catalysts of the biochemical reactions on which living organisms depend. As such, enzymes function best under conditions that are compatible with life: neutral pH, mild temperature and pressure, and an aqueous (water-based) environment.

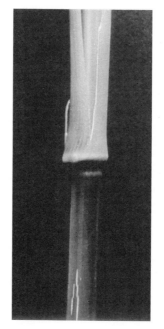

Figure 1.12 (Left) Corn plant genetically engineered to produce Bt toxin; (right) control showing European corn borer damage. (Photograph courtesy of Novartis, Inc.)

In certain manufacturing processes where catalysis by enzymes might prove useful, the conditions are too harsh for enzymes to function. Most enzymes literally fall apart at high temperatures, in very acidic or basic solutions, or when exposed to organic (non-water-based) solvents. Currently, we are modifying enzymes, both directly through chemical manipulations and indirectly by specifically mutating genes that code for enzymes, to increase their stability under harsh manufacturing conditions, to broaden their substrate specificity, and to improve their catalytic power.

Abzyme Engineering

Although most of the protein engineering work has been directed at changing the catalytic properties of existing enzymes, scientists have also invented a way to synthesize novel catalysts: antibodies with catalytic ability, or **abzymes**. Antibodies resemble enzymes because both are proteins that bind to specific molecules. But the similarity ends there. Antibodies bind for the sake of binding; enzymes bind to make reactions happen.

We are learning to custom design abzymes to catalyze reactions for which there are no known enzymes, opening up a new world of exciting possibilities. For example, we might be able to create antibodies that can function as proteases, the enzymes that down proteins. If we design the proteolytic

1. An Overview of Biotechnology

so that they break peptide bonds with great specificity, they will be the protease equivalent of a restriction endonuclease, a type of enzyme that is an invaluable tool of the genetic engineer. These "restriction endoproteases" might allow us to cut protein molecules with great specificity.

Antisense technology

Antisense technology is being used to block or decrease the production of certain proteins. Scientists accomplish this blocking by using small nucleic acids (oligonucleotides) that prevent translation of the information encoded in DNA into a protein (Figure 1.13). For more details on the molecular aspects of antisense technology, please see chapters 5 and 8.

The potential applicability of this technology is enormous. In any situation in which blocking a gene would be beneficial, antisense technology provides a valuable approach to the problem. An obvious example is a situation in which you want to block the production of a protein product. Currently, researchers are using this technology to slow food spoilage, control viral diseases, inhibit the inflammatory response, and treat asthma, cancers, and thalassemia, a hereditary form of anemia common in many parts of the world.

Metabolic Engineering

A less obvious but very exciting use of antisense technology will be in **metabolic engineering.** Many compounds in nature that have great commercial applicability are not proteins. For example, most compounds produced by plants to deter insect feeding could be useful as crop protectants but are not proteins. We could increase their production in crop plants by using antisense technology to block the production of enzymes in certain pathways, thus rerouting the plant's metabolism to favor production of these compounds.

On the other hand, we might want to decrease the production of a substance that is not a protein. An excellent example is cholesterol. Given the right **antisense molecule,** perhaps we could significantly decrease production of cholesterol by blocking key enzymatic steps in its synthesis.

To visualize how metabolic engineering might work, refer to Figure 1.5A, which depicts some of the valuable products that result from the sequential enzymatic breakdown of a single starting substance, glucose. None of these useful products is a protein, but all have commercial value. If you wanted to maximize production of isopropanol in a microbial fermentation process, which steps, and therefore enzymes, would you block with antisense technology?

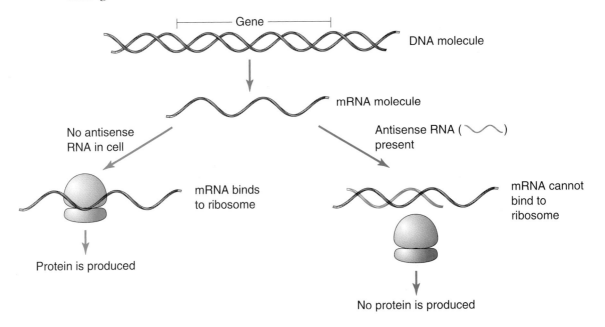

Figure 1.13 Schematic representation of antisense technology. In this example, the antisense oligonucleotide is an RNA molecule that blocks protein production by preventing the binding of mRNA to the ribosome.

DNA chip technology

DNA chip technology, a marriage of the semiconductor industry and molecular biology research, will transform genetic analysis because it allows tens of thousands of genes to be analyzed on a single "microchip." The manufacturing process of microchips and **DNA chips** is similar in principle, but instead of shining light through a series of masks to etch circuits into silicon, automated DNA chip makers use a series of masks to create a sequence of **DNA probes** on a glass slide.

For analysis of genetic information, DNA is removed from cells, tagged with fluorescent markers, and placed on the chip. Hybridized sequences attach to the probes, and unmatched bits of DNA are then washed away. Using a laser reader, computer, and high-powered microscopes, scientists can analyze thousands of sequences at one time and determine where the tagged DNA found a match with a chip-mounted DNA probe.

DNA chip technology is already being used to detect mutations in disease-causing genes, monitor gene expression in yeast and cancer cell lines, diagnose infectious diseases, and tell us whether a pathogen is resistant to certain drugs. In the next decade, DNA chips will contribute to crop biotechnology, improve screening for microbes used in bioremediation, hasten drug discovery, and provide answers to questions about gene function and the significance and clinical manifestations, if any, of polymorphisms in the human genome.

Bioinformatics technology

Bioinformatics is the use and organization of information about biology. In biotechnology applications, bioinformatics technology exists at the interface of computer science, mathematics, and molecular biology.

The pace of discovery in molecular biology is astronomical. We already have huge amounts of sequence data for genes and proteins; information on the three-dimensional structure of proteins, carbohydrates, and **glycoproteins;** and genetic maps for many species. Without methods for organizing the information, however, we will simply drown in it. To analyze and interpret the information, we need tools and methods for organizing, entering, processing, storing, accessing, and integrating data from different sources. And these tools and methods must be used consistently by laboratories involved in molecular biology research all over the world.

Bioinformatics technology uses computational tools such as algorithms, graphics, artificial intelligence, statistical software, simulation, and database management to help us map and compare genomes, determine protein structure, simulate molecular binding, pursue structure-based drug design, identify genes, assess the effects of virtual mutations, and determine phylogenetic relationships.

The Applications of Biotechnology

In the coming years, the diverse collection of technologies labeled "biotechnology" will yield an even broader array of products. Most of the commercial applications of biotechnology will be in three markets: human health care, agriculture, and environmental management.

Medical biotechnology

Biotechnology has already provided us with quicker and more accurate diagnostic tests, therapeutic compounds with fewer side effects, and safer vaccines.

Diagnostics

We can now detect many diseases and medical conditions more quickly and with greater accuracy because of the sensitivity of new diagnostic tools and techniques developed through biotechnology, such as monoclonal antibodies, biosensors, DNA probes, DNA chips, **restriction fragment length polymorphisms (RFLPs),** and the **polymerase chain reaction (PCR).** More information about DNA probes, RFLPs, and PCR is provided in chapter 5 and in Classroom Activities, chapters 16 and 18. For example, the time required to diagnose strep throat, gonorrhea, and chlamydia has dropped from days to minutes. Certain cancers are now diagnosed by simply taking a blood sample, thus eliminating the need for invasive and costly surgery. Through the Human Genome Project, scientists are also making remarkable progress in identifying and sequencing genes. These advances will greatly assist doctors in diagnosing hereditary diseases, many of which we cannot detect currently.

Biotechnology-based diagnostics are also altering healthcare provision. Advances based on biosensors are allowing physicians to carry out tests and interpret results right at the patient's bedside. These point-of-care tests allow doctors to make decisions immediately rather than waiting hours or days for test results to come from a centralized hospital laboratory.

Therapeutics

Biotechnology will provide improved versions of today's therapeutic regimens as well as treatments that would not be possible without these new techniques. Here are just a few examples of the novel therapeutic advances biotechnology now makes feasible.

Natural products as pharmaceuticals. Many plants produce compounds with human therapeutic value. For years, we have used a chemical derived from foxglove (digitalis) for treating heart conditions. A newly discovered chemical extracted from yew trees is being used to treat breast and ovarian cancers. We are investigating ticks and leeches for potential anticoagulant compounds and poison arrow frogs for painkillers (Figure 1.14).

We have recently turned our attention to the extraordinarily diverse ecosystems found in the sea and have discovered compounds that heal wounds, destroy tumors, prevent inflammation, relieve pain, and kill microorganisms. As exciting as these developments are, we could not turn pharmaceutical potential into reality without plant cell culture, microbial fermentation, and animal cell culture. Researchers collected 2,400 kg of sponges to obtain 1 mg of an anticancer drug! Using these sponges as sources of a pharmaceutical would not be feasible economically; more important, it would be an ecological disaster. If the sponge cells cannot be maintained in cell culture, we might iden-

Figure 1.14 For centuries, native tribes in South America have used the poison from certain frogs, such as this dendrobatid frog species, on the arrows they use to kill prey. Scientists are studying these poisons, which are neurotoxins, for their potential use as painkillers. (Photograph courtesy of VU/©Ken Lucas.)

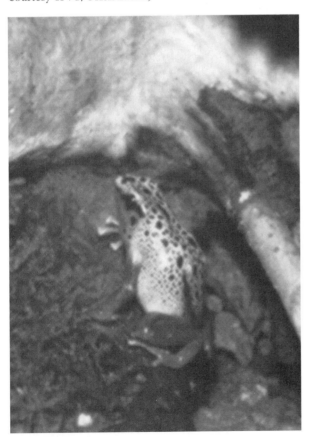

tify the genes required to produce the compound and move them into organisms that do well under culture conditions. By developing sophisticated cell culture and bioprocessing technologies, as well as recombinant DNA technologies, we will broaden our ability to use many more compounds from nature.

Endogenous therapeutic agents. The human body produces many of its own therapeutic compounds, and many of them are proteins. As proteins, they are prime candidates for possible production by genetically engineered bacteria. Such production would provide quantities that would allow us to better analyze their functions and make their commercialization economically feasible. As we increase our understanding of these and other endogenous therapeutic agents, we will be able to capitalize more on the body's innate healing ability. Following are examples of some endogenous therapeutic compounds:

- **Interleukin-2** activates **T-cell** responses (Figure 1.15).
- **Erythropoietin** regulates red blood cell production.
- **Tissue plasminogen activator** dissolves blood clots.

Biopolymers as medical devices. Nature has provided us with substances that are useful medical devices. Some are superior to inorganic, manmade substances, because, being biological materials, they are more compatible with our tissues and are degraded and absorbed when their job is done. The manufacture of biological polymers is also more environmentally benign. Following are examples of naturally occurring biopolymers used as medical devices:

Figure 1.15 Mice with lung cancer were treated with activated killer T cells plus recombinant interleukin-2. More than 250 tumor foci were reduced to fewer than 12 in mice receiving this treatment. (Photograph courtesy of Dr. Steven Rosenberg, National Cancer Institute.)

- The carbohydrate **hyaluronate** is a viscous, elastic, plasticlike, water-soluble substance that is used to treat arthritis, to prevent postsurgical scarring in cataract surgery, and for drug delivery.
- Adhesive protein polymers derived from living organisms are replacing sutures and staples in wound healing. They set quickly, produce strong bonds, and are absorbed.
- **Chitin,** a carbohydrate found in the exoskeletons of insects and crustaceans, combined with a natural fiber, polynosic, creates a material that limits bacterial and fungal growth.

Designer drugs. Using principles of protein engineering and computer molecular modeling, a scientist may be able to design effective therapeutic compounds before ever stepping into a laboratory (Figure 1.16).

Replacement therapies. Many disease states result from defective genes that cause the total lack or inadequate production of substances the body normally produces. Following are examples of compounds being produced (thanks to genetic engineering and bioprocess technologies) in sufficient quantity to provide a reliable source of replacement therapies.

- **Factor VIII** is a protein involved in the blood-clotting process. Some hemophiliacs lack this protein.
- **Insulin** is a protein hormone that regulates blood glucose levels by affecting cellular uptake of glucose. Diabetes results from an inadequate supply of insulin.

Figure 1.16 Drug development researchers are increasingly using computer imaging to help them relate data on gene and amino acid sequences to protein three-dimensional structure and, more important, protein function. (Photograph courtesy of Inpharmatica Company, London, England.)

Manufacturing pharmaceuticals with the use of transgenic plants and animals. Earlier we described methods for producing large amounts of proteins by using microbial fermentation and mammalian cell culture. While these production systems have allowed us to produce sufficient amounts of therapeutic proteins, the costs of building and maintaining the manufacturing facilities are very high. Through genetic engineering, we are now able to use **transgenic** plants or milk from transgenic animals as sources of pharmaceuticals.

Scientists have inserted the genes that code for therapeutic proteins into a variety of commonly grown crops such as tobacco, corn, and soybeans. The proteins have maintained their stability for over 2 years in dried seeds of the transgenic plants. However, some of the human proteins produced in plant systems are not therapeutically effective. By using transgenic animals such as goats and sheep, complex, therapeutically active human proteins can be expressed with great fidelity. Using goats rather than mammalian cell culture for production of a therapeutic protein could decrease the cost of production of 100 kg from $50–70 million to $16–25 million. However, obtaining purified proteins from milk is difficult, because milk contains fats, many proteins, minerals, sugars, and whole cells from mammary glands. Other concerns involve the ethics of using animals in this fashion.

Because both transgenic animal and plant production methods would require relatively small capital investments and minimal costs of production and maintenance, they may provide the only economically viable option for independent production of therapeutic proteins in developing countries.

Gene therapy. The ability to isolate and clone specific genes gives us the power to do far more than simply replace missing or dysfunctional proteins. Through gene therapy, we will be able to treat genetic disorders by giving patients functional genes in place of defective ones. Only certain inherited genetic diseases are amenable to correction via replacement gene therapy. Primary candidates for replacement gene therapy are hereditary diseases like adenosine deaminase (ADA) deficiency, or the "bubble boy" disease. This disorder results from a mutation in the gene that codes for the enzyme ADA. When ADA is not produced, intracellular toxins build up and kill cells of the immune system. Successful gene therapy experiments have been conducted on children who suffer from ADA deficiency (Figure 1.17) and cystic fibrosis, another inherited genetic disorder.

Gene therapy has proved to be simpler in theory than it is in practice. Our initial enthusiasm for the extraordinary promise of replacement gene therapy has

Figure 1.17   Both of these children were born with a rare genetic defect that results in a defective immune system. The boy on the left, the "bubble boy," was born prior to advances in biotechnology. (Photograph courtesy of James DeLeon, Jr., Children's Hospital, Texas Medical Center, Houston, Texas.) The girl on the right received gene therapy treatment, supplemented with ADA injections, and leads a relatively normal life. (Photograph courtesy of Jennifer Coate, March of Dimes.)

been tempered with a large dose of reality. Enormous technical barriers must be overcome before replacement gene therapy can live up to its potential to cure inherited genetic disorders: getting replacement genes into the appropriate cells, inserting them into the proper site within the genome of those cells, and getting them to function and respond to normal physiological signals.

Even though these technical impediments may have tempered our optimism for using replacement gene therapy for inherited genetic disorders, our enthusiasm for the potential of more transient gene therapy as a tool for treating other diseases has grown. Leading scientists are investigating the use of briefly introduced genes as therapeutics for a variety of cancers, autoimmune diseases, acquired immune deficiency syndrome (AIDS), hemophilia, and other diseases (Table 1.4).

Cell therapy. When gene therapy is not an option, we may be able to correct certain diseases by providing patients with healthy versions of malfunctioning

Table 1.4 Disorders for which gene therapy is being tested

Brain tumors
Hemophilia
Liver cancer
Prostate cancer
Hemoglobin disorders
Hyperlipidemia
Metabolic storage disorder
AIDS
Colon cancer
Melanoma
Solid tumors
Graft vs. host in bone marrow transplants for leukemia
Point mutations in bone marrow cells
Head and neck cancer
Asthma
Focal muscle atrophy
Graft vs. host disease
Breast and ovarian cancer
Cardiovascular disease
Non-small-cell lung cancer
Liver cancer
Hypercholesterolemia
Infectious diseases
Neurodegenerative diseases

cells. Cell therapy, like gene therapy, would come closer to curing the disease than do repeated injections of missing proteins. For example, cell therapy may be a more favorable treatment for diabetics than daily insulin injections. Diabetes is an autoimmune disease in which the body's immune system attacks the pancreatic cells that produce insulin, the **islet cells**. If we could provide diabetics with healthy islet cells, coated in a protective material to shield them from immune system attacks, there would be no need for insulin injections.

Researchers have achieved successful results with animals by giving them cardiac muscle cells to replace dead cells and neurons to correct neurological damage. Muscular dystrophy patients who received muscle cells have also shown some improvement.

Immunosuppressive therapies. Without a responsive immune system, we would face death as a real possibility every day. Sometimes, however, our vigilant immune system works against us. In organ transplant rejections and autoimmune diseases, suppressing our immune system would be in our best interest. Currently, we are testing the feasibility of using MCAb to accomplish this suppression. Here's how it works.

When exposed to foreign tissue in an organ transplant, the T cells of the immune system go to work to rid the body of this nonself component. By injecting the organ recipient with an MCAb that binds to a protein found on the surface of T cells, the T cells are selectively incapacitated. Patients injected with this MCAb show significantly less transplant rejection than do those given an immunosuppressive drug such as cyclosporin. Because immunosuppressive drugs suppress all immune function, they leave organ transplant patients vulnerable to infection. By using a specific MCAb in place of an immunosuppressive drug, we selectively knock out only one aspect of the immune response, the T cells, leaving other aspects intact. Once again, because of the extraordinary specificity of this technique, we can zero in on the problem and cause few, if any, side effects.

In autoimmune diseases, such as rheumatoid arthritis and multiple sclerosis, our immune system turns against our own tissues, leading to the progressive degeneration of those tissues. MCAb are proving to be a successful therapy for slowing the progression of autoimmune diseases.

Cancer therapies. In addition to chemotherapeutic agents derived from plants and MCAb that selectively deliver toxins to tumors, biotechnology research tools have permitted progress in treating cancer on a

variety of fronts (Figure 1.18). One is related to our new understanding of the genetic basis of cell growth and differentiation. When genes involved in certain critical events of cell growth and development mutate, they become **oncogenes,** or tumor-producing genes. MCAb are being used to bind to and inactivate the proteins produced by these genes. Antisense technology may also provide us with a way to block expression of these genes.

All of us normally have genes called tumor suppressor genes. When these genes function correctly, they suppress cell growth. When both copies of these genes become inactive or are absent, the genes then act as oncogenes. By introducing normal copies of these genes into tumor cells through gene therapy, the tumor may regress. (See chapter 33 for more information on the genetics of cancer.)

A second front for treating cancer involves stimulating the immune system of cancer patients through

Figure 1.18 Biotechnology provides a number of novel approaches for treating cancer. Physicians used gene therapy to treat a woman with melanoma by providing her with tumor-infiltrating lymphocytes genetically engineered to contain the gene for tumor necrosis factor. Before treatment, we see multiple live melanoma cells. The posttreatment biopsy shows massive necrosis of the tumor. (Photograph courtesy of Dr. Steven Rosenberg, National Cancer Institute.)

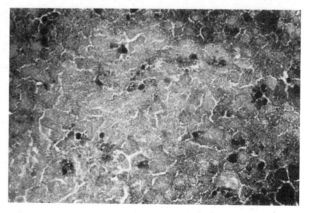

cancer vaccines. Unlike other vaccines, cancer vaccines are given *after* the patient has contracted the disease. Antigens from tumor cells stimulate a weak, ineffective attack by the immune system. Cancer vaccines help the immune system find and kill the tumor by intensifying the reactions between the immune system and the tumor antigen. Some cancer vaccines work by increasing the antigenicity of the tumor; others work by teaching the immune system to recognize the tumor as foreign.

Other candidate cancer therapies involve genetically engineering the tumor cells to make them more sensitive to drugs or to reduce their ability to evolve resistance to chemotherapeutic agents. Still other therapies use endogenous small proteins such as the interleukins or **tumor necrosis factor** to stimulate an immune system attack. Another interesting new approach is to use **endostatin** or other **antiangiogenesis** peptides to starve tumors by blocking the formation of blood vessels needed to nourish growing tumors.

Vaccines

There is much truth to the adage "an ounce of prevention is worth a pound of cure." The best way to battle diseases is not to develop new therapeutics but to prevent the diseases. Through biotechnology, we are developing better preventative agents by improving on a practice that has been with us since the 19th century: vaccination.

Vaccine design and production. The vaccines we have produced to prevent smallpox and other diseases (polio, diphtheria, tetanus, measles) are based on the use of either killed viruses or live but weakened viruses. When vaccinated with such a nonvirulent virus, your body produces antibodies to that organism, but you don't get the disease. If you are exposed to that virus again, your body has a ready supply of antibodies to defend itself. Vaccines are analogous to the "threat of war" that incites us to build up a supply of weapons. For the most part, vaccines cause no serious problems, but they do have side effects: allergic reactions, aches and pains, and fever. In a very few individuals, the vaccine has caused the disease it was intended to prevent.

A second problem with this method of vaccination is consistent production. Growing large amounts of some human pathogenic viruses outside of the human body is not easy. Viruses are quasi-living organisms; they need the biochemical machinery of a living cell to reproduce.

Finally, developing vaccinations for some deadly, infectious diseases, such as human immunodeficiency virus/AIDS and malaria, is problematic and risky. Because of this shortcoming, we have been unable to develop vaccines that successfully protect people from these diseases.

Using genetic engineering, scientists are improving existing vaccines and developing new ones. Usually, only one or a few proteins on the surface of the pathogen trigger the production of antibodies. By isolating the gene for the pathogen's cell surface protein(s) and inserting it into a yeast, or a bacterium such as ***Escherichia coli,*** bioprocess scientists can produce large quantities of this protein to serve as the vaccine. Because no live animals are needed for vaccine production, there are virtually no limitations on the amount of vaccine that can be produced. When the protein is injected, the body produces antibodies that can recognize the pathogen but suffers none of the adverse side effects that sometimes go hand in hand with vaccination. Using these new techniques of biotechnology, scientists have developed vaccines against the life-threatening diseases hepatitis B and meningitis.

Much to the surprise of many scientists, injecting naked DNA into muscles or skin cells also elicits an immune response. Researchers had assumed that DNA alone would not trigger an immune response of sufficient strength to impart protection against infectious diseases. However, in early gene therapy trials, the immune response against the therapeutic protein was too strong to make the gene therapy effective. While this result was disappointing to the gene therapy researchers, other scientists saw the positive side. The exciting discovery of **DNA vaccines** could lead to more advances in vaccine production, improved vaccines with fewer side effects, and more organisms for which we can develop effective vaccines.

How do DNA vaccines work? Researchers insert genes for one or more of the pathogen's proteins into a small, circular, noninfectious piece of DNA called a **plasmid**. After introduction into the host, the host cells synthesize the pathogen's protein(s) (Figure 1.19). Recognizing the protein as foreign, the immune system produces both antibodies and T cells specific for that antigen. DNA vaccines against AIDS, malaria, herpes, hepatitis B, and influenza are currently in clinical trials.

Vaccine delivery systems. Whether the vaccine we are developing is a live virus, coat protein, or a piece of its DNA, the production of vaccines requires very elaborate and costly facilities and procedures, both of which are heavily regulated by the federal government. And then there's the issue of painful injections. How might the tools of biotechnology help us lessen these problems?

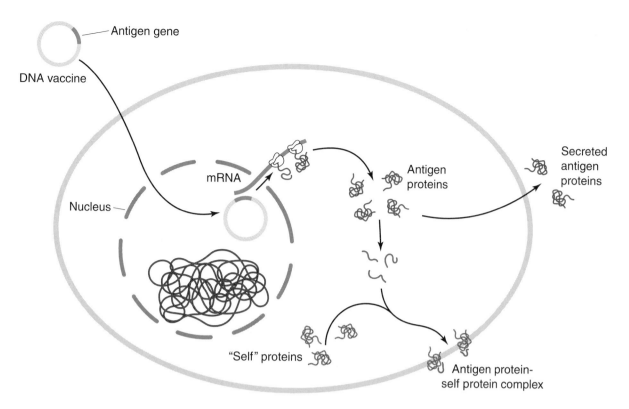

Antigen gene

DNA vaccine

Nucleus

mRNA

Antigen proteins

Secreted antigen proteins

"Self" proteins

Antigen protein-self protein complex

Figure 1.19 Plasmids altered to carry a gene for a protein (antigen) produced by a pathogen are injected into muscle cells. The antigen-coding gene is transcribed into messenger RNA (mRNA) in the nucleus. The mRNA moves from the muscle cell nucleus to the ribosomes, where the antigenic protein is produced. Some copies of the antigen are excreted from the cell into the blood. Other copies are chopped into smaller pieces to enable them to fit into grooves on other proteins, unique to each individual, that are synthesized by the muscle cell. These "self" proteins bind to the pieces of antigen and move to the cell surface. In response, the immune system synthesizes T cells that will recognize the pathogen's antigen in the future.

Earlier we mentioned the prospect of using genetically engineered plants and animals for the production of therapeutic proteins. Transgenic organisms could also produce antigens for vaccines made of a pathogen's cell surface proteins. Industrial and academic researchers are using biotechnology to develop edible vaccines. A company has genetically engineered goats to produce a malaria antigen in milk. University researchers have obtained positive results using human volunteers who consumed hepatitis vaccines in bananas and *E. coli* and cholera vaccines in potatoes. If edible vaccines become a reality, being vaccinated by drinking a glass of milk and eating a banana is much more appealing than a shot in the arm, isn't it? In addition, because these vaccines are genetically incorporated into food plants and need no refrigeration, sterilization equipment, or needles, they may be especially important for developing countries.

Medical Research Tools

One way that biotechnology will help us prevent diseases is less obvious than the ready example of vacci-

nations. For medical researchers, some of the most important outcomes of advances in biotechnology are not commercial products but the powerful research tools biotechnology provides.

The technologies of biotechnology discussed earlier are not simply techniques for producing vaccines, pharmaceuticals, and diagnostic kits. As research tools, these technologies have given us a much more valuable product: a deeper understanding of biological systems. They have provided us with new answers to old questions and have prompted us to ask questions we never would have dreamed of asking. Learning more about healthy biological processes and what goes wrong when they fail will enable us to develop even better diagnostics, therapeutics, and preventative agents.

One of the most powerful research tools provided by advances in biotechnology is targeted mutations, or **gene knockouts**. By disrupting specific genes, we gain valuable information about the role that the gene plays in healthy individuals. For example, we have used knockout mice that lack normal copies of DNA repair enzymes to investigate the role these enzymes may play in cancer induction.

Agricultural biotechnology

The opportunities biotechnology will create for agriculture are as impressive and extensive as those for human health. We will witness progress in improving the quality, nutritional value, and yields of our agricultural products and decreasing production costs. A related aspect of agriculture that will benefit from biotechnology is food processing.

Plant Agriculture

Because plants are genetically complex, plant agricultural biotechnology lagged behind medical advances in biotechnology. Another, perhaps equally important factor that may explain the different rates of progress is that over the years, animal research has received much more federal funding than plant research.

Nonetheless, we have made remarkable progress in plant biotechnology, largely because of improvements in two fundamental techniques of biotechnology: genetic engineering and plant cell and tissue culture. In the early 1980s, around the time we were already using genetically engineered bacteria to produce human insulin, we discovered a way to genetically engineer plants by using recombinant methods. To insert novel genes into plants, scientists exploited a genetic engineer that occurs in nature, **Agrobacterium tumefaciens**. A common soil bacterium, *A. tumefaciens* infects plants by injecting a portion of its DNA into plant cells (Figure 1.20). Since the mid-

1980s, plant geneticists have used *A. tumefaciens* to genetically engineer many important agricultural crops (see chapter 26) and have developed a number of alternative methods for adding new genes to plants.

The traits agricultural scientists are incorporating into our crops through genetic engineering are the same traits they have incorporated into crops through selective breeding: improved nutritional content; delayed ripening; resistance to diseases caused by bacteria, fungi, and viruses; better taste; the ability to withstand harsh environmental conditions such as freezes and droughts; and resistance to pests such as insects, weeds, and nematodes.

Crop production and protection. The first generation of genetically engineered crops to come to market was engineered for better pest management capabilities and included a number of insect-resistant, herbicide-tolerant, and disease-resistant crop varieties. Since they first became available to farmers in 1996, the number of acres devoted to these crops has increased markedly every year. In fact, the rate of adoption of this new farming technology is unprecedented in American agriculture. The limiting factor in determining the number of acres planted in genetically engineered crops has been seed availability and not market demand.

According to the U.S. Department of Agriculture (USDA), the number of acres planted in genetically engineered crops in the United States has increased from 8 million acres in 1996 to more than 50 million acres in 1998. For the 1998 growing season, approximately 25% of the corn, 40% of the soybeans, and 45% of the cotton grown in the United States were genetically engineered varieties. This rapid rate of increase is not unique to U.S. agriculture. In its 1999 annual report on the global use of transgenic crops, the nonprofit organization International Service for the Acquisition of Agri-Biotech Applications (ISAAA) (www.isaaa.org) notes that "between 1996 and 1999, twelve countries, 8 industrialized and 4 developing, have contributed to more than a twenty fold increase in the global area of transgenic crops" (Table 1.5).

Why has the rate of adoption of this new farming technology increased so dramatically, not only in the United States but also globally? According to ISAAA's 1999 re-

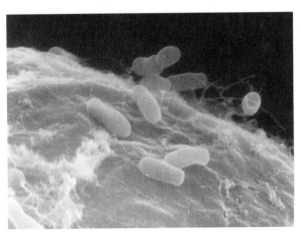

Figure 1.20 Scanning electron micrograph of *Agrobacterium tumefaciens* infecting plant cell. (Photograph courtesy of Ann Matthysse, Biology Department, University of North Carolina, Chapel Hill, N.C.)

Table 1.5 Global area of transgenic crops in millions of hectares

1996	1997	1998	1999
1.7	11.0	27.8	39.9

port: "High adoption rates reflect grower satisfaction with the products that offer significant benefits ranging from more convenient and flexible crop management, higher productivity or net returns/hectare and a safer environment through decreased use of conventional pesticides, which collectively contribute to a more sustainable agriculture." A survey of farmers conducted by the Economic Research Service (ERS) of the USDA reveals that U.S. farmers adopted genetically engineered crops hoping "to increase yields through improved pest control" and "to decrease pesticide costs." In most cases, the crop yields have increased and/or the use of pesticides and herbicides has decreased, but, as expected, results vary with the year, the region of the country, and the crop.

According to the USDA-ERS 1999 study (www.econ.ag.gov), the use of insect-resistant (genetically engineered to contain the Bt gene) cotton and corn is associated with significantly higher yields in most years for some regions and also with fewer treatments of insecticides aimed at killing those pests susceptible to Bt. Using Bt crops has no effect on the use of insecticides to treat other insect pests. Herbicide tolerance in soybeans is associated with significantly higher yields in some regions. Herbicide-tolerant technology has significantly reduced herbicide treatments for soybeans and, to a lesser extent, for cotton. (For information on how the use of herbicide-tolerant crops can decrease the amount of herbicide applied to a crop, see chapter 36.)

However, genetically engineered seeds cost more than non-genetically engineered varieties, and some companies also require farmers to pay a technology user's fee. As a result, weed and insect problems must be serious enough to compensate for this increased cost. This threshold level of infestation must be met before farmers can obtain the economic benefits of reduced pesticide and herbicide inputs and higher yields. According to historical data collected by the USDA on pest infestations, in 7 out of 10 years, pest levels should be high enough for farmers to benefit economically from using genetically engineered insect-resistant seeds.

Using biological methods to protect crops. Just as biotechnology is allowing us to make better use of the natural therapeutic compounds our bodies produce, it is also providing us with more opportunities to work with nature in plant agriculture.

We have discovered that plants, like animals, have endogenous defense systems, the **hypersensitive response (HR)** and **systemic acquired resistance (SAR)**. We are searching for environmentally benign chemicals that we can use to trigger these two means

of defense so that plants can better protect themselves against insects and diseases.

To deter crop pests, we will also be able to rely more heavily on biological methods of pest control. Biological control, or biocontrol as it is often called, is the suppression of pests and diseases through the use of biological agents. For example, a virus may be used to control an insect pest, or a fungus may deter the growth of a weed (Figure 1.21).

A truly extraordinary variety of alternatives to the chemical control of insects is available. All have this in common: they are biological solutions, based on understanding of the living organisms they seek to control.
Rachel Carson, Silent Spring, 1962

Biological control has been used in agricultural systems in the United States since the 1800s, when we began using bacteria and a type of virus called a **baculovirus** to control pests, but the potential for biocontrol has been limited by the constraints that exist in nature. It is no simple task to find a specific microbe that kills a specific insect pest but not others, a certain fungus that causes diseases in weeds but not in crop plants, or an insect predator that preys only on "bad" and not "good" insects.

On the other hand, sometimes the unique characteristics of an organism or the ecological relationship between an organism and a crop would make the organism an excellent biocontrol agent, but it lacks the "ammunition." A perfect example of this situation is

Figure 1.21 Parasitic wasps are effective biocontrol agents. Female wasps deposit eggs inside caterpillars where larvae develop and feed. After larvae emerge, they form the white, ovoid pupal cases attached to the caterpillar's body. (Photograph courtesy of VU/©W. Mike Howell.)

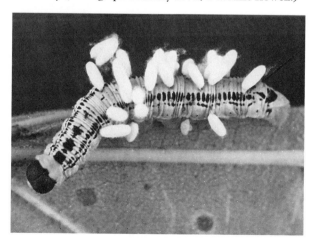

provided by the bacterium that lives in the stalks of corn plants and was discussed earlier under genetic engineering and Bt.

The flexibility that comes from genetic engineering removes those constraints. If we find a gene for insect resistance in an oak tree, we could incorporate it into our crops. If a very rare mold that occurs only in showers in homes in Australia has a gene that codes for a protein that can kill weeds, we can isolate that gene, engineer it into bacteria, and have the bacteria produce the herbicidal protein.

Exploiting cooperative relationships in nature.
In addition to capitalizing on nature's negative inter-actions—predation and parasitism—to control pests, we might also use existing positive relationships that are important to plant growth. One example is the symbiosis between plants in the bean family and cer-tain nitrogen-fixing bacteria (Figure 1.22). By provid-ing the crop plant with a usable form of nitrogen, the bacteria encourage plant growth. Scientists are work-ing to understand the genetic basis of this symbiotic relationship so that we can give nitrogen-fixing capa-bilities to crops other than legumes.

Figure 1.22   (A) Alfalfa nitrogen-fixing nodules (photograph courtesy of VU/©C. P. Vance); (B) alfalfa nodule cells containing actively nitrogen-fixing bacteroids. ×640. (Photograph courtesy of VU/©C. P. Vance.)

A

B

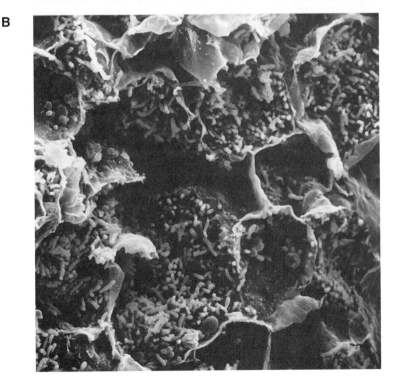

A second and less well-known example of a positive symbiotic relationship we might exploit for increasing crop production involves fungi and plant roots and is known as a **mycorrhiza.** In mycorrhizal associations, fungi extract nutrients from the soil and make them available to plants.

At this time, plant genetic engineers (or any genetic engineers, for that matter) are limited to one- or two-gene traits, those controlled by one or two genes. Most of the agronomically valuable traits are **multi-genic;** that is, they are controlled by many genes. One example is nitrogen fixation, which is controlled by at least 15 different genes. So while the idea of having our major crops obtain useful nitrogen from the air via a bacterium that lives on its roots is an appealing one, it is unrealistic in the near future.

Nutritional value of crops. The first generation of genetically engineered crops primarily benefited farmers. Although there are benefits to consumers in growing these crops, the benefits are largely invisible. For example, studies have shown that because Bt corn sustains relatively little insect damage, it is also infected significantly less often by fungi and molds that produce toxins that are fatal to livestock and harmful to humans.

The benefits of the next wave of agricultural biotechnology products to hit the market will be more obvious to consumers. Some of those benefits will involve improvements in food quality and safety, both of which we discuss later in this section. Biotechnology will also provide consumers with plant products that are designed specifically to be healthier and more nutritious.

A variety of healthier cooking oils derived from biotechnology are currently on the market. Using genetic engineering, plant scientists have decreased the concentration of saturated fatty acids in certain vegetable oils. They have also increased the conversion of linoleic acid to alpha linoleic acid in these oils. Not only does this conversion decrease the saturated fatty acid concentration, but it also increases the type of oil, found mainly in fish, that is associated with lowering cholesterol levels.

Another nutritional concern related to edible oils is the negative health effects produced when vegetable oils are hydrogenated to increase their heat stability for cooking or to solidify oils used in making margarine. We have given soybean oil these same properties, not through hydrogenation but by using genetic engineering to increase the amount of the naturally occurring fatty acid stearic acid.

Future nutritionally beneficial products are grains engineered to have a complete protein profile; fruits, vegetables, and grains with improved vitamin and mineral content; and potatoes with higher solid content, to reduce the amount of oil absorbed when fried.

Biotechnology also promises to improve the health benefits of **functional foods**. Functional foods are foods containing significant levels of biologically active components that impart health benefits or desirable physiological effects beyond our basic need for sufficient calories, the essential amino acids, vitamins, and minerals. Familiar examples of functional foods include compounds in garlic and onions that lower cholesterol and improve the immune response, antioxidants found in green tea, and the glucosinolates in broccoli and cabbage that stimulate anticancer enzymes. We are using biotechnology to increase the production of these types of compounds.

Animal Agriculture

The field of animal agriculture will continue to progress as biotechnology provides new ways to improve animal health and increase productivity. As is true of human health, improvements in animal health have come from advances in diagnostics, therapeutics, and vaccines. According to the U.S. Congress Office of Technology Assessment, animal diseases cost U.S. agriculture over $17 billion each year, so progress in detecting, curing, and preventing animal diseases will have considerable economic impact.

Animal health. Farmers are using MCAb on their livestock for early and more accurate diagnosis of certain diseases. The same principles discussed under human health apply here: because of their specificity, these diagnostics are faster, more accurate, and more sensitive than traditional diagnostics. Many of these diagnostics are also portable and allow veterinarians to diagnose livestock diseases on site. Diagnosing a disease sooner and with greater accuracy means that appropriate therapy can be chosen and started sooner, thus decreasing the spread of the disease. Brucellosis, pseudorabies, scours, foot-and-mouth disease, and trichinosis are a few of the economically important infectious diseases we can now diagnose thanks to advances in biotechnology.

Another parallel exists in the type of therapies biotechnology is providing for both humans and animals. Genetic engineering is being used to produce sufficient quantities of endogenous therapeutic proteins found in animals. Cattle naturally make **interferon** and interleukin-2, proteins produced by the immune system to fight viruses. The genes for these proteins have been cloned and placed into bacteria for mass production. Injections of interferon and interleukin-2 can help decrease the incidence of shipping fever in cattle, a disease that costs the beef industry more than $250 million a year.

Great progress has also been made in disease prevention. Techniques being used to improve human vaccines are finding their way into veterinary medicine as well. Vaccines are now available for a number of diseases and parasites, including pseudorabies, foot-and-mouth disease, coccidiosis in poultry, tapeworms, ticks, and tick-borne diseases. Scientists are also exploring edible vaccines for livestock.

A second route to disease prevention is to produce livestock that are resistant to diseases, just as we have created disease-resistant plants. Some breeds are naturally resistant to some bacterial diseases, such as mastitis, and this resistance has a genetic basis. If only one or a few genes are responsible for disease resistance, creating transgenic animals resistant to bacterial diseases may be possible.

Increasing productivity. Farmers are always interested in improving the productivity of agricultural animals. Their goal is to obtain the same output (milk, eggs, meat, wool) with less input (food), or increased output with the same input. Increasing muscle mass and decreasing fat in cattle and pigs has long been a goal of livestock breeders. These goals can be reached sooner by using genetic engineering to create transgenic animals with these traits. A number of industrial and academic researchers are investigating genetic engineering as a method for increasing the muscle mass in livestock (Figure 1.23).

Figure 1.23 Researchers at Johns Hopkins University created a strain of mice that is twice as muscular as its parent strain by deleting a naturally occurring gene that encodes a growth factor that limits skeletal muscle growth. Similar genetic engineering of livestock could significantly increase meat production. (Photograph courtesy of Se-Jin Lee, Johns Hopkins University.)

Another method we are using to increase the productivity of our livestock is a variation on the theme of selective breeding. We first choose those individuals that possess desirable traits. Then, instead of breeding the animals, we collect their gametes (eggs and sperm) and allow fertilization to occur in a laboratory dish. This in vitro fertilization is followed by embryo culture, a form of mammalian cell culture in which the fertilized egg develops into an embryo. At an early stage in development, the embryo is taken from the laboratory dish and implanted into a female of the same species but not necessarily of the same breed. This is known as **embryo transplant**. Using this method, a farmer can improve the genetic makeup of the herd more quickly than by simply relying on a single female who produces one calf per year.

Scientists have also tested ways to increase production efficiency by applying the same principle to some very different animals: cows, pigs, and fish. The principle involves increasing the level of endogenous growth hormone found in these animals. Protein synthesis in all animals is dependent upon growth hormone. As a result, growth hormone in cows, which is also known as **bovine somatotropin (BST)**, is important in growth and milk production. By increasing a cow's blood level of growth hormone, farmers can increase milk output or growth rate through stimulating protein synthesis.

One of the first products of biotechnology that became available to farmers was BST, which is also known as bovine growth hormone (BGH). Scientists isolated the gene for BST and inserted it into bacteria. Companies use these genetically engineered bacteria to produce BST. The BST is extracted from the fermentation tanks, purified, and injected into dairy cows, causing a 10 to 20% increase in milk production.

The safety of BST has been studied extensively by regulatory agencies in the United States, Canada, Europe, and many other parts of the world. Although all of these agencies have determined that BST in not active in humans and that meat and milk taken from cows injected with BST contain the same amount of BST as the milk of uninjected cows, some public concern about the safety of BST still exists. As a result, in some grocery stores you may see milk labeled "rBGH free" ("r" for recombinant).

On the other hand, the product **porcine somatotropin (PST)** has applications that are more attractive to many consumers. Injections of PST, the growth hormone in pigs that is the functional equivalent of BST, cause a dramatic increase in protein and a decrease in the amount of pork fat (Figure 1.24). Finally,

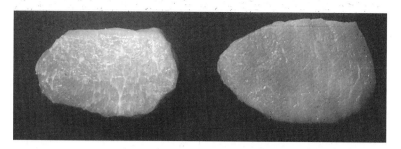

Figure 1.24 Because porcine somatropin (PST) decreases fat deposition in pigs, consumers will have access to leaner cuts of pork. (Left) Boneless pork chop control; (right) boneless pork chop from pig treated with PST. (Photograph courtesy of VU/©Ken Prusa.)

scientists have introduced into salmon a gene that codes for the production of fish growth hormone to boost the natural production of growth hormone. Scientists are hopeful that farm-raised, transgenic fish will reach a marketable size sooner than fish with lower levels of growth hormone.

Food Processing

An aspect of agriculture that people often forget is food processing: what happens to the food when it leaves the farm. The food processing industry will be affected by a variety of developments in biotechnology. Some applications will involve the food product itself, but just as many, if not more, will be directed toward improving food additives.

Traditionally, microbes have been essential to the food processing industry not only for the role they play in the production of fermented foods but also as a rich source of food additives and enzymes used in food processing (Table 1.6). Through advances in genetic engineering and bioprocessing technology, their importance to the food industry will only increase in the future.

Product quality. In plant agriculture we are altering crops to have more desirable processing qualities. Many companies that produce soup, ketchup, and tomato paste now use tomatoes that were derived from a biotechnology technique, **somaclonal variant selection**. The new tomatoes contain 30% less water and are processed with greater efficiency. A 0.5% increase in the solid content is worth $35 million to the U.S. processed-tomato industry.

The first product of plant genetic engineering approved by the U.S. Food and Drug Administration was a tomato that is allowed to ripen on the vine instead of being picked while it is green. Calgene, the company that commercialized the Flavr Savr tomato, de-

veloped this tomato by biochemically separating the ripening process from the spoiling process. These processes usually go hand in hand, but in fact, some different enzymes and therefore different genes are involved. Using antisense technology, Calgene blocked a gene that codes for one of the enzymes involved in spoiling while leaving the ripening enzymatic pathways intact.

Table 1.6 Microbial fermentation is essential to the production of foods, food additives, and enzymes used in food processing

Foods	
Cheese	Vinegar
Yogurt	Bologna
Buttermilk	Salami
Sour cream	Tofu
Soy sauce, tamari	Miso
Bread and other baked goods	Sauerkraut
Wine, beer	Pickles
Cider	Olives

Food additives
  Acidulants (citric and lactic acids)
  Amino acids (glutamine, lysine)
  Vitamins (beta-carotene, riboflavin)
  Flavor enhancers (monosodium glutamate)
  Thickeners (xanthan gum)
  Stabilizers (dextran)
  Flavors (methyl salicylate [wintergreen], benzaldehyde [almond])

Enzymes
  Proteinase (Gouda and Edam cheeses)
  Peptidase (cheddar cheese)
  Lipase (blue cheese)
  Amylase (high-fructose corn syrup)
  Invertase (soft-centered candies)
  Pectinase (fruit juices)

Food additives are substances used to increase nutritional value, retard spoilage, change consistency, and enhance flavor. The compounds food processors use as food additives are substances nature has provided and are usually of plant or microbial origin, such as xanthan gum and guar gum which are produced by microbes.

Through genetic engineering, food processors will be able to produce many compounds that could serve as food additives but that now occur in scant supply or are found in microbes that are difficult to maintain in fermentation systems. We will also be able to capitalize on the extraordinary diversity of the microbial world and obtain new enzymes that will prove important in food processing.

Food safety. The most important advance in food processing will be in food safety. Through genetic engineering, we will be able to decrease natural plant toxins and food allergens. In addition, MCAb, biosensors, and DNA probes are being developed that will be used to determine the presence of harmful bacteria, such as *Salmonella* species, *Clostridium botulinum*, and *E. coli* 0157:H7, the strain of *E. coli* that was responsible for a number of deaths in recent years. Once again, these tests will be quicker and more sensitive to low levels of contamination than were previous tests because of the increased specificity of molecular techniques. Biotechnology-based diagnostics have also been developed that will allow us to detect toxins produced by fungi that grow on crops and to determine whether food products have inadvertently been contaminated with peanuts, a potent allergen for a significant portion of the population.

Environmental biotechnology

Few people doubt that we are in the midst of an environmental crisis. The air, soil, and water of the earth are contaminated as a result of human activities. We are depleting the earth's nonrenewable resources and generating massive amounts of wastes that don't degrade readily. Many people hope that biotechnology will help solve some of our environmental problems.

Most people believe that in addition to solving environmental problems, biotechnology will cause fewer environmental problems than previous physical or chemical technologies. This optimism is based on the use of biological, renewable resources in place of chemical, nonrenewable ones and the greater specificity, precision, and predictability that characterize biologically based technologies. Because we will be working with the biology of organisms in more specific and targeted ways, we should be able to develop technological solutions that generate fewer side effects and have fewer or less severe unintended consequences.

In addition, just as biotechnology is providing us with new tools for diagnosing health problems and detecting harmful contaminants in food, it is yielding new methods of monitoring environmental conditions and detecting environmental pollutants.

Cleaning Up Pollution through Bioremediation

Using biotechnology to treat pollution problems is not a new idea. Communities have depended on complex populations of naturally occurring microbes for sewage treatment for many years. Microbes help purify water by breaking down solid organic wastes before the water is recycled. But solid organic wastes are not the only type of pollutants that need to be removed from our water supplies. More and more, we are finding that our water is contaminated by chemical pollutants; more and more, we are turning to microbes for help in removing pollutants both from water and from soil (Figure 1.25).

Why microbes? Over the billions of years they have been on earth, microbial populations have adapted to every imaginable environment. No matter how harsh the habitat, some microbe has found a way to make a living there. We have found microbes in hot springs that are the source of the geysers in Yellowstone, thousands of feet underwater in hydrothermal vents, in salty seas and lakes, and living off inorganic materials such as copper sulfide in copper mines. Microbes that are able to live in such extreme environments are called **extremophiles** (Figure 1.26).

Life in unusual habitats makes for unique biochemical machinery, so the range of compounds microbes can degrade is enormous. We are capitalizing on this biochemical diversity to repair a number of pollution problems. We are using a fungus to clean up a noxious substance discharged by the papermaking industry. Indigenous microbes that contribute to the natural cycling of metals, such as mercury, offer prospects for removing heavy metals from water. Other naturally occurring microbes that live on toxic waste dumps are degrading wastes such as polychlorobiphenyls to harmless compounds.

Recently acquired information from microbial genomic studies is helping us better utilize the wealth of genetic diversity in microbial populations. We are using DNA probes to fish, on a molecular level, for microbes with specific capabilities. Once caught, such organisms can be identified and cultured for their ability to carry out particular reactions.

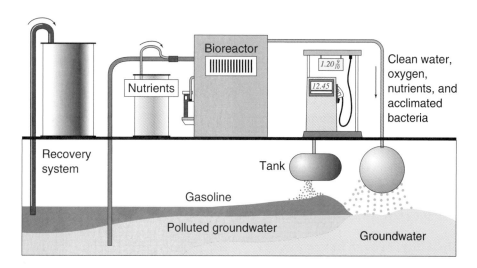

Figure 1.25 Gasoline from an underground storage tank seeps through the soil to the water table. After the leak is stopped, the free-floating gasoline is pumped out to a recovery tank, and polluted groundwater is pumped into a bioreactor with oxygen, nutrients, and hungry microbes. After the microbes eat the gasoline, the mixture of clean water, nutrients, and microbes is pumped back into the ground so that more of the pollutant can be degraded. (From *Carolina Genes,* 1994, North Carolina Biotechnology Center Education and Training Program, Research Triangle Park, N.C.)

Not all pollution problems that threaten human health result from human activities. During the hot summer months in Australia, aquatic cyanobacterial populations increase rapidly and secrete carcinogenic chemical compounds into the water supply. To counter these population blooms, copper sulfate, which kills the cyanobacteria but doesn't break down the toxins, is dumped into the water. As you might expect, microbes that coexist with the cyanobacteria have evolved a better solution to the problem. Recently, scientists discovered a bacterium that produces three enzymes that sequentially break down the cyanobacterial toxins until they are harmless.

Figure 1.26 Microbes living in extreme environments, such as hot pools in Yellowstone National Park, will provide industrial manufacturing companies with enzymes that retain their function under extreme manufacturing conditions. (Photograph courtesy of VU/©Hal Beral.)

Conveniently for scientists, the three genes that code for the enzymes are clustered together in a single functional unit, an **operon**.

In these examples, naturally occurring organisms are performing the cleaning duties. Using only naturally occurring microbes can be somewhat limiting, however. Microbes that degrade hazardous wastes in clay soils of North Carolina may not work in the silty soils of the Mississippi delta. Genetic engineering gives us the flexibility we need to maximize our use of the biochemical capabilities of microbes and at the same time circumvent problems such as habitat specificity.

If useful enzymes are discovered in microbes that can survive only in certain habitats, the gene that codes for that enzyme may be movable into microbes known to prosper in other habitats. For example, we have discovered a number of species of aquatic microbes that degrade some of the 200 naturally occurring halogenated hydrocarbons found in the ocean. Some of our primary soil pollutants belong to this class of chemical compounds. If we isolate the genes that code for the enzymes that degrade these compounds, we could insert these genes into soil microorganisms. These genetically engineered microorganisms might then be able to clean up hazardous waste sites.

William Reilly, the former head of the Environmental Protection Agency (EPA), the World Wildlife Fund, and The Conservation Foundation, is enthusiastic about the extraordinary potential of bioremediation. Not only is it a natural process, it is also inexpensive and effective and requires little energy input. After the *Exxon Valdez* oil spill, Reilly created a new office at EPA to investigate and develop biologically based solutions to some of our environmental problems.

Preventing Environmental Problems

If an ounce of prevention is worth a pound of cure in human health, surely the same can be said for environmental health. Rather than cleaning up environmental problems, wouldn't it be better if we did not create them? Biotechnology is opening up a number of avenues for preventing environmental problems.

Industrial sustainability through biotechnology. **Industrial sustainability** includes processes and products that meet the current consumer demand for products without compromising the resources and energy supply of future generations. How can industrial manufacturing achieve sustainability? The key words are "clean" and "efficient." Any change in production processes, practices, or products that makes production cleaner and more efficient per unit of production or consumption is a move toward sustainability. In practical terms, industrial sustainability means utilizing technologies and know-how to

- reduce material and energy inputs while maximizing renewable resources as inputs
- minimize the generation of harmful pollutants or waste during product manufacture and use
- produce recyclable or biodegradable products

Companies in a variety of industry sectors are practicing proactive environmental protection through industrial sustainability, and biotechnology is helping them achieve that objective.

Material and energy inputs. Manufacturing processes have long relied on petroleum, a nonrenewable resource and major contributor to pollution and solid waste generation, as a source of materials and energy. We are using biotechnology to remove sulfur from fossil fuels, which significantly decreases their polluting power. More important, biotechnology will increase the use of renewable resources for feedstock chemicals and energy. The environmental advantages of using biomass for feedstocks and energy are that production will be cleaner, in most cases, than petroleum-based production and will generate less waste (Figure 1.27). When the biomass source is agricultural refuse, our gains will double: we will enjoy all the advantages of renewable resources while reducing wastes generated from another human endeavor, agriculture. A final advantage of using plant biomass as feedstock is that as our crop

Figure 1.27 To decrease pollution problems, scientists are developing diesel fuels and lubricants based on vegetable oils. Seventy to 80% of the vegetable-based products biodegrade to harmless products, compared with 20 to 40% of conventional, petroleum-based fuels and lubricants. (Photograph courtesy of West Central Cooperative, Ralston, Iowa.)

of feedstock grows, it consumes CO_2, one of the greenhouse gases.

Today, at least 5 billion kilograms of commodity chemicals that use plant biomass as the primary feedstock are produced annually in the United States. Government laboratories have devoted significant resources to research on genetic engineering and bioprocess engineering to improve the economic feasibility of biomass-derived energy.

Process applications of biotechnology. Biotechnology's versatility as a tool for industrial sustainability is most apparent in the improved manufacturing processes being implemented by a wide variety of industries. The chemical, textiles, pharmaceutical, pulp and paper, food and feed, metal and minerals, and energy industries have all benefited from cleaner, more energy-efficient production made possible by incorporating biotechnology into their production processes. Almost always, these improvements are due to biocatalysts, which are living organisms and/or their enzymes. Because biocatalysts are more specific than chemical catalysts, they produce fewer unwanted by-products. In addition, the environment benefits from industry's use of biocatalysts because they are water soluble and catalyze reactions most efficiently at low temperatures, when compared with nonbiological catalysts.

Many of these advantages can become limitations in certain industrial processes, however. Biocatalysts function in a narrow temperature range, and most fall apart at temperatures above 100°F. Their relatively low productivity is made even less acceptable by their relatively high cost. Because of the many environmental advantages offered by biocatalysts, however, industrial scientists are using biotechnology to circumvent these limitations.

To decrease the costs of biocatalysts, scientists are using genetic engineering to increase enzyme production in microorganisms that are commonly used in production or to give these cheap and easy to grow microbes the ability to make enzymes usually made by microbes that are too expensive to maintain or impossible to culture. In addition, protein engineering and directed protein evolution have been used to modify the substrate specificity of an enzyme, improve its catalytic properties, or broaden the reaction conditions under which it can function so that it is more compatible with existing industrial processes. Scientists also hope to discover novel biocatalysts that will function optimally at the relatively high levels of acidity, salinity, temperature, or pressures found in some industrial manufacturing processes.

Biodegradable products. Biotechnology also offers us the prospect of replacing petroleum-derived polymers with biological polymers. Cotton that is genetically engineered to contain a bacterial gene produces a polyesterlike substance that is biodegradable and has the texture of cotton but is warmer. Other biopolymers with the potential to replace synthetic fabrics and fibers are under development in Japan and the United States. Industrial scientists have genetically engineered plants to produce **polyhydroxybutyrate (PHB),** a naturally occurring bacterial polyester that could be used as a feedstock chemical for manufacturing biodegradable plastics. Finally, genetic engineering provides us with the opportunity to produce abundant amounts of natural protein polymers, such as spider silk and adhesives from barnacles, through microbial fermentation processes.

In summary, no matter what stage of industrial production you choose—inputs, manufacturing process, or final product—biotechnology is providing industry with tools, techniques, and know-how to move beyond command-and-control regulatory compliance to proactive pollution prevention and resource conservation strategies that are characteristic of industrial sustainability.

Monitoring the Environment

The techniques of biotechnology are providing us with novel methods for diagnosing environmental problems and assessing normal environmental conditions so that we can be more informed environmental stewards in the future.

Companies have developed methods for detecting harmful organic pollutants in the soil by using MCAb (Figure 1.28) and PCR, while scientists in government laboratories have produced antibody-based biosensors that detect explosives at old munitions sites. Not only are these methods cheaper and faster than the current laboratory methods, which require large and expensive instruments, but they are also portable. Rather than gathering soil samples and sending them to a laboratory for analysis, scientists can measure the level of contamination on site and know the results immediately.

The remarkable ability of microbes to break down chemicals is proving useful not only in pollution remediation but also in pollutant detection. A group of scientists at Los Alamos National Laboratory work with bacteria that metabolize a class of organic chemicals called phenols, many of which are considered pollutants by the EPA. When the bacteria ingest phenolic compounds, the phenols attach to a receptor. The phenol-receptor complex then binds to DNA, triggering the activation of the genes involved in phenol metabolism. The Los Alamos scientists added to the bacteria a reporter gene that, when activated by a phenol-receptor complex, produces an easily detectable protein, thus indicating the presence of phenolic compounds in the environment. A variation on this theme involves linking fluorescent reporter genes to operons involved in microbial breakdown of environmental pollutants.

Figure 1.28 On-site monitoring of environmental pollutants made possible through advances in biotechnology. (Photograph courtesy of EnSys, Inc.)

1. An Overview of Biotechnology • 31

Laying the Foundation

Summary

Biotechnology is a set of very flexible and powerful tools that offer great potential for improving human health, increasing the quality and yield of our agricultural products, and improving our relationship with the environment. Some of the most important biotechnologies include monoclonal antibody technology, cell culture, genetic engineering, bioprocessing technology, and protein engineering. The common thread that joins these technologies is that they are based on the use of cells and biological molecules.

An essential advantage of biotechnology over other technologies is that it is based on biology. It can work with the biology of organisms in very specific, predictable ways to solve biological problems or make products. Of all the technologies developed so far, biotechnology has the potential to be more compatible with sustainable life on this planet.

Because biotechnology is in the earliest stages of development, we are at a critical stage in deciding how best to utilize its power. We can use the tools of biotechnology to answer scientific questions, make new products, solve problems, and achieve goals deemed desirable by society. The issue before all of us is determining which questions, products, and problems are our highest priorities and deciding what sort of society we want in the new millennium. In the next chapter we discuss some of the societal issues raised by technology in general and biotechnology in particular.

Selected Readings

Abelson, Philip H., and Pamela Hines. 1999. The plant revolution. *Science* 285:367–389. This issue of *Science* contains a number of excellent articles on plant agricultural biotechnology.

Anderson, W. F. 1995. Gene therapy. *Scientific American* 273:124–128. In 1997 *Scientific American* (276 [5]) also produced a special edition on gene therapy.

Bains, William B. 1993. *Biotechnology from A to Z*. Oxford University Press, Oxford, England. A concise reference for some of the basic terms used in biotechnology and molecular biology.

Bayley, H. 1997. Building doors into cells. *Scientific American* 277:62. A number of interesting protein engineering examples.

Benton, David. 1996. Bioinformatics—principles and potential of a new multidisciplinary tool. *Trends in Biotechnology* 14:261–272.

Cohen, E. P., E. F. de Zoeten, and Morton Schatzman. 1999. DNA vaccines as cancer treatment. *American Scientist* 87:328–334.

Glick, Bernard R., and Jack J. Pasternak. 1998. *Molecular Biotechnology*, 2nd edition. ASM Press, Washington, D.C. An excellent general reference on both the science and applications of biotechnology.

James, Clive. 1999. *Global Review of Commercialized Transgenic Crops: 1999*. The International Service for the Acquisition of Agri-Biotech Applications, Ithaca, N.Y. Contains data on global growth of plant agricultural biotechnology. For information on ordering this and other ISAAA publications on agricultural biotechnology in developing countries, visit their website (www.isaaa.org).

Lander, Eric S. 1996. The new genomics: global views of biology. *Science* 274:536–539.

Lewin, B. 1997. *Genes VI*. John Wiley and Sons, Inc., New York, N.Y. The title speaks for itself. Excellent general reference, as are Lewin's earlier editions, *Genes I* through *Genes V*.

Lysaght, Michael J., and Patrick Aebischer. 1999. Encapsulated cells as therapy. *Scientific American* 280:76–82.

Marshall, E. 1995. Gene therapy's growing pains. *Science* 269:1050–1055.

Marx, Jean. 1989. *A Revolution in Biotechnology*. Cambridge University Press, Cambridge, England. Though somewhat dated, this remains an excellent reference book, particularly for information on bioprocessing and the history of industrial use of microorganisms.

Moffat, Anne Simon. 1998. Toting up the early harvest of transgenic plants. *Science* 282:2176–2178.

Mooney, David J., and Antonios G. Mikos. 1999. Growing new organs. *Scientific American* 280:60–65.

Organization for Economic Cooperation and Development. 1998. *Biotechnology for Clean Industrial Products and Processes: Towards Industrial Sustainability*. OECD, Paris, France. A thorough treatment of biotechnology's potential to improve environmental effects of industrial processes. To order, visit the OECD website (www.oecd.org).

Pederson, Roger A. 1999. Embryonic stem cells for medicine. *Scientific American* 280:68–73.

Roush, R. 1997. Antisense aims for a renaissance. *Science* 276:1192–1193.

Smith, John E. 1996. *Biotechnology*. Cambridge University Press. Cambridge, England. Provides an excellent overview of the breadth of biotechnology applications and is especially useful for information on bioprocessing, cell culture, and industrial uses of microorganisms.

Watson, James, M. Gilman, J. Witkowski, and M. Zoller. 1992. *Recombinant DNA*. Scientific American Books, W. H. Freeman & Co., New York, N.Y. An excellent reference focused on the science and techniques of gene-based biotechnology.

Weiner, Donald B., and Ronald C. Kennedy. 1999. Genetic vaccines. *Scientific American* 281:50–57.

Biotechnology and Society

Introduction

We often hear that advances in biotechnology will give rise to problems, issues, and concerns humans have never before faced. But what exactly are they? Are the societal issues said to be rooted in biotechnology unique to the new powers that biologically based technologies give us, or are they the old, unresolved, sometimes unacknowledged problems associated with earlier technologies?

Many of the issues attributed to the development of biotechnology have actually been with us in slightly different forms for many years, but we haven't been paying attention. We have enjoyed the fruits of technology without fully considering its implications. Now we recognize that the short-term gratification technology provides is sometimes accompanied by long-term costs. As a result of this recognition, we are focusing much more attention on biotechnology at an early developmental stage, and much of that attention seems to be negative. Why is biotechnology receiving so much negative attention? Is the negativity deserved, or is biotechnology paying the price for the mistakes of previous technologies?

We can trace some of the negativity to an antitechnology sentiment that is not uncommon in today's society. Our sentiments about some technologies seem to have swung from inattentive acceptance of the benefits to cynical disillusionment with the costs. Unfortunately, evaluating technology is not an either-or issue. Forming a rational and fair opinion of any technology depends on looking at its costs and benefits side by side.

Considering the Benefits of Technology

Take a minute and try to imagine the world in the absence of any modern technology. That is very difficult, for technology is woven tightly into the fabric of our lives and ourselves. The essence of our species seems inseparable from technology. We cannot objectively analyze technology and our relationship to it any more than a fish can step back and observe water. Nonetheless, try and imagine a world without technology.

Would you really want to go back to the "good old days" of no running water, no electricity, sewage in the streets, an unpredictable food supply, and an expected life span of 40 years? These conditions existed relatively recently in the United States and continue to exist in many parts of the world. According to the World Bank, 1 billion of the 3 billion people living in underdeveloped countries cannot count on finding food or shelter from one day to the next. Every year, 14 million children under the age of 4 years die from starvation or starvation-related diseases.

Agriculture was one of the earliest technologies developed and used by humans. It allowed us to move from nomadic hunter-gatherer societies to geographically stable, self-supporting communities. All other technological, cultural, and political progress presupposes and depends on having adequate food supplies. As recently as 100 years ago, almost 50% of the U.S. population lived on farms. That number of farm workers was required to provide enough food to feed our population. As a result, most children had no schooling beyond the first few years because they were needed as laborers. Now all of our food is produced by less than 1% of the population. We can trace this remarkable change to the development of new farm technologies that allowed machines and chemicals to take the place of people (Figure 2.1). The people who used to work on farms now sell insurance, play professional sports, watch professional sports, and write books on biotechnology. Much of the growth of this country's economy is linked to the movement of people away from the farm and into the workforce. How many of us truly want to get rid of machines and chemicals and go back to the "good old days" of hard manual labor and no schooling? Some might, but most probably don't.

If you look around the room, what do you see that is "natural," and what is a product of technology? How many of these products would you give up readily?

Figure 2.1 As new farm technologies replace hand labor with machines, people leave farms and move into the industrial workforce, stimulating a country's economic growth. In industrialized countries, farmers use a combine (A) to both harvest and thresh wheat. In other parts of the world, the separate operations of harvesting (B) and threshing (C) are done by hand. (Photographs courtesy of VU/©Inga Spence [A] and Charles Preitner [B and C].)

On the other hand, how often do you really come face to face with nature? Not often, because humans rejected the idea of living with nature long ago.

In evaluating any technology, we must, first and foremost, be honest with ourselves. We depend on technology, and few of us could survive without it.

Considering the Costs of Technology

Much of the negative sentiment surrounding technology stems from the perception that technology is destroying the environment. Technology has a profound effect on our relationship with the natural world. That, after all, is its intent. We use technology to insulate us from the often harsh realities of nature, and some of those technologies have harmful effects on the natural environment. For example, burning fossil

fuels to generate energy for heating and cooling our homes pollutes our air supply and produces acid rain.

But does technology have to be incompatible with nature? Not necessarily. If we look at technology as engineering solutions to the problems of life on earth, then technology exists in nature. Other species have evolved technologies that are harmonious with nature. An obvious example is the conversion of solar radiation into chemical energy (glucose) by plants. This is a solution to a problem all living organisms must face: obtaining energy.

Biologists call the technological achievements of other organisms "adaptations." The forms of adaptations that have evolved have been shaped by the indifferent force of natural selection. Natural selection cares nothing about politics and economics, so these factors have neither driven nor shaped the technological solutions evolution has designed. So maybe it's not technology per se but the technological solutions we have chosen that have had detrimental environmental effects. Knowing what we do now about the potential costs of technology, perhaps we will aim for applications of biotechnology that are energy, resource, and capital efficient as well as minimally polluting.

But even if we do develop technologies that are more environmentally benign, that will not rectify the relationship between technological advance and environmental health. No matter what ingenious engineering solutions we observe in other organisms, only one species has been able to use technology to increase its population size beyond ecologically sustainable limits, and that's us. Even if we, like other organisms, develop technologies that are harmonious with nature, if our population continues to increase at its current rate, we will only increase the strain we are placing on the planet and ourselves.

This, in fact, is one way that advances in biotechnology may contribute to our current problems. Biotechnology will give us what we want: lives that are long and healthy. As appealing as that is to us as individuals, the result will be more people living on an already overcrowded planet, using its resources and generating wastes. As a species we should remember the adage taught in the myth of Midas and the Golden Touch: Be careful what you wish for—you may get it.

Technology alters our relationship with nature in ways even more subtle than allowing an increase in population size. It has insulated us from nature, and increasingly from one another, so effectively that we have forgotten we are part of the natural world. We are no longer conscious of the very real biological ties we have with other species and with each other.

We are in the midst of an environmental crisis and must begin to change our ways if we and countless other species are to survive. Loss of this sense of connection with the natural world ultimately makes the tasks ahead of us even more difficult. The less we feel we are a part of nature, the more we feel we can play outside its rules. The less we know about nature, the harder it is for us to respect it. Without respect for nature, and each other, will we be able to make the necessary choices to balance our wants and needs with the needs of other species?

Our evolutionary heritage has provided us with a conscious mind. As thinking beings, we question and explore and then use our discoveries to control our environment and change things to our liking. For the most part, we have directed these abilities toward things *outside* of ourselves. That's what science and technology are all about. Turning inward and questioning, exploring, controlling, and changing ourselves does not come quite as easily to us. Now we must stretch the minds evolution has provided us and learn the more difficult behaviors of questioning our sense of values and our place in the scheme of things, exploring ourselves and our motives, and accepting responsibility for our actions.

We can do better than muddling through problems we create, trying to clean up our mess as we go along. We know tomorrow's problems can be predicted, in part, from our past and our present. Rather than simply react, we can anticipate and plan. More than any technology, biotechnology places this challenge before us. We will not meet this challenge by relying solely on the cleverness of our science. This technology raises deeper philosophical questions that science alone can't answer. For these we need wisdom.

How can we act wisely in directing biotechnology? Neither blind acceptance nor wholesale rejection qualifies as acting wisely. By attempting to completely block the development of biotechnology, we ultimately deny ourselves possible choices. What wisdom is there in limiting our options? On the other hand, ignoring the implications of this technology and passively allowing it to develop on its own will more than likely guarantee we will do no better with this technology than we have with others. Instead, it is in our best interest to attend to biotechnology's development by consciously making choices based on critical evaluation of real issues. How do we do this?

Analyzing the Issues of Biotechnology

A person does not have to be a scientist to analyze the societal issues attendant to the development of biotechnology. A productive and user-friendly approach to analyzing any issue associated with biotechnology is simply to view modern biotechnology within the context of technological change. Remember, biotechnology is one of many technologies we have developed. Medicine, agriculture, and energy production are age-old attempts to control our environment with technology. Biotechnology now places at our disposal a new set of technologies for shaping the world. The tools may be different, but the goals remain the same: to improve human health and alter our environment so that our lives are as long and easy as possible. Will biotechnology be better than previous technologies in providing us with tools for reaching these goals, or will the costs of developing and applying these tools be greater than the benefits?

In evaluating societal issues raised by biotechnology, placing biotechnology in the historical context of a continuum of technological change can help us:

- understand the changes biotechnology might bring and determine whether the changes are significant when compared with current technologies and their societal implications;
- use our past experiences to separate real problems from remote possibilities;
- assess the costs and benefits of this technology compared with those of past activities. (Quantitative cost-benefit models that allow us to evaluate the impact of technologies much more rigorously are discussed in chapter 35.)

When presented with a societal issue related to developments in biotechnology, we can look at our history of using technologies and ask ourselves whether the issue is new or unique to biotechnology. Does the concern or issue derive from biotechnology's novel powers that provide capabilities we never before have had, or do we currently engage in practices that raise similar or even identical issues? If the practice and issues it raises are unique to biotechnology, do all aspects of the issue raise concerns? If not, we should specifically define those areas that trigger concern and focus our attention there.

We may determine that the practices that elicit concerns are not at all unique to biotechnology but are accepted as part of life. Does their commonness mean they are not worth analyzing? Absolutely not. Just because we *are* doing something does not mean we *should* be. For example, some people have expressed concern that genetic engineering allows us to genetically modify food by moving genes between different species. But before the development of genetic engineering, plant breeders genetically modified much of our food by interbreeding different plant species. After learning this, some people may still believe we should not move genes between species, whether we accomplish this through genetic engineering or crossbreeding. In other words, what we are doing, not how we are doing it, concerns them.

Below are just a few examples of issues said to be derived from advances in agricultural and medical biotechnology and suggested ways of analyzing them. Most of the concerns expressed about biotechnology revolve around a particular subset of biotechnology, gene-based biotechnology. Others are linked to our ability to grow human cells in culture. We will begin with agricultural biotechnology, because many of the applications of agricultural biotechnology, and therefore the resulting issues, are identical to past agricul-

tural practices and issues. As a result, we have historical data that can be used in analyzing the issues of agricultural biotechnology. Beginning with issues for which we have facts, not simply speculations, will allow us to become familiar with the method of analyzing complex issues before we move into a more difficult area, that of medical biotechnology.

In Part III of this book, we provide more information on societal issues associated with gene-based biotechnology, as well as detailed lesson plans on conducting classroom discussions and debates.

Agricultural biotechnology

Issue: Environmental Introductions of Genetically Engineered Organisms

Concern: Many people have expressed concerns about the environmental risks associated with introducing genetically engineered organisms into our agricultural ecosystems.

Their concerns focus primarily on

- the transfer of the gene(s) to other, related species
- harm to nontarget species
- the creation of exceptionally hardy crops that will escape from agricultural systems and displace indigenous species

Are these concerns new or unique to biotechnology? No. We have been introducing genetically altered organisms into the environment for thousands of years. We have changed the genetic makeup of all the crops we plant and livestock we raise. Every year, seed companies introduce many new varieties that differ genetically from those they've sold before. All of the genetically modified organisms we have released differ from genetically engineered organisms only in the methods used to alter their genetic makeup. Traditionally, we have used selective breeding, both within and between different species, and mutagenesis. Both are less specific methods of genetic alteration than genetic engineering.

When assessing the environmental impact of genetically modified organisms, we should focus not on the method of genetic modification but on the trait that results from the modification. For the most part, the traits we have incorporated and will continue to incorporate into crops and livestock through genetic engineering are those we have incorporated into organisms through breeding, such as increased yields, drought resistance, insect resistance, disease resistance, nematode resistance, and herbicide tolerance. As a result, we have decades of experience to draw from in assessing the concerns associated with the environmental introduction of genetically engineered organisms. We can look at our extensive agricultural history to see if environmental problems have been caused by the movement of genes from crops to other organisms. Have nontarget species been harmed? Have crop plants displaced indigenous species?

However, some of the genetically engineered traits we are incorporating into crops will be novel. As described earlier, we are genetically engineering crops such as tobacco to serve as sources of pharmaceutical compounds or industrial chemicals such as biodegradable plastics. In addition, we are genetically engineering crops to have different oils, decreased allergenicity, cold tolerance, more vitamins and minerals, and more complete proteins. Because we have no previous experience with these types of crops, we have no body of experience to use for predictive purposes. In the future, the novel traits introduced through genetic engineering must be assessed thoroughly to determine whether or not they pose a risk to the environment.

We have very little experience with certain species, such as fish, that have been genetically engineered for increased productivity or certain insect predators and parasites that are being investigated as targets for genetically improved biological control capabilities. In the absence of existing data on their uses, these species must also be carefully evaluated for their potential environmental impact.

Many independent groups of scientists have studied the environmental risks of genetically engineered organisms released into agricultural ecosystems, and all have reached the same conclusion: in principle, the risks associated with genetically engineered crops do not differ from those of traditionally bred crops with similar traits. According to information provided at the 1999 risk assessment workshop "Ecological Effects of Pest Resistance Genes in Managed Ecosystems," this hypothesis is supported by scientific data gathered on North American crop species coexisting with wild weedy relatives. (See Traynor and Irwin in *Selected Readings* to obtain the workshop proceedings.)

As of January 2000, government-approved field tests of genetically engineered organisms have occurred in at least 30 countries worldwide. Over 10,000 field tests of over 100 species of genetically engineered organisms have been approved by regulatory agencies in the United States, Canada, and the European Union alone (Table 2.1), and farmers throughout the

Table 2.1 Field test notifications received by the United States, Canada, and the European Union as of January 1, 2000[a]

Country/region	No. of field tests	No. of species tested
United States	>5,800	88
Canada	>3,600	22
European Union	>1,500	58

[a]To track the international growth in the number of field tests, genetically engineered traits, and species tested, review the BioTrack database at the OECD website (http://www.oecd.org//ehs/service.htm).

world have grown over 100 million acres of genetically engineered crops. There have been no reports of any unanticipated adverse environmental consequences. More information on environmental introduction of genetically engineered organisms is provided in Part III. For now, suffice it to say that our extensive experience in releasing genetically altered organisms serves as a rich information base from which to discuss and evaluate any concerns about environmental introduction of genetically engineered organisms.

Issue: Crossing Genetic Boundaries Established by Nature

Concern: People have expressed concern that genetic engineering allows us to cross the normal reproductive and genetic barriers that exist in nature.

Is this ability new or unique to biotechnology? No. When we selectively bred many agricultural species, we forced fertilizations that would never have occurred without human intervention. So we have already crossed genetic barriers that would not have been breached in nature. Many of these crosses involved organisms belonging to different species but the same genus (Table 2.2). A few have been forced between organisms of different genera. By using increasingly sophisticated laboratory techniques, we have successfully crossed organisms and produced fertile offspring from organisms that cannot interbreed naturally (Table 2.3).

If the concern is that we should not force crosses between organisms that would not breed in nature, what if the two organisms are in the same species? Sometimes, for a variety of reasons, members of the same species will not mate or will not produce viable offspring. Is it acceptable for us to intervene in these cases? If so, why?

If we decide that crossbreeding organisms in different species is acceptable but we object to moving genes between species through genetic engineering,

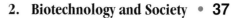

Table 2.2 Examples of interspecific and intergeneric hybridizations in common crop plants and the traits transferred into the crop species from donor species through the hybridizations

Crop species	Donor species	Trait
Avena sativa (oat)	*Avena sterilis*	Increase yield 25–30%
Beta vulgaris (sugar beet)	*Beta procumbens*	Nematode resistance
Gossypium hirsutum (cotton)	*Gossypium tomentosum*	Decrease boll rot
	Gossypium raimondii	Fungal disease resistance
Lycopersicon esculentum (tomato)	*Lycopersicon peruvianum*	Nematode resistance
		Virus resistance
		Harvesting traits
	Lycopersicon pimpinellifolium	Fungal disease resistance
Oryza sativa (rice)	*Oryza nivora*	Virus resistance
Solanum tuberosum (potato)	*Solanum demissum*	Fungal disease resistance
		Virus resistance
	Solanum stoloniferum	Fungal disease resistance
		Virus resistance
Triticum aestivum (wheat)	*Aegilops comosa*	Fungal disease resistance
	Aegilops ovata	Increase protein
	Aegliops speltoides	Fungal disease resistance
	Agropyron elongatum	Drought tolerance
	Secale cereale	Winter hardiness
		Fungal disease resistance
Zea mays (corn)	*Tripsacum dactyloides*	Fungal disease resistance

we must try to delineate our concerns precisely. What is the heart of the issue? Is it the method, genetic engineering, that causes the concern? Why? Is it that we shouldn't move genes between organisms that are "too different" from each other? How will we define "too different?" Where will we draw the line, and why will we draw it there?

Does it matter that, left to their own devices in the absence of human intervention, organisms do not respect the genetic boundaries we have erected to delineate species? Some don't. This natural exchange of genetic material between species is called *horizontal gene transfer,* and it is more common than you might imagine.

Table 2.3 Examples of natural physiological barriers to reproduction and recombination and techniques for overcoming the barriers

Barrier	Techniques for overcoming the barrier
Prefertilization barriers	
Failure of pollen germination	Remove pistil, then pollinate exposed end of the style
	Use recognition mentor pollen
Slow pollen tube growth	Use recognition mentor pollen
	Chemical treatment with organic solvents or growth regulators
Pollen tube stops in style, ovary, and ovule	In vitro fertilization
	Use of plant growth hormones and chemicals like chloramphenicol and acriflavin
Failure to obtain sexual hybrids	Protoplast fusion
Differences in ploidy level	Chemically induce chromosome doubling
	Reduce chromosome number of cultivated polyploid species before hybridization
Postfertilization barriers	
Embryo abortion (immediate)	In vivo/vitro embryo rescue/implantation
Embryo abortion (early stages of development)	Ovary culture
	In vitro fertilization
Lethality of F_1 hybrids	Graft hybrids
	Regenerate plants from callus
Chromosome elimination	Alter genomic ratios of species
	Induce chromosomal exchanges before onset of elimination
Hybrid sterility	Chemically induce chromosome doubling
Lack of recombination	Induce chromosomal exchanges through tissue culture or irradiation
	Induce homologous recombination through manipulation of chromosome pairing system

Most of us are familiar with the ready transfer of genes between bacteria of different species. That is one of the factors responsible for the rapid spread of resistance to antibiotics in the microbial world. But bacteria are not the only organisms that exchange genetic material between species. So do many other species, and they are not "lowly" organisms like bacteria. In fact, primate ecologists have described natural hybrids between two different monkey species. In plants, outcrossing between different species has been crucial to the evolution of the higher plants, the angiosperms.

The evolutionary distance over which gene transfer occurs in nature can be very great. According to molecular evidence, it seems that viruses have transferred genetic material between distantly related insect and mammalian species. Similar evidence shows that a parasitic mite has transferred genes between two *Drosophila* species and kept a few *Drosophila* genes for itself.

Of course, significant obstacles block the frequent transfer of genetic information across species boundaries. Blocking interspecific gene flow is much more prevalent than permitting it. Nonetheless, it is important for us to realize that the issue of natural versus unnatural gene transfer is not as clear-cut as we might like to think.

ISSUE: The Evolution of Resistance to Pesticides

Concern: People have expressed concern that pests will evolve resistance to the pesticidal products of biotechnology just as they have evolved resistance to other chemical pesticides. In particular, they are concerned that insects will evolve resistance to *Bacillus thuringiensis* (Bt) and its toxic protein.

Is this evolutionary response new or unique to biotechnology? No. The evolution of resistance is not a problem specific to biotechnology. We can expect all pests—insects, weeds, fungi, nematodes, bacteria, viruses—to evolve resistance to *any* control measure we use against them (Figure 2.2). Weeds have evolved to physically resemble crop plants to avoid being pulled up by hand. Insect pests have lengthened their diapause from 1 year to 2 to escape control by crop rotation and have even skirted biological control mechanisms.

If we continue to try and control insects, weeds, and disease, then there can be no doubt that certain pests will evolve resistance to certain pesticidal products produced through biotechnology. No matter what methods we use to kill pests, they will do their best to evolve resistance. We are powerless to stop evolution or change the mechanism through which it acts—natural selection. To stop the evolution of resistance to

control mechanisms, we would have to stop trying to control our crop pests. We can, however, slow evolution down by decreasing the selective pressure we place on pest populations.

Some say that evolution of resistance to Bt is a unique problem because Bt is a rare, natural resource that shouldn't be squandered. Again, this problem and the solution are not specific to biotechnology. We have selectively bred certain crops to be resistant to certain insect and microbial pests. The genetic basis of this resistance is also a natural resource that should not be squandered. Knowing what we do about the extraordinary adaptability of pests, we should use all products judiciously so that we can maximize their effective life span by retarding the rate of the evolution of resistance. In other words, we should use products designed to control problem organisms in ways that lessen the selective pressure we place on a population from any single control measure—whether the control measures are genes or chemicals for protecting our crops or antibiotics for protecting our health.

ISSUE: The Safety of Genetically Engineered Foods

Concern: People have expressed concern over the safety of foods we have genetically engineered.

There is something new and unique to biotechnology here, but it is only a portion of the concerns that have been expressed.

By now you know we have genetically altered everything we consume except for nondomesticated organisms like the wild fish we catch or the blackberries we gather. None of our food has escaped our genetic tinkering via selective breeding, a less precise means of changing an organism's genetic makeup than genetic engineering. In addition, the traits we are introducing through genetic engineering are similar to those produced through selective breeding. So the overarching concern about the safety of genetically engineered foods compared to our current foods is misplaced. We have been eating insect-resistant, disease-resistant, and herbicide-tolerant plants for decades.

One aspect of public concern that may be appropriate relates to food allergens. A handful of foods are known to be common allergens: soybeans, peanuts, wheat, eggs, and shellfish. These foodstuffs are so common in our diet that people who are allergic to them become aware of their allergies early on. They then avoid those foods. We now have the ability to move genes from common allergens to foods not typically seen as allergenic. If we are interested in transferring certain traits in an allergenic food, we might

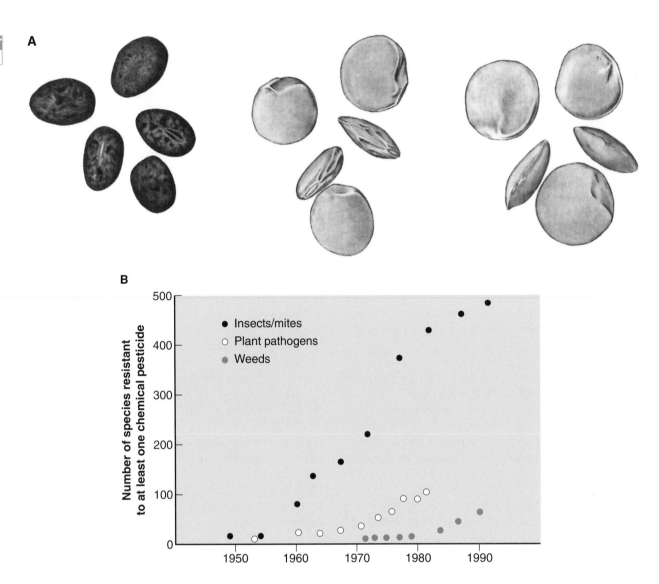

Figure 2.2 Crop pests will evolve resistance to whatever selective pressures we place upon them. (A) Seed mimicry in common vetch (*Vicia sativa*), a weed that sometimes infests lentil (*Lens culinaris*) fields. Typical vetch seeds from plants that have not adapted to lentil field habitats are shown on the far left, and lentil seeds are shown on the right. Between the two are vetch seeds whose shape and color have evolved to resemble lentil seeds. This adaptation tricks the farmer into planting vetch seeds along with lentil seeds. (Adapted from Virge Kask, *American Scientist,* volume 79, 1991.) (B) The evolution of resistance to chemical pesticides. (Data from Green, LeBaron, and Moberg, 1990, and National Research Council, 1986; see *Selected Readings*.)

move the gene that codes for the protein that triggers the allergic reaction to the nonallergenic food. People with food allergies should and will be alerted to that fact. In its 1993 announcement regarding the regulation of genetically engineered foods, the U.S. Food and Drug Administration announced that it will require labeling in those cases where a gene from a known allergen has been moved to a nonallergen.

But what about unknown allergens or foods that are allergenic in very rare cases? Should the government require labeling for a genetically engineered product that might cause an allergic reaction in a small percentage of the population? Non-genetically engineered products currently on the market might cause allergic reactions in some people. Should they be labeled? When kiwi fruits were first introduced into grocery stores in the United States, many people learned they were allergic to kiwi fruits only after eating them. Should the fruits have carried warning labels? What should the labels have said?

Medical biotechnology

The most difficult issues raised by biotechnology deal with medical biotechnology and bioethics. Here are just a few examples. For a more thorough treatment of these issues, see chapters 37 to 39.

ISSUE: Genetic Testing

Concern: People have voiced a number of interrelated and complex concerns associated with genetic testing. To facilitate increased understanding, we will consider each issue separately. Analyzing each issue apart from the others makes the task more manageable and also may reveal any hidden—and perhaps incorrect—assumptions. This method is particularly useful when analyzing the complex societal issues raised by biotechnology.

ISSUE: Genetic Screening for Inherited Disorders

Concern: Some believe that our ability to identify genetic defects will cause more harm than good by placing individuals in psychological, emotional, and social quandaries without easy answers.

They fear that people with information about personal genetic disorders could be psychologically harmed by having this knowledge. They believe that knowing you have a genetic disorder that causes an incurable disease, such as Huntington's chorea, would be emotionally devastating. Others say that knowing you are pregnant with a child suffering from a genetic disorder that causes an incurable disease would force parents into a psychologically and emotionally difficult decision regarding abortion.

What about this issue is new or unique to biotechnology? First, let's consider what's not new, because people, including physicians, grapple with many of the same issues today.

If a routine physical examination reveals you have an incurable cancer, do you expect your physician to share that information with you? Probably so. But just a few decades ago, physicians did not tell their patients they had cancer because they thought not knowing was best for the patient's emotional well-being. The tension between withholding and disclosing information is not new to medicine and varies with the disease, the physician, the patient, and the prevailing practice. It is not specific to new genetic testing capabilities.

The same can be said about prenatal testing for inherited disorders. Today, many obstetricians routinely advise women over 35 to have amniocentesis or chorionic villi sampling to test fetuses for genetic dis-

orders. Biochemical tests on fetal cells can detect more than 200 metabolic genetic disorders and a number of chromosomal abnormalities. A few of these disorders can be treated once the baby is born. For example, the course of the inherited disorder phenylketonuria can be changed by placing the infant on a special diet early in life. The mental retardation associated with phenylketonuria will not result if the individual maintains a diet low in the amino acid phenylalanine.

However, most of the genetic disorders these tests can detect are incurable. Their progress cannot be subverted by any known treatments. Parents today must decide what to do with that information. If we are concerned about the emotional trauma caused by sharing fetal genetic information with parents, we should reassess our current practices.

Here's what is new about genetic screening and the issues it creates. The first issue relates to methodology: how the genetic basis of a disorder is discovered and how screening methods for this disorder are developed.

Before knowing the exact location of a gene that causes a disorder, scientists first develop "markers" that are associated with the defective genes. The process of developing markers depends on acquiring blood samples from many members of a family and subjecting their DNA to analysis. As a part of this discovery process, a person's genetic makeup becomes known to researchers and may be published in a scientific journal. Many researchers who have studied family trees to trace disease-causing genes have discovered incest, mistaken paternities, and abortion histories. On the other hand, many lives have already been saved by these studies. For an individual, a dilemma is posed between the right to "genetic privacy" and a responsibility to family members. Where would you draw the line in balancing your need for privacy with their need to know?

The second new issue raised by widespread availability of genetic testing is more a question of a change in degree than a true novelty. But that change in degree presents exceptionally difficult ethical issues.

We have had the power to conduct fetal testing for genetic disorders for many years. Parents have had to face some very difficult decisions as a result, because most of those tests have been for incurable diseases or serious medical problems that leave little or no room for human intervention. In the future, we will be able to detect many more defective genes than those that cause incurable illnesses. Some genetic defects will directly cause serious problems that can be reversed with appropriate treatment (hemophilia). Other defective genes will simply indicate a propensity to develop a certain disease (emphysema). Still others will cause defects that are in no way life threatening (color blindness).

Where will we draw the line when defining a genetic defect and determining how to handle it? When is a defect truly a defect? At what point will eliminating genetic defects inch toward selecting desired characteristics, such as large height or attractive physical appearance? Will each couple be free to make decisions regarding the fate of offspring carrying a "defective" gene?

Currently, genetic alteration in human germ cells (sperm and eggs) is not allowed. If the government ever changes its position on germ line gene changes, the door for altering the species' gene pool will be opened. Will we do our best to decrease genetic diversity? We may praise genetic diversity in the abstract, but historically, our behaviors are not consonant with that viewpoint. If we decrease genetic diversity in the human gene pool, will we ultimately be limiting our species' capacity to evolve?

You may be asking why we see these issues as potential problems for individuals seeking information about the genetic makeup of their offspring. Can't people avoid these difficult decisions by simply not subjecting themselves to genetic testing? After all, no one requires them to take the tests, do they?

If we look at history, we see that often technologies that are supposed to increase our options may also increase our obligations. A choice at one time may become a requirement later on. For example, most parents probably would be surprised to learn that their newborn babies were tested for as many as nine inherited disorders immediately after delivery. Such tests are required by law in most states. So although certain genetic tests are an option today, tomorrow we may have no choice but to be tested.

ISSUE: The Meaning of Genetic Information
Concern: What will this information really tell us?

People object to the large-scale use of genetic testing because, by their nature, the test results are often ambiguous. The presence of certain genetic defects does not guarantee that an individual will actually develop the disease. In many cases, the presence of a defective gene means only that the person is a carrier. In other, more ambiguous cases, the presence of a defective gene means only that you have a genetic predisposi-

tion to that disease. In other words, it tells you that you are at risk.

But how does this differ from current practices? Every time we have our blood pressure taken or our cholesterol measured, we are assessing risk factors. When we describe our family's medical history to a physician, we are providing data on genetic predisposition. Is the information gained through genetic testing any more ambiguous than the information physicians already gather?

Both bodies of information are ambiguous, but while most people are prepared to accept the ambiguity of blood pressure readings and cholesterol measurements, they do not know that genetic testing data can be equally ambiguous. The nature and power of genes are thoroughly misunderstood by the public. An aura of predestination surrounds a gene. People believe that if you have a gene for a given trait, then all is determined, nothing is left to chance.

Biologists know that this is not true. The path from genotype to phenotype is an indirect one filled with opportunities for myriad factors other than a single gene to exert an influence (Figure 2.3). The final expression of a gene is affected by many variables: the allelic forms of the gene in question, other genes, and environmental influences.

Figure 2.3   This diagram illustrates the complex relationship between genes and observable traits. The path from the gene to its primary protein product to visible traits is not straightforward and linear as many people assume. One gene can affect a number of traits, and any one trait may be the result of interactions between many genes. In addition, gene products have feedback effects on the activities of other genes. Although the diagram is complex, it is an oversimplification of the actual processes. Consider how complicated the diagram would be if we listed all intermediate products and regulatory steps between the gene and the trait, and all the environmental influences that affect the observable trait.

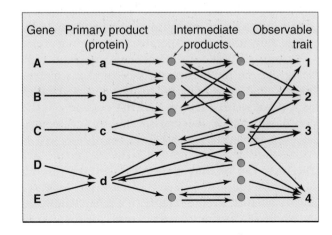

These facts are often ignored when people are assessing the meaning of genetic data. Members of the public who are in positions of power might assume that genetic information is much more meaningful than it really is. Such unenlightened viewpoints can be particularly pernicious when held by individuals in a position to exercise control over our lives or by institutions and organizations concerned with issues other than humanitarianism. This leads us to the final issue related to genetic testing.

ISSUE: Genetic Privacy

Concern: Who will have access to genetic testing information and for what purposes?

This concern involves the confidentiality of the results of the genetic tests and the interpretation and dissemination of those results. People fear that the results will be used unjustly by employers to withhold jobs and promotions and by insurance companies to deny health coverage.

What is unique about this concern? Again, let's start by describing what's not new, because a precedent for some of these concerns already exists. Many American companies go to great lengths to screen potential employees for health-related factors. At least 50% of the employers in the United States require that job applicants have medical examinations. In those examinations, many of the tests conducted are indirect measures, that is, phenotypic manifestations of genetically based traits. As early as 1982, 23 of 366 companies surveyed by the Office of Technology Assessment had begun to use genetic tests in screening applicants (and another 57 expressed an interest in using them).

Companies are interested in the health of workers for reasons related to decreasing costs:

- Time lost due to sickness or disability costs money.
- Hiring healthy people allows corporations to hold down health care costs.
- By screening workers susceptible to certain occupational hazards, companies protect themselves against lawsuits.

Currently, insurance companies do not require genetic testing, but historically, insurers have used medical technologies to identify preexisting conditions as a basis for denying coverage or raising premiums. They have also excluded people from coverage because their family medical history put them in a high-risk category.

Even though these concerns about privacy and the use of genetic information are not specific to new ge-

netic technologies, others can be traced to these newer technologies. In general, concerns specifically related to genetic technologies are derived from the true nature of genes and people's misunderstandings about their nature as described above.

When insurance companies require medical examinations to determine the health of their applicants, the physicians measure physiological and morphological correlates of health and disease. That is, if the person has begun to develop cardiovascular disease, the physician will be able to discern the problem through blood tests, electrocardiograms, and blood pressure measurements. The same is true of a lung disease such as emphysema. Measurable factors such as lung capacity and force of exhalation indicate that the applicant-patient is developing or has developed the disease.

Genetic information is unlike this type of data. Genes foretell the likely future. They might allow physicians to predict the possible appearance of cardiovascular or lung disease in the absence of symptoms. Will these telltale markers then be classified as determinants of preexisting conditions, thus allowing insurance coverage to be denied?

A second problem regarding the nature of genetic information is linked to the general public misunderstanding of what genes are, what they do, and what genetic information means. If employers and insurance companies view genes as final arbiters of traits rather than one of many variables that determine the final product, any genetic defect may be seen as a preexisting condition.

This leads us to another important and difficult question that must be answered. What exactly will be considered a defect? There is extraordinary genetic variation for most traits in all populations. What variants of what traits will be labeled normal and which will be labeled defective? Geneticists estimate that probably everyone has 5 to 10 genes that could cause an illness of some sort. As a result, all of us may be seen as genetically defective. How could all of us be uninsurable and unemployable? Again, where will we draw the line, and who will draw it?

Human embryonic stem cell research

All of the applications of biotechnology discussed above involve gene-based technologies. Recently, applications that rely on a different type of biotechnology, mammalian cell culture, have triggered societal concerns. The first of these, human embryonic stem cell research, raises a number of novel questions to which there are no simple answers.

Background

In November 1998, researchers reported that they had established human embryonic stem cell (ESC) lines capable of existing indefinitely in laboratory dishes and of giving rise to any human cell type. The availability of a source of undifferentiated human cells opens up new avenues for treating diseases with cell therapy, studying human development, discovering new drugs, testing drug safety, and engineering replacement tissues. However, many people object to any research that involves growing human embryos in laboratories, even if it is limited to the earliest stages of development.

To assess the implications of ESC research, we need to understand the methodologies used in human ESC culture. After fertilization, cells of the embryo divide until they form a blastocyst, a hollow ball of approximately 140 cells, in approximately 5 days. Blastocysts consist of an outer layer of cells, called trophoblast cells, and an inner cell mass. The trophoblast cells differentiate into placental tissue, while the inner cell mass is made up of the cells that will divide and develop into the individual. The inner cell mass cells give rise to the ESC lines (Figure 2.4).

These cells can differentiate into any cell type. Mouse ESC in culture have spontaneously differentiated into **hematopoietic** cells and heart muscle cells; when transplanted into adult mice, they have differentiated into organized structures such as neurons, muscle, bone, and teeth. Despite their remarkable developmen-

Figure 2.4 Human blastocysts are potential sources of ESC, which are undifferentiated cells capable of developing into any type of cell in the human body. Blastocysts consist of an outer layer of cells and an inner cell mass, which gives rise to the ECS lines. (From D. K. Gardner, W. B. Schoolcraft, L. Wagley, T. Schlenker, J. Stevens, and J. Hesla, *Hum. Reprod.* **13**:3434–3440, 1998. © European Society of Human Reproduction and Embryology. Reproduced by permission of Oxford University Press/Human Reproduction. Photograph courtesy of David K. Gardner.)

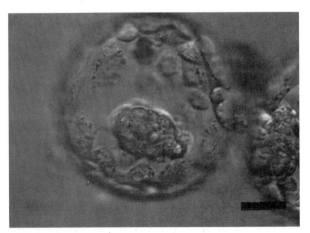

tal potential, these ESC will not develop into a mouse if inserted into a female, because they lack the cells required for implantation into the uterine wall. However if ESC are added to trophoblast cells, the resulting embryo will implant in the uterus, and a normal mouse, derived completely from cultured ESC, will develop.

Work on human ESC demonstrates similar developmental potential. Researchers obtained 14 blastocysts that were produced through in vitro fertilization and donated, with informed consent, by the couples. Scientists removed the inner cell masses and successfully established five ESC lines that have shown remarkable plasticity. When implanted in mice, these human ESC have differentiated into tissues resembling neural epithelium, bone, cartilage, gut, striated muscle, and kidney. Of course, no one has conducted an experiment to determine whether these human ESC have the potential to develop into a human being. However, in discussing the legal and ethical implications of this work, it is best to assume that the results of the mouse studies apply to human ESC as well. In other words, the ESC by themselves would not have the ability to implant and form a placenta; however, if placed inside trophoblast cells, the ESC, if implanted into a woman's uterus, could develop into a normal human being.

Issues associated with ESC research

A number of people are concerned about the source of ESC. Blastocysts from which the inner cell masses are removed could develop into humans if given the chance. On the other hand, blastocysts derived from in vitro fertilization and not implanted in the mother are frozen and, more than likely, are discarded in 2 or 3 years. Which of these fates is more respectful of the potential human life, or are both equally disrespectful? If both are unethical applications of technology, should in vitro fertilization be banned, or should all blastocysts resulting from in vitro fertilization be implanted? If we decide we should not use blastocysts as sources of ESC in the future, is it acceptable to use the immortal ESC lines researchers have already established? Or should they be discarded?

Laws, Politics, and Science: The Role of the Federal Government

What should be the role, if any, of politicians in determining which paths of scientific inquiry are pursued and therefore which medical breakthroughs are most likely to occur? This is a question that is not unique to biotechnology research.

The U.S. government has banned the use of federal funds for human embryo research for over 20 years.

As a result, almost all research on human reproduction, including in vitro fertilization and the ESC work described above, has been financed by private companies. In 1994, the National Institutes of Health (NIH) convened a panel of scientific and ethical experts who recommended that some embryo research, including the derivation of human ESC, was ethical and should be considered for federal funding. Nonetheless, in 1995, the U.S. Congress again denied federal funds for "research in which a human embryo or embryos are destroyed, discarded, or knowingly subjected to risk of injury or death greater than that allowed for research on fetuses in utero." Other countries, including the United Kingdom and Denmark, have approved research on human embryos up to 14 days old.

In response to advances in human ESC research, Dr. Harold Varmus, then director of the NIH, asked for a legal opinion regarding whether or not research using human ESC lines was banned from federal funding as a result of the 1995 law banning funding for human embryo research. In 1999, the Office of the General Counsel of the Department of Health and Human Services (DHHS), the cabinet-level agency that oversees the NIH, ruled that human ESC should be excluded from the ban because they lack the natural "capacity to develop into human beings" (i.e., lacking trophoblast cells, they would not be able to implant) and therefore "are not embryos." As a result of this ruling, the NIH may provide funds for research on human ESC as long as the research does not involve *deriving* these cell lines from the inner cell masses. Because human ESC lines have already been established and are immortal, researchers will rarely need to derive new ESC lines.

Does this ruling violate the spirit of the law even if it upholds the letter of the law? Is the line that has been drawn between deriving and using ESC arbitrary, or is it defensible on scientific grounds? A group from Congress has objected to the DHHS ruling and may overturn it. On the other hand, Senator Tom Harkin, a Democrat from Iowa, applauded the ruling and said that the government should not issue "blanket bans" on large areas of medical research. Why? Because doing so places control of the research methods, results, and products firmly in the hands of the private sector.

Government-Funded Research or Private Sector Research?

Many people, including Dr. Varmus, have publicly stated that they are concerned that ESC research has been conducted solely by private companies. Should research that is potentially very lucrative, but that some find ethically questionable, be left to compa-

nies? Is that better or worse than having government involvement? Again, this issue is not unique to the new powers provided by biotechnology.

Research funded by NIH is overseen by an institutional review board. No such legally mandated oversight is guaranteed in private sector research, although private companies do have their own review boards. Some people may question whether a company's internal review board would do anything to ensure that the company abides by high ethical standards. However, in the United States, another powerful incentive governing corporate behavior exists: the fear of litigation. The U.S. government can be sued only if it *allows* itself to be sued. Private companies, concerned with patent protection, do not tend to publish their research, so independent verification of findings is often difficult for research financed by the private sector. On the other hand, NIH also patents discoveries and enters into cooperative research agreements with companies.

Ethical Questions

To circumvent the ban on human embryo research, some researchers have transplanted nuclei from human cells (sometimes their own) into cow eggs, claiming that the ban refers only to products of human fertilization. If these human-cow chimeric embryos developed to the blastocyst stage, would the inner cell masses be a source of cells for tissue transplants, thereby avoiding the use of human embryos? Is this possibility more or less ethically problematic than using human embryos?

The potential applications of human ESC are truly extraordinary. Disorders that involve the loss of normal cells, such as Alzheimer's disease, diabetes, and Parkinson's disease, could perhaps be corrected with cell therapy. ESC offer the prospect of eliminating genetic diseases. Organ transplants that do not trigger rejection could be produced by using ESC. Although these applications are at least a decade away, the potential is there. What are the ethical and legal implications of denying living human beings these lifesaving cures while the government focuses on the rights of potential human beings? Would your answer change if you suffered from a disease that could be cured by ESC research and the government denied funds for that research?

Who will own tissues derived from the ESC lines? The parents of the embryo conceived by in vitro fertilization? The reproduction technology company that worked with the couple and actually carried out the in vitro fertilization laboratory work and nurtured the cells to the blastocyst stage? The person receiving the tissue? The company that produced the ESC lines from discarded blastocysts? Although this issue may be similar to the question of ownership of organs in organ transplants, there are also important distinctions because of the source of the tissue.

These questions are only some of the legal, political, scientific, economic, and ethical issues associated with ESC research. While much public attention and concern directed at biotechnology and medical ethics has focused on gene-based technologies, the above discussion clearly shows that some of the most difficult questions may arise from applications of other biotechnologies. Another such example, animal cloning, is provided below.

A case study for analyzing biotechnology issues: cloning

In February 1997, scientists and nonscientists alike were shocked by the news that an adult mammal had been cloned. The responses of these two populations were similar on the surface, but the reasons underlying their responses had little in common. The existence of this clone, a sheep named Dolly, unleashed a global media frenzy. The story, which was essentially scientific in nature, received unprecedented coverage, but the key elements of the scientific discovery were lost in the storm. For excellent background information on the cloning of Dolly and related topics, visit the website of the Roslin Institute, the research center in Scotland where the cloning of Dolly occurred (www.ri.bbsrc.ac.uk).

Studying the media's coverage of Dolly offers an excellent opportunity for evaluating how reporters and editors handled the story and whether they fulfilled their responsibility to inform the public. More important, the issue of cloning presents us with an opportunity to use the strategy described earlier in this chapter, and further elaborated in Part III, for analyzing a societal issue raised by biotechnology. That is, mentally walk through the issue to clarify the core concern; determine whether the application introduces novel issues; and assess the benefits and costs of the technology and its potential applications. What other steps can help us follow this strategy for issue analysis?

- Define scientific terms in order to focus the analysis.
- Gather facts.
- Put the topic in context by determining what is new about it and what is not new.
- Clarify the issues or concerns. When appropriate, analyze issues separately. For example, when discussing the "ethics of cloning," we should be sure to separate the ethics of cloning plants, animals, and people. Some people may feel that the issues are the same irrespective of the organism involved; others may not.
- Assess the costs and benefits, either informally or by using a more quantitative model of cost-benefit analysis as described in chapter 35.

Defining the Terms
What is a clone? A clone is a collection of genetically identical individuals, all derived from a single parent. Cloning is the process of propagating those genetically identical individuals. Biologists use the word "clone" in a number of ways, depending on what is being cloned.

In molecular cloning, the cloned "individual" of interest is a gene or a length of a DNA molecule. A clone, in this case, usually refers to both the piece of DNA being copied and, by extension, the collection of organisms (usually bacteria) containing the same piece of DNA. In recombinant DNA work, the clone is a recombinant organism, and the piece of DNA is a recombinant molecule.

Cells can also be the cloned entity. Cellular cloning results in "cell lines" of identical cells. For example, in monoclonal antibody technology, the B cell used to manufacture the antibody of interest is separated from a large population of other B cells. The B cells producing the antibody are clones of each other.

Finally, sometimes the clone is a multicellular organism. This is the type of clone most nonscientists mean when they use the word. Many plants are propagated through cloning by rooting plant parts taken from fully developed plants. Cloning animals is more difficult. We cannot simply break off a part of an animal and grow an entirely new organism. In animals, the starting material must be reproductive cells, either eggs or cells taken from an embryo at the earliest stage of development, such as the 16- to 32-cell stage. For production of clones, the embryo is split into single cells, or the nuclei are removed from the embryonic cells and placed in eggs. When development is complete, each organism derived from the original embryo is a clone.

After reading these different definitions of "clone" and "cloning," it is easy to see why defining terms is essential for an informed discussion. If someone says "clone" and means a mammal that's genetically identical to its parent, but the listener pictures a genetically identical population of bacteria, misunderstandings will exist from the outset and the discussion will not prove to be very productive.

For purposes of our discussion, we will use "clone" to mean a collection of genetically identical multicellular organisms.

What mammalian cloning is not. Cloning mammals is not a matter of creating a copy of an organism instantaneously in the laboratory. If the goal of the work is to produce a whole organism and not simply maintain cells in culture, the clone must go through normal mammalian gestation and development. The researcher must implant the eggs or embryonic cells into the uterus of a surrogate mother very early in the life of the embryo; then the organism must go through the gestation period typical for that organism. For example, the embryo that became Dolly was implanted into a ewe after having been cultured within sheep oviduct tissue for only 6 days. Dolly was born approximately 6 months later, the typical gestation period for sheep.

Nor is a mammalian clone a precise copy of the organism that served as the source of the genetic material. As we discussed earlier, environmental influences, including the uterine environment during gestation in mammals, provide powerful input into the ultimate organism. To assume clones are perfectly identical copies of one another is to ignore the "nurture" component of nature + nurture = an organism.

Gathering Facts

What are some of the facts about Dolly's beginnings? Dolly was the first animal to be cloned from a cell from an adult animal. Scientists removed 277 cells from the udder of an adult sheep, then fused those cells with 277 unfertilized egg cells from which the nuclear genetic material had been removed. After culturing the resulting embryos for 6 days, scientists implanted the 29 embryos that had developed normally into surrogate mothers. Only one produced a live lamb, Dolly, approximately 5$\frac{1}{2}$ months later.

What makes Dolly special? Since the mid-1980s, researchers have used *embryonic* cells as sources of genetic material to clone sheep, cows, and other mammals. Until Dolly was born, attempts to use *adult* cells as sources of genetic material for cloning had ended in failure. Scientists assumed that cells containing adult genetic material could not develop into complete organisms because the adult cells had become specialized into a certain cell type and could not shed their specific role and "remember" how to give rise to a complete organism. In other words, scientists thought that when certain sections of the cell's DNA "turned off" during cell specialization, it could not be turned on again. Dolly proved that differentiation was not irreversible and, in being born, challenged a fundamental truth of developmental biology.

That is the scientific achievement that got lost in the hoopla: the despecialization of genetic material that had been committed to a special function and its ultimate reprogramming into embryolike genetic material capable of directing the development of a complete organism. When you read in news reports that scientists are excited about Dolly, it's this scientific discovery that excites them, not the prospect of making genetically identical copies of an organism. That's old news.

Nuclear transfer of genetic material from adult cells is also important because it introduces the possibility of producing organisms from cultured cells grown in culture. We will explain the implications of this breakthrough later.

Putting Cloning in Context with Other Technological Developments

What "unnatural" reproductive technologies have been used in the past?

Livestock breeding. Animal breeders began to use techniques that interfered with natural methods of reproduction approximately 50 years ago. Some of the techniques listed below are not cloning techniques, but all are "not natural." Animal breeders developed these techniques to incorporate desired genetic changes into a herd more quickly. As a result, livestock animals are healthier, and the food derived from them is cheaper and of higher quality.

ARTIFICIAL INSEMINATION. Artificial insemination, which began to be widely used 45 years ago, has allowed U.S. farmers to produce more milk with 10 million cows than they did in 1945 with 25 million. Semen from the best bulls is frozen and used repeatedly to inseminate cows anywhere in the world. Because of efforts to keep the bull studs healthy, calves produced through artificial insemination are healthier. In addition, its low cost—approximately $10 per breeding—does not place small farmers at a disadvantage when compared to large farmers.

MULTIPLE OVULATION EMBRYO TRANSFER. Commercial use of embryo transfer as a method for genetically improving livestock herds began in the early 1970s, and now 100,000 calves are produced annually by this method. To produce calves by embryo transfer, a valuable cow with superior genetic attributes is inseminated after being treated with hormones that stimulate ovulation. The resulting four or five embryos, which are not genetically identical to each other or to either parent, are removed from the mother and placed in surrogate mothers who are less valuable than the genetic mother. The genetic mother does not become pregnant, and thus another reproductive cycle is triggered. Not only has embryo transfer increased the speed with which superior animals can be produced and bred, but it is also an important means for transporting superior genetic material to other countries. Embryos carry fewer diseases than semen or whole organisms, and transportation costs are much lower than for adult animals. Multiple ovulation embryo transfer is not a cloning technique.

EMBRYO SPLITTING. Embryo splitting, or embryo twinning, has been used in cattle breeding for approximately 15 to 20 years and is similar in concept to embryo transfer. At a very early stage in development, an embryo is split in two, and each half is implanted in a surrogate female. In this case, the resulting embryos are genetically identical, so embryo splitting is a cloning technique.

NUCLEAR TRANSFER FROM EMBRYONIC CELLS. Nuclear transfer from embryonic cells is a relatively recent development that bears a resemblance to both embryo splitting and adult cell cloning. It involves taking the nucleus from an embryonic cell of one individual and placing it in the egg (from which the nucleus has been removed) of another individual (Figure 2.5). In the mid-1980s a number of research teams successfully produced calves by transferring nuclei from 64-cell-stage embryos into enucleated eggs and implanting these eggs in surrogate females. A variation on this theme is fusing individual cells taken from a 64-cell-stage embryo with 64 eggs whose nuclei have been removed, then implanting these into surrogate mothers. This technique is called nuclear fusion. The distinction between nuclear transfer and nuclear fusion is relevant, because in nuclear transfer, mitochondrial DNA, which carries approximately 1% of our genetic information, is provided by only one individual, the egg donor. In nuclear fusion, mitochondria from both the embryo cells and the eggs may contribute genetic material to the resulting organisms.

NUCLEAR TRANSFER FROM ADULT CELLS. The trait that distinguishes nuclear transfer from adult cells from the one described above is the source of the genetic material: adult cells versus embryonic cells. Nuclear fusion is also used to transfer adult genetic material into egg cells. This is the technology used to produce Dolly. Both the nuclear transfer and nuclear fusion techniques are cloning techniques.

Laboratory animals. The application of cloning by nuclear transfer is not unique to mammals. Since the 1950s, developmental biologists have used nuclear transfer from embryonic cells to eggs to study amphibian development.

Humans. In vitro fertilization has been used since 1978 as a method for conceiving children in infertile couples. Eggs and sperm are placed into a test tube, where fertilization occurs. The resulting embryos (usually more than one) are then implanted into a female who is usually, but not always, the biological mother. Clearly, in vitro fertilization is not a form of cloning, because genetic material from two sources is combined, creating a genetically unique individual. It is, however, a reproductive technology that is "not natural."

Assessing Benefits and Costs of Animal Cloning

What are some of the benefits of animal cloning?

1. Animal cloning should allow great progress in understanding what turns genes on and off.
2. By using genetically identical animals in experiments, scientists are able to get results quicker and use fewer animals, because the variation in experimental results due to genetic variation is eliminated.
3. Improvements in livestock can be incorporated into herds much more rapidly.
4. Animal cloning could help to save endangered species. In August 1998, a rare ungulate species was successfully cloned at a zoo in the United States.

Figure 2.5 Nuclear transfer. (Photograph courtesy of VU.)

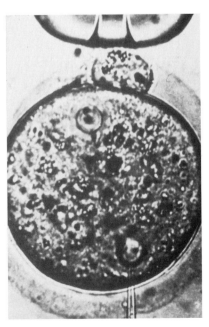

5. When donor cells are in culture, scientists can introduce desirable genetic changes before the cell is fused with an egg to form the embryonic clone. Therefore, animal clones of a consistent quality can be reproduced in large numbers quickly and cheaply. As a result, animal cloning, used in conjunction with genetic engineering, could provide

- an alternative and more efficient way of producing transgenic animals that can be used in the production of human therapeutic proteins or for organs to be used in xenotransplantation
- excellent animal models for studying genetic diseases, aging, and cancer, for discovering new drugs, and for evaluating other forms of therapy, such as gene and cell therapy

What are the concerns people have voiced? The news of Dolly's birth triggered widespread fear and, in many countries, a call for a total ban on cloning humans. A few have suggested that all animal cloning should be banned. Those who want to ban cloning have justified such bans with vague statements that cloning "poses new ethical problems" or "is contrary to nature." To assess the validity of these claims, we need to determine if cloning does pose new ethical problems and, if so, we need to define what they are (see below). Above, we provided information for evaluating the naturalness or unnaturalness of cloning and compared cloning with other widely used unnatural means of producing offspring, both human and non-human. If we ban cloning because it is "contrary to nature," then we should reconsider the other unnatural reproductive technologies we have used for decades.

Others have sought to ban cloning because of the horrible things they feel could be done if this technology is extended to humans. They have speculated about armies of Hitler clones and brainless cloned duplicates serving as private stocks of replacement parts. Such sensationalist speculation is counterproductive, at best. To assess the costs and benefits of cloning, we must put concerns in perspective by aligning them with the more immediate and realistic effects of this breakthrough.

What are the current problems associated with animal cloning research based on the use of nonembryonic nuclear transfer? Clearly, using the Dolly research as an example, the success rate for animal cloning is very low (1 in 277). Few implantable embryos result from the fusion of an adult cell with an enucleated egg cell. Of those that are implanted, most have developmental abnormalities and do not survive. Of those, most die early in the pregnancy, and a few die in late pregnancy or soon after birth. Since Dolly's birth, a number of other species have been cloned by

nuclear transfer from adult cells—mice, cows, goats—and the success rates were equally dismal.

Scientists are not sure about the long-term health implications of this type of cloning. Because "old" DNA is used to produce the clone, the clone may inherit mutations caused by ultraviolet light or other environmental influences, or chromosomes with shortened telomeres. This could predispose the clone to premature aging or a higher incidence of environmentally induced cancer.

Other issues related to livestock management may surface, because a herd composed of clones would be genetically homogeneous. This lack of genetic variation could result in health problems that affect the entire herd and not just a few individuals. For example, all individuals might be susceptible to the same infectious diseases.

As for ethical issues associated with this research, it is difficult to think of ethical issues that are not associated with animal research in general but are unique to the cloning research.

What are the ethical issues involved in cloning humans? In discussing the question of cloning humans, it is essential that we make a clear distinction between cloning to produce 32- to 64-cell embryos and cloning to make babies. In other words, we must recognize that if the type of cloning being discussed involves implanting an embryo into a woman in hopes of producing a cloned offspring, the ethical issues will differ (for most people) from those related to simply growing a mass of undifferentiated mammalian cells in culture. In our discussion, we will focus only on the ethical issues associated with cloning to produce a baby. It is also important to point out that we do not know whether human cloning from specialized cells with adult genetic material is even possible.

Assuming that the problems associated with cloning in sheep would be similar to those in human cloning, then the great majority of the embryos implanted would not survive. Of those that survive, shortened telomeres and environmentally induced mutations may pose the same potential risks as they do for cloned sheep. Even if our experience with cloned sheep identifies ways of reducing these risks, is it ethical to perform similar experiments with women to determine whether the risks in humans are significant?

Under what circumstances, if any, might the cloning of humans be acceptable? Some people believe that there may be situations in the future when nuclear transfer of DNA into an egg cell in hopes of producing a baby could be beneficial. One example often cited involves people with mitochondrial DNA genetic diseases. Some types of mitochondrial DNA diseases cause blindness and epilepsy. Other evidence suggests that a form of Alzheimer's is transmitted via mitochondrial DNA. By removing the nucleus from an embryo produced by in vitro fertilization and inserting it into an egg from a donor, the resulting baby would not have the disorder inherited through mitochondrial DNA, but 99% of its DNA would come from its two parents.

Another example that some find acceptable is those cases of infertility that cannot be overcome with in vitro fertilization. The nucleus from an adult cell of one of the parents could be added to an enucleated egg taken from the mother and then implanted. If the donor of the nuclear genetic material is the father, the baby would be the biological offspring of both parents, but 99% of its genes would come from only one parent—the father.

The third example sometimes given as a potentially acceptable use of cloning technology does not deal with cloning an embryo with the intent of producing a baby, but simply with cloning human cells in culture. This involves producing cells for cell therapy or tissue engineering. The value is that the cells could be a perfect match genetically if the nucleus inserted into the enucleated egg was taken from a cell of the recipient of the new cells or tissue.

Summary

The societal issues raised by advances in biotechnology are often not unique to biotechnology but are the same issues raised by other technologies. However, other issues result solely from the new powers biotechnology gives us. By analyzing the issues of biotechnology within the continuum of technological change, we gain information that will help us determine how best to use the extraordinary capability biotechnology provides. By carefully overseeing its development, we are better able to ensure that the decisions shaping its future are informed by past mistakes and current understandings. We have an opportunity to avoid the mistakes of the past, to act a little wiser, to use what we have learned.

But sometimes careful oversight of technological development becomes excessive vigilance that encourages stagnation, and moving cautiously may turn into not moving at all. Should we ask biotechnology to pay the price for previous technologies by scrutinizing it so thoroughly that we stifle its development and effectively limit our options?

Today many people are antitechnology because of what they perceive as past mistakes. They equate technology with problems, as did many people in each generation that was presented with new technologies. Unless we are prepared to turn our backs on progress—to say "no thank you" to a cure for AIDS, to return to the days when 50% of the population worked on farms to ensure an adequate food supply, to stop pursuing new and better ways to treat cancer—we must honestly admit to ourselves that we want technologies that make our life on earth better. No matter how antitechnology people may claim to be, would they be willing personally to pay the price for deterring advances in biotechnology?

Each individual has a responsibility to see that biotechnology is developed and used wisely. We must do whatever we can to avoid mistakes and ensure that biotechnology benefits not only us but also the other organisms that share this planet with us.

Selected Readings

Alpern, Kenneth, D. (ed.). 1992. *The Ethics of Reproductive Technology*. Oxford University Press, Oxford, England.

Beardsley, Tim. 1999. Stem cells come of age. *Scientific American* 281:30-31.

Berg, Paul, and Maxine Singer. 1998. The regulation of human cloning. *Science* 282:410-412.

Bronkowski, J. 1956. *Science and Human Values*. Harper & Row, New York, N.Y. A classic work.

Burley, J. (ed.). 1999. *The Genetic Revolution and Human Rights*. Oxford University Press, Oxford, England.

Congressional Office of Technology Assessment. 1990. *Genetic Monitoring and Screening in the Work Place*. U.S. Government Printing Office. Washington, D.C. To order publications from the U.S. Government Printing Office, write to the Superintendent of Documents, Mail Stop: SSOP, Washington, DC 20402-9328.

Frankel, Mark, and Albert Teich (ed.). 1995. *The Genetic Frontier: Ethics, Law and Policy*. AAAS, Washington, D.C.

Gasser, C. S., and R. T. Fraley. 1992. Transgenic crops. *Scientific American* 266:62-65.

Gearhart, John. 1998. New potential for human embryonic stem cells. *Science* 282:1061-1062.

Gray, P. E. 1989. The paradox of technological development. In Jesse Ausubel and Hedy Sladovich (ed.), *Technology and Environment*, National Academy Press, Washington, D.C. This edited volume includes a number of good articles on the relationship between technological advance and environmental issues.

Green, N. G., H. M. LeBaron, and W. K. Moberg. 1990. *Managing Resistance to Agrochemicals: From Fundamental Research to Practical Strategies*. American Chemical Society, Washington, D.C.

Griffiths, Anthony J. F., Jeffrey H. Miller, David Suzuki, Richard C. Lewontin, and William M. Gelbart. 1996. *An Introduction to Genetic Analysis*. W. H. Freeman & Co., New York, N.Y. An excellent general genetics textbook that discusses not only molecular genetics but also quantitative and population genetics.

Holland, A., and A. J. Holland (ed.). 1997. *Animal Biotechnology and Ethics*. Chapman & Hall, Ltd., London, England.

Holtzman, Neil. 1999. Are genetic tests adequately regulated? *Science* 286:409.

Kahn, Paula. 1996. Coming to grips with genes and risk. *Science* 274:496–498.

Kevles, Daniel, and Leroy Hood (ed.). 1993. *The Code of Codes. Scientific and Social Issues in the Human Genome Project*. Harvard University Press, Cambridge, Mass.

Knobil, E., and J. D. Neil (ed.). 1998. *Encyclopedia of Reproduction*. Academic Press, San Diego, Calif.

National Research Council. 1986. *Pesticide Resistance: Strategies and Tactics for Management*. Committee on Strategies for the Management of Pesticide Resistant Pest Populations. National Academy Press, Washington, D.C.

National Research Council. 1989. *Field-Testing Genetically Modified Organisms: Framework for Decisions*. National Academy Press, Washington, D.C. One of many studies by broadly based scientific advisory committees dealing with environmental releases of genetically engineered organisms.

Organization for Economic Cooperation and Development. 1994. *Field Releases of Transgenic Plants, 1986–1992: an Analysis*. OECD, Paris, France. This report, by an international agency, is available by contacting the OECD website (www.oecd.org).

Pederson, Roger A. 1999. Embryonic stem cells for medicine. *Scientific American* 280:68–73.

Richardson, H. 1999. *How to Clone a Sheep*. Oxford University Press, Oxford, England. Appropriate for young students.

Thompson, James A., J. Itskovitz-Eldor, S. S. Shapiro, M. A. Waknitz, J. J. Swiergiel, V. S. Marshall, and J. M. Jones. 1998. Embryonic stem cell lines derived from human blastocysts. *Science* 282:1145–1147.

Tiedje, J. M., R. K. Colwell, Y. L. Grossman, R. Hodson, R. E. Lenski, R. N. Mack, and P. J. Regal. 1989. The planned introduction of genetically engineered organisms: ecological considerations and recommendations. *Ecology* 70:298–315. This excellent paper presents the position of the Ecological Society of America on the ecological issues of field releases of genetically engineered organisms.

Traynor, Patricia, and Ruth Irwin (ed.). 1999. *Ecological Effects of Pest Resistance Genes in Managed Ecosystems*. Virginia Polytechnic Institute and State University, Blacksburg, Va. These proceedings are available for free by contacting the Information Systems for Biotechnology at isb@vt.edu.

Wenk, Edward. 1986. *Tradeoffs and Imperatives of Choice in a High-Tech World*. Johns Hopkins Press, Baltimore, Md.

Wilmut, Ian. 1998. Cloning for medicine. *Scientific American* 279:58–64.

Wright, Shirley. 1999. Human embryonic stem-cell research: science and ethics. *American Scientist* 87:352–361.

Zorpette, Glenn, and Carol Ezzell (ed.). 1999. *Your Bionic Future*. Scientific American Books, W. H. Freeman & Co., New York, N.Y.

Genes, Genetics, and Geneticists   3

The Centrality of Genes

The remainder of this book is about genes: what they are, what they do, how they do it, and how we are using them. But before we delve into the workings of genes on a molecular level, let's step back and look at these extraordinary particles from a broader perspective.

Everyone recognizes that we must learn about genes to understand genetics and heredity, the transmission of inherited traits from one generation to the next. What you may not realize is that an understanding of genes is central to an understanding of all of biology. Each of the diverse branches of biology can be seen as a way of investigating questions about genetics.

- Developmental biology is, in large part, the study of gene regulation: what turns genes on and off.
- Physiology and its companion science, morphology, focus on the structural and functional manifestations of gene expression.
- Ecologists study the genetic adaptation of organisms to their environments.
- Taxonomy is the study of the genetic differences and similarities within and between species.
- Evolutionary biologists investigate changes in gene frequencies in populations over time.

You can take any biological phenomenon of interest and explore it from the angle of the gene.

First, using your mind's eye as a microscope, put the phenomenon under maximum magnification and ask about the molecular details of its genetic basis. Lower the magnification to the physiological level and ask how the function of this particular gene integrates with the functions of other genes in producing a living organism.

Now look at the whole organism. How much of what you see has a genetic basis? How much can be explained by environmental influences? How do the two interact?

Now take a giant step back and use your mind's eye like a telescope to explore the same gene over time and space. How has that gene's structure changed over time? Has the change in structure brought about a concomitant change in function? How much of that change has been meaningful in an evolutionary sense? How does the gene vary within the population? Between populations, but within the species? Between species?

To an outsider, biology appears to be a fragmented field of study with many loosely related subdisciplines, and textbooks that dissect the branches of biology into separate, apparently unrelated chapters do little to correct this misconception. Yet all of the various threads of biological thought can be woven into a coherent whole with genetics as the unifying concept. Keep returning to the central landmark genes provide, and you will be better able to integrate the field of biology into a coherent whole.

Genetics is central to biology, the study of life, because genes are central to life. The fundamental characteristic of living organisms is reproducibility. Living organisms produce more living organisms like themselves. For organisms other than those that reproduce by simple division, reproduction involves both the transmission of hereditary information from parent to offspring and a developmental process that transforms that information into a living organism. Both of these hallmarks of living organisms reflect the functioning of genes: faithful replication and information transmission. If life depends on genes, it only makes sense that the study of life, biology, would have genetics at its core.

A Brief History of Genetics

In the beginning

The relationship between genes, reproduction, and development seems obvious to us now, but it was not always so. In fact, for thousands of years—99% of human history—we were totally ignorant of the biological basis of reproduction. Sperm and eggs are invisible to the naked eye, and we could understand only those phenomena we could see. The invisible details of reproduction were left to our imaginations, and imagine we did!

Some truly great scientists developed theories of reproduction that seem silly to us now. Hippocrates, Aristotle, and, much later, even Charles Darwin believed that offspring resulted from the blending of male and female genital secretions that contained "seeds" that had been shed by each body part: eye seeds, liver seeds, heart seeds. During reproduction, these seed-filled fluids coagulated and were transformed into an organism as each organ seed developed.

It seems to me that among the things we commonly see there are wonders so incomprehensible that they surpass all the perplexity of miracles. What a wonderful thing it is that this drop of seed from which we are produced bears in itself the impressions not only of the bodily form, but also the thoughts and inclinations of our fathers! Where can that drop of fluid contain that infinite variety of forms?

Michel de Montaigne, 1570

The invention of the microscope in the 1600s permitted observation of human sperm and eggs but did not quell the development of imaginative theories of reproduction. Many scientists believed that tiny humans resided in each sperm cell and that these minuscule humans took root and grew in the female's uterus. In this scenario, the male provided all of the genetic material, and the female provided a nurturing womb for safe development. Some scientists even claimed that, using microscopes, they could see little chickens and horses in sperm taken from those animals. Other scientists were skeptical, however. In fact, one scientist ridiculed this theory of reproduction by saying he had seen a tiny human enclosed in a sperm take off its even tinier coat!

Even though our ancestors were ignorant of the precise details of reproduction at the microscopic level, they could easily see consistent hereditary patterns in the natural world. Life begets life, and offspring look like their parents.

As early as 10,000 years ago, the first agriculturists put these observations to work by systematically crossbreeding the plants and animals they had begun to domesticate. As soon as we domesticated living organisms, we began to genetically manipulate them through this selective breeding. Our agrarian ancestors accomplished this genetic manipulation by trial and error, relying only on minimal understanding acquired through observation. Their knowledge was incorrect and incomplete but nonetheless sufficient to enable them to domesticate and genetically improve virtually all of the crops we use today for food and fiber.

As successful as the early agriculturists were, their attempts at selective breeding were still hit-or-miss ventures. Sometimes the "magic" worked, sometimes it didn't. Such uncertainty is not very satisfying to humans. We want to explore and understand things from the inside out so that we can exert more control over our lives.

The mysteries of heredity provided us with a fertile field for such exploration. Inheritance patterns often seem unpredictable. Traits disappear in one generation and reappear in another. Some observable traits have a genetic basis and therefore can be subjected to successful selection; others don't. Some traits are controlled by one gene, while others are controlled by hundreds of genes.

For centuries, most people invoked supernatural powers to understand the apparently inexplicable nature of biological inheritance. A few, however, persisted in trying to understand. Through their efforts we have come to understand so much about genes that now not only do we know what genes are made of and what they do but we can also manipulate them to our advantage. In other words, we know genes inside out.

How did we come to learn so much? Who were the pioneers in the field of genetics? What questions did they ask, and how did they answer them?

Placing our current understanding of genetics in a historical perspective is a valuable exercise. Tracing the development of this field not only informs us about the workings of biological systems at a basic level but also provides insights into how science advances. Sometimes the next critical discovery in a field is obvious, and scientists race each other in an attempt to be the one who captures the prize. At other times, the most important discoveries are ignored because the timing isn't right. The field of genetics was born at such a time. No one could hear what the "Father of Genetics" was saying, because he was years ahead of his time.

Discovery: the discrete nature of genes

One evening in February 1865, a monk named Gregor Mendel presented a lecture to the local scientific society in Brno, Czechoslovakia. He described in great detail the results of 8 years of data on crossbreeding thousands of peas. No one was impressed, but the results he shared that night were some of the most important findings in the history of biology. We are fortunate that his paper, like all papers of that society, was published in a journal kept in libraries across Europe. Thirty-five years later, his work was rediscovered, and only then did scientists understand its importance.

Every biology textbook discusses Mendel's work in detail. All biology teachers and students know of Mendel's work and could probably describe the methods and results in their sleep. Yellow peas, green peas, round peas, wrinkled peas. Do the crosses; see the results; now do more crosses; count more offspring. But what was so significant about Mendel's work? Why do biologists speak of him in reverential tones?

Stated simply, Mendel changed the way we view the natural world. Although he never used the word "gene," his work revealed the inherent nature of genes and gave birth to a new branch of biology: genetics, the study of heredity. He demonstrated that the hereditary substance that passes from one generation to the next is organized as discrete packets of information. Heredity does not involve blending together fluidlike contributions. Instead, heredity depends upon combining discrete particles from both parents (Figure 3.1). When egg and sperm fuse during fertilization, the maternal and paternal hereditary particles do not become joined together but retain their distinct identities.

What is so important about having hereditary information packaged as discrete particles? Bundling up genetic information into separable chunks provides a constant means of generating genetically variable offspring, even in the absence of **mutation.** Without genetic variation, life on earth could not have evolved. Now, that's important.

Figure 3.1 Diagram of the different models of inheritance: fluid blending and discrete particle.

Fluid blending   Discrete particle

Because genes are discrete particles, the maternal and paternal genes for a given trait (**alleles**) separate from each other during the production of gametes. To understand the importance of this separation in producing genetic variation, contrast the two models of heredity: fluid blending and discrete particle.

The Fluid-Blending Model of Inheritance

Imagine pouring together milk (maternal genes) and chocolate syrup (paternal genes) to make chocolate milk (offspring). Now take the chocolate milk and divide it into four equal parts (gametes) and pour two of them together (fertilization). You still have chocolate milk. No matter how many times you repeated this dividing and mixing, crossing offspring gametes to produce the next generation, you would still end up with chocolate milk offspring and only chocolate milk offspring. You would never again have a glass of pure chocolate or a glass of pure milk. In other words, the offspring would always be genetically identical.

Since genetic variation is the key to the evolution of life on earth, such a system of genetic blending would be an evolutionary dead end. For evolution to proceed, genetically variable offspring must be generated constantly. When used in conjunction with sexual reproduction, the discrete nature of genes provides this variation.

The Discrete Particle Model of Inheritance

To demonstrate the discrete particle model, use separate objects—jelly beans, M&Ms, marbles, index cards—as genes, and mimic sexual reproduction: produce gametes through meiosis, and join them in fertilization. Now mimic a second round of reproduction with the gametes of the offspring. For example, if you use two red (maternal) and two green (paternal) jelly bean genes, the first-generation offspring will all have a single genotype: red/green. Cross those individuals, and you will have three genotypes: pure red, red/green, and pure green—genetically variable offspring.

Mutation was not a factor in creating these genetically variable offspring. The variation results from reassorting existing alleles through meiosis and fertilization, the constituent elements of sexual reproduction.

Mendelian principles. This separation of the maternal and paternal genes for the same trait (male and female alleles) during gamete formation is known as Mendel's Principle of Segregation (Figure 3.2).

A second important Mendelian principle also depends upon having hereditary information organized as discrete particles. This second principle, the Principle of

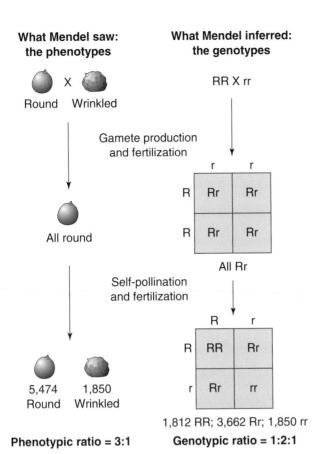

What Mendel saw:
the phenotypes

Round X Wrinkled

Gamete production
and fertilization

All round

Self-pollination
and fertilization

5,474 Round   1,850 Wrinkled

Phenotypic ratio = 3:1

What Mendel inferred:
the genotypes

RR X rr

	r	r
R	Rr	Rr
R	Rr	Rr

All Rr

	R	r
R	RR	Rr
r	Rr	rr

1,812 RR; 3,662 Rr; 1,850 rr

Genotypic ratio = 1:2:1

Figure 3.2 The Principle of Segregation. Mendel observed that experimental crosses between pea plants with round or wrinkled seeds yielded plants with round seeds in the first generation and a 3:1 mix of plants with round and wrinkled seeds in the second generation. From this result and observations of other characteristics, he inferred that maternal and paternal hereditary information is packaged as discrete particles and that maternal and paternal particles separate from each other during gamete formation.

Independent Assortment, involves separation of male and female genes for different traits (Figure 3.3).

Its essence is this: on average, any gamete has an equal number of maternal and paternal alleles. On a purely mechanical level, this means that during meiosis, the paternal **chromosomes** do not line up on one side of the divide and the maternal chromosomes on the other. Maternal and paternal chromosomes do not stick together as a group during gamete production. In humans, each new gamete has 23 chromosomes. On average, 11 to 12 are of maternal origin, and 11 to 12 are paternal. (Of course, crossing over between homologous chromosomes makes the situation not as tidy as just described when one looks at the specific maternal and paternal genes on chromosomes.)

Dominant and recessive alleles. The discrete nature of the gene permitted Mendel to observe and de-

scribe the concept of dominant and recessive characteristics. Think back to the chocolate milk versus jelly bean analogy. For the chocolate × milk cross, all offspring in all generations are chocolate milk. For jelly beans, assume that one color is dominant and the other is recessive. The first cross, red × green, tells us very little, because all of the offspring look the same and have the same genetic makeup. The revealing cross is the next one, in which red/green offspring are crossed with each other. This cross results in both red and green offspring as well as three genotypes. A trait that disappeared in the first generation reappears in pure form in the second. The traits do not blend into each other. One trait (**dominant allele**) simply overpowers the other (**recessive allele**) in the first-generation offspring.

Genotypes and phenotypes. Because Mendel chose traits that exhibited clear dominance relationships, he was able to elucidate another concept fundamental to our understanding of inheritance: the relationship between **genotype** and **phenotype**. The outward appearance of an organism (phenotype) may or may not directly reflect the genes that are present (genotype). When you want to understand the details of an organism's genetic makeup, sometimes the phenotype is helpful, and other times it can be very misleading.

Genetics, statistics, and probability. Another ancillary but very important feature of the discrete nature of genes involves not biological inheritance but the ease with which genetic traits can be scientifically investigated. Because genes are discrete entities (like coins, dice, or cards), the observed results of crosses can be scientifically analyzed by using the laws of statistics and probability (as with coins, dice, and cards). Using observable differences, scientists can quantify patterns of inheritance and infer the genetic basis of traits without knowing the details of the molecular biology of the gene of interest.

It would be difficult to overstate the importance of Mendel's findings to our understanding of the workings of biological systems from DNA to evolution. We could not learn all we know about the natural world until we could appreciate and understand his remarkable work on the particulate nature of inheritance. He could see patterns where everyone else saw disarray. From these patterns he made inferences and established the fundamental concepts of genetics before anyone knew genes existed.

And yet, no one was impressed the night he presented his results or for 35 years afterward. Mendel

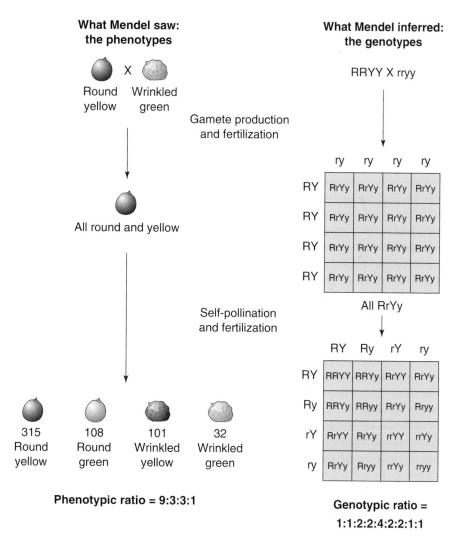

**What Mendel saw:
the phenotypes**

Round   X   Wrinkled
yellow   green

Gamete production
and fertilization

All round and yellow

Self-pollination
and fertilization

315   108   101   32
Round   Round   Wrinkled   Wrinkled
yellow   green   yellow   green

Phenotypic ratio = 9:3:3:1

**What Mendel inferred:
the genotypes**

RRYY X rryy

	ry	ry	ry	ry
RY	RrYy	RrYy	RrYy	RrYy
RY	RrYy	RrYy	RrYy	RrYy
RY	RrYy	RrYy	RrYy	RrYy
RY	RrYy	RrYy	RrYy	RrYy

All RrYy

	RY	Ry	rY	ry
RY	RRYY	RRYy	RrYY	RrYy
Ry	RRYy	RRyy	RrYy	Rryy
rY	RrYY	RrYy	rrYY	rrYy
ry	RrYy	Rryy	rrYy	rryy

**Genotypic ratio =
1:1:2:2:4:2:2:1:1**

Figure 3.3   The Principle of Independent Assortment. By observing the hereditary patterns of two separate traits, seed shape and color, Mendel inferred that during gamete formation, the alleles for one trait segregate independently of the alleles for a trait located on a different chromosome. What would Mendel have seen if the genes for seed color and shape were adjacent on a chromosome?

revolutionized biology 16 years after he died. Science works like that more often than you might guess.

Discovery: the chromosomal nature of inheritance

Once Mendel established that the hereditary material is organized into packets of information that separate from each other during gamete formation, scientists had to localize these particles within the cell.

Improvements in microscopy permitted the next set of discoveries, because we could actually see, on a cellular level, the phenomena Mendel had described. Using microscopy, we began to uncover the underlying mechanism of the principles Mendel had inferred

from observing phenotypic variation. In other words, technological advances enabled the next generation of scientists to answer the next question: Where are these discrete particles located?

In 1902, 2 years after Mendel's work was rediscovered, Walter Sutton observed chromosomes behaving in ways that resembled the segregation of hereditary material Mendel had postulated. Studying meiosis in grasshoppers, Sutton, a graduate student at Columbia University, observed that chromosomes occur in morphologically similar pairs and that the two members of a chromosome pair separate from each other during gamete formation. He used this cytological evidence in conjunction with Mendel's findings to hypothesize that the hereditary material is associated with chromosomes. When he discovered this critical relationship, he ran excitedly to his major professor and mentor, E. B. Smith, who was one of the foremost biologists of his time. This is how Dr. Smith described what happened.

> *I well remember when in the early spring of 1902, Sutton first brought his main conclusion to my attention. I also clearly recall that at the time I did not fully comprehend his conception or realize its entire weight.*

Smith may not have understood what Sutton discovered, but Sutton was right on target. His theory was confirmed and elaborated by a remarkable group of biologists, also at Columbia University. In a 5-year period, they uncovered the major details of the chromosomal basis of heredity in Thomas Hunt Morgan's laboratory, an infamous room known to biologists as The Fly Room.

Thomas Hunt Morgan once joked that God made the fruit fly just for him. With its relatively simple genetic makeup of only eight chromosomes (four pairs) and a very short generation time (12 to 14 days), the fruit fly, *Drosophila melanogaster,* is an excellent experimental animal for studying genetics. Much of our basic understanding of classical genetics comes from fruit fly studies; the great bulk of that early work came from Morgan's laboratory.

In the early 1900s, Morgan began to use *Drosophila* to address problems in evolution. He was not particularly interested in studying the cellular mechanics of inheritance. Instead, as an experimental scientist, he wanted to test existing theories about the role of mutation in evolution. Over and over he bombarded fruit flies with X rays or exposed them to chemicals in an attempt to mutate an easily observable trait.

One day, he finally got lucky. In the midst of thousands of red-eyed flies was a mutant at last: a white-eyed male. This single mutant mated with a red-eyed female, left its mutated gene, and in that act provided the raw material for a series of experiments that provided the foundation for classical genetics.

By all accounts, Morgan was quite an interesting character. On the one hand, he was quite informal, humorous, and a stereotypical absent-minded scientist. He was often mistaken for a beggar because his clothes were so ragged. His lab was just as disorderly as his clothes. Many desks and tables were crammed into a tiny room, leaving little room to move. Hundreds of half-pint milk bottles filled with rotting bananas and thousands of flies were everywhere. An equal number of flies lived free in the room, hovering around

ubiquitous stalks of bananas in various stages of decay. He recorded data on scraps of paper and the backs of envelopes, which promptly got buried under the piles of paper on his desk.

Despite his appearance, he was from one of the finest families in Kentucky and an heir to the fortune of industrialist J. P. Morgan. He was known for giving money to graduate students who had a hard time making ends meet. He was also a brilliant investigator and an inspiring leader. He led an extraordinary group of young students to key findings that served as the footings on which to build our understanding of genetics. In a very few years, Morgan and his students, Alfred Sturtevant, Calvin Bridges, and Herman Muller,

- proved genes are chromosomally located
- introduced the concept of **sex-linked inheritance**
- introduced and elaborated the concept of genetic **linkage**
- originated the idea of **gene mapping**
- constructed the first genetic map of a chromosome
- demonstrated crossing over between homologous chromosomes

Discovery: the chemical nature of genes

Now that scientists knew where genes were located, they began to question the molecular makeup of genes.

In 1869, a German chemist, Frederick Miescher, had isolated a novel substance from the nuclei of white blood cells. He gave it the name "nuclein." Unlike proteins, nuclein had a high concentration of phosphorus. His colleagues tried to convince Miescher that nuclein was just another protein and phosphorus was a contaminant, but Miescher persisted in his belief that nuclein was another type of molecule. Eventually, chemical analysis revealed that chromosomes are made of both protein and Miescher's nuclein, which we now call deoxyribonucleic acid (DNA).

Of these two substances, which carries hereditary information? Almost everyone thought that proteins must be the hereditary material. Chemically, proteins are much more complicated than DNA, and everyone knew that the molecule of heredity had to be able to contain an extraordinary amount of information. It did not seem possible that a simple molecule like DNA could be the hereditary material.

Of course, we now know that they were wrong. The heredity material is not protein but DNA. But how did we discover this?

The Transforming Factor

The first chapter of the story occurred in 1928 in the laboratory of a man interested not in heredity or chromosomes but in vaccines. Frederick Griffith was trying to understand the differences between the strains of *Diplococcus pneumoniae* that cause pneumonia (virulent strains) and the strains that are harmless (nonvirulent). The virulent forms had polysaccharide capsules that gave them a smooth appearance. The nonvirulent bacteria had no capsules and were rough.

Griffith hoped that either heat-killed virulent strains or live nonvirulent strains could be used as a vaccine. He mixed heat-killed virulent and live nonvirulent strains together in hopes of making an effective vaccine. He injected the mixture into mice, and they died. How could dead bacteria be virulent?

He was able to retrieve virulent bacteria from blood samples taken from the dead mice. Something had been transferred from the dead virulent bacteria to the live nonvirulent bacteria. Griffith called this substance the "transforming factor." We now know the transforming factor was DNA, but Griffith didn't know that (Figure 3.4).

The Griffith story also illustrates how science progresses. Often, findings in one branch of science shed light on a different realm.

In 1943, O. T. Avery and his colleagues at the Rockefeller Institute purified the transforming factor and announced that it was DNA. Many scientists doubted this announcement and persisted in believing the hereditary material was made of protein. They doubted that DNA, with only four subunits, could carry enough information to turn a fertilized egg into a human being. Proteins, which have 20 subunits, can carry much more information.

In a way, it seems incredible now that scientists doubted that four subunits could provide sufficient information, when every day we transfer and process tremendous amounts of information via computer languages that have only two subunits!

DNA or Protein?

The final answer to the DNA versus protein debate was provided by definitive experiments conducted in 1952 by Alfred Hershey and Martha Chase. Because of the work done on the atomic bomb in World War II, scientists had access to the radioactive substances that have proved invaluable in scientific investigation. These isotopes were crucial to the Hershey-Chase series of experiments on the molecular nature of the hereditary material.

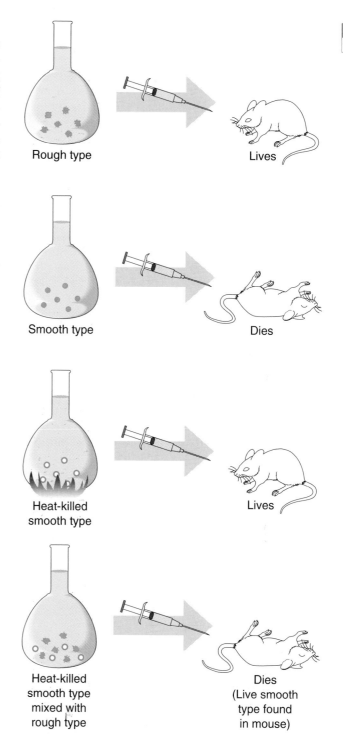

Figure 3.4 Discovery of the "transforming factor" by Griffith. The strain of pneumonia-causing bacteria (smooth) is virulent because of a gene that codes for a protective outer covering. A different strain (rough) does not cause pneumonia because it lacks the gene for the protective covering. When nonvirulent bacteria are mixed with heat-killed virulent bacteria, the nonvirulent bacteria are "transformed" into virulent bacteria. The gene for the protective outer layer moves from the dead smooth bacteria to the living nonvirulent bacteria.

By the 1950s, researchers had begun to use experimental organisms that were even simpler than bacteria, the bacteriophages. Bacteriophages are viruses that infect bacteria. Viruses consist of a coat protein with a small amount of genetic material, usually made of DNA, inside. They infect bacteria by injecting their genetic material into the bacterium while leaving the coat protein outside. Hershey and Chase exploited this bit of viral molecular biology to settle once and for all the question of the molecular nature of the genetic material.

Proteins and DNA differ chemically in many ways, but one simple chemical difference was critical to the Hershey-Chase experiment. DNA has phosphorus but not sulfur, and proteins have sulfur but not phosphorus.

Hershey and Chase grew viruses labeled with the radioactive isotopes of sulfur and phosphorus, ^{35}S and ^{32}P. These viruses were used to infect *Escherichia coli* cells. Then the solutions containing the infected bacteria and the viruses were agitated in a blender and centrifuged to separate the viral coats from the infected bacteria. The supernatant containing the viral coats was rich in ^{35}S, while the pellet of infected bacterial cells contained ^{32}P. These infected *E. coli* cells produced phage progeny that contained ^{32}P and no ^{35}S (Figure 3.5).

Thus, the DNA versus protein debate was finally resolved. Because the phages injected their DNA into the bacteria while leaving their protein coats outside, Hershey and Chase concluded that DNA was the genetic material.

Again, note the recurring themes of scientific progress:

- Technological advances promote scientific discoveries.
- Findings in one branch of science shed light on others.

Discovery: the structure of DNA

As soon as the DNA versus protein debate was settled, another question was immediately born: What was the structure of the DNA molecule? Whatever the structure, scientists knew DNA's function depended on it. They also knew that whoever described the structure would be guaranteed an important place in history. And so the race was on.

Before scientists proved conclusively that DNA is the genetic material, we actually knew a great deal about its biochemistry. We knew all of its components (phosphate, **deoxyribose**, and the four nitrogenous bases) and their molecular structures (see chapter 4). By 1930, we knew that each subunit (**nucleotide**) of

Figure 3.5   The experiments of Alfred Hershey and Martha Chase. One group of viruses containing protein labeled with the radioactive isotope ^{35}S and a second group of viruses containing DNA labeled with the radioactive isotope ^{32}P infected bacterial cells by injecting their genetic material into the cell. Hershey and Chase separated the viral coats from the bacterial cells and found ^{35}S in the viral coats and ^{32}P within the bacterial cells. Viral progeny that resulted from the infection also contained radioactive ^{32}P. Hershey and Chase thus concluded that the genetic material of the virus was DNA and not protein.

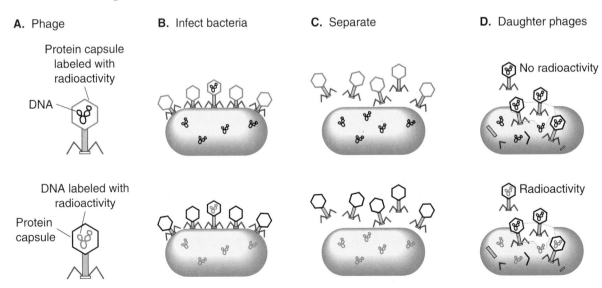

DNA consists of one phosphate, one deoxyribose, and one nitrogenous base. Finally, in 1952, we discovered that the phosphate and deoxyribose moieties are linked together to form a chain and that the nitrogenous base is attached to this chain.

So what else was there to know? The three-dimensional structure. We could not deduce the three-dimensional structure from the information available, and we knew that the key to explaining DNA's function resided in its three-dimensional structure.

Findings from a number of branches of science coalesced and provided us with sufficient information to discover this next piece of the puzzle.

- The physicists Maurice Wilkins and Rosalind Franklin provided data from X-ray diffraction analysis of DNA crystals.
- The chemist Erwin Chargaff discovered that DNA molecules always have equal amounts of **adenine** and **thymine** and equal amounts of **cytosine** and **guanine**.
- Another chemist, Linus Pauling, described the rules governing the formation of chemical bonds and published details of a novel chemical structure certain proteins could assume, the alpha helix.

Watson, Crick, and the Double Helix

Frances Crick, a graduate student in physics at Cambridge, and James Watson, an American geneticist at Cambridge on a postdoctoral fellowship, both young and brash, were determined to be the ones who unearthed DNA's structure, and they did. Their success stemmed not only from their determination but also from a mutual distaste for sloppy thinking; their blunt, open, passionate style of communication; and a bit of luck.

They set out to discover DNA's three-dimensional structure by building a model. In 1953, when they happened upon the structure—a helix composed of two strands running in opposite directions with internal pairing between the bases, adenine, thymine, cytosine and guanine—they knew they had it! They could immediately see one DNA function, information replication, in its structure. Each strand could be used as a template for making another strand. (See chapter 4.)

In their brief, understated report in the journal *Nature* in which they described DNA's structure, Watson and Crick state:

> *This structure has novel features which are of biological interest . . . .It has not escaped our notice that the specific pairing we have postulated im-*

> *mediately suggests a possible copying mechanism for the genetic material.*

In his book *The Eighth Day of Creation*, Horace Freeland Judson beautifully describes their discovery in this way:

> *That morning, Watson and Crick knew, although still in mind only, the entire structure: It had emerged from the shadows of billions of years, absolute and simple, and was seen and understood for the first time.*

Ever since Watson and Crick discovered the structure of DNA, scientist after scientist has described DNA's structure as "beautiful." In reply, nonscientists usually ask why we see beauty in a molecule. DNA's beauty lies in its structural simplicity. It is a simple, straightforward molecule: simple in structure but astoundingly complex in its function.

The elegance of DNA has inspired awe in some nonscientists as well. A quote from Salvador Dali eloquently describes the feeling many scientists share with the painter regarding DNA:

> *And now the announcement of Watson and Crick about DNA. This, for me, is the real proof of the existence of God.*

In another report in *Nature* 2 months later, Watson and Crick provided insights into the second DNA function: the elaboration of the information it contains into a living organism:

> *It therefore seems likely that the precise sequence of the bases is the code which carries the genetic information.*

So Watson and Crick were able to deduce how DNA carries out its two functions, replication of the information it contains and elaboration of that information into an organism, by simply looking at its structure. A fundamental principle of architecture—"form follows function"—is also a fundamental truth in biology.

Theory: the central dogma

In 1957, Frances Crick proposed what was to become known as the central dogma describing the process of information translation:

$$DNA \rightarrow RNA \rightarrow Protein$$

Information flows from DNA through RNA to proteins. The information in DNA is encoded in the linear sequences of nucleotides, just as the information on this page is encoded in linear sequences of letters. The information in DNA is translated into the language of proteins, which is composed of different "characters": amino acids in place of nucleotides. This translation is facilitated by a middleman, RNA, which, like DNA, is a nucleic acid composed of a sequence of nucleotides. Translating the information in DNA into a protein is somewhat like copying pages from a book and taking those sheets of paper to someone who can translate it into Japanese. (For more details, see chapter 4.)

Further studies supported the central dogma and led to the entrenched and very satisfying view of the gene as a stable, continuous segment of DNA whose linear nucleotide sequence corresponds to a linear sequence of amino acids. It was not long before this tidy picture of gene structure and function had to be revised, however. The more one learns about biological systems, the more one realizes how untidy they can be. If you look long enough, you can find exceptions to any rule or pattern in biology. Biological systems are too complex to always follow rules. This lack of predictability can be a source of both joy and frustration for biologists.

Revising the Central Dogma

As we have learned more about gene structure and function, we have fine-tuned and elaborated our original concept of DNA translation. First, we found that the flow of information from DNA to RNA is not one way. Information can flow from RNA to DNA. We also discovered that not all DNA is translated into proteins. Often a portion of a gene's nucleotide sequence does not contain information that prescribes an amino acid sequence, and this section is excised before DNA's information is translated into protein. The translated sequence of nucleotides in a stretch of DNA is known as an **exon;** the excised sequences are called **introns.** (See chapter 4.)

We also found sequences of nucleotides that repeat themselves over and over and over, like a phonograph record that's stuck in a certain place. (Does anyone remember phonograph records?) Because each person's pattern of repeats is unique, these sequences are important in DNA typing. Scientists assumed such sequences were irrelevant to biological functions and labeled them "junk DNA." Now, however, there is evidence implicating these DNA repeats in some cancers and also in a number of neuromuscular disorders.

We then found that some nucleotide sequences in a gene don't stay put but hop around. They're known as transposable elements, **transposons,** or "jumping genes." (See the sidebar.)

Perhaps most surprising and intriguing, we found that our perception of RNA as a passive middleman, relaying information between the "real" molecules, DNA and pro-

teins, is not appropriate. RNA, like an enzyme, has the ability to catalyze chemical reactions. This discovery may help us explain the origin of life on earth and, in particular, the evolution of nucleic acids and proteins.

Who knows what we will find next?

Understanding Genes

Because

- genes occupy a central place in the study of all biological sciences,
- the structure of the DNA molecule is so simple and yet so powerful,
- the mechanism of gene action is so awe inspiring,
- genetic variation is crucial to the evolution of life on earth,

biologists sometimes sound as if they are talking about a supernatural force when they discuss genes.

There's no doubt that genes are potent particles. If information is power, then genes are the powerhouses of the planet, for they contain the information of life. They are the repositories of the information that makes us who we are, that binds us to our species, and that separates us from other species. Genes forecast the future and reflect the past. Your genes contain information on your genealogical relationships and ethnic background and a chapter on the evolution of life on earth.

Yes, genes are powerful, but they are not omnipotent. We sometimes forget that fact, and it is important that we don't, particularly as we assess the applications and societal issues of gene-based biotechnologies.

Misunderstandings about genes

Biologists add to the misunderstanding surrounding genes by talking in shorthand and using insider's jargon that is easily misinterpreted by those not as familiar with the workings of genes.

Recall from chapter 2 that many of the fears and concerns surrounding biotechnology can be traced to public misunderstanding of the nature of genes. It is important that we do whatever we can to correct these misunderstandings and be more disciplined with the words and phrases we use to describe genes. Here are a few common misstatements made about genes and more accurate, albeit lengthier, explanations:

Genes make proteins. Genes do not make proteins. Genes contain the information for making proteins. Each nucleotide sequence in a gene specifies the

Scientific Models: Gateways or Barriers?

If you think that what happened to Mendel wouldn't happen today, think again. The story of another great geneticist, Barbara McClintock, parallels Mendel's so closely that it is eerie. Dr. McClintock is the scientist who discovered "jumping genes," or transposons, in the 1950s.

Like Mendel, she studied an organism so familiar it was almost dull—corn. Like Mendel, she was blessed with great powers of observation, an extraordinary intellect, and an uncommon amount of patience and persistence. For over 30 years she carried out the slow, laborious work of cross-pollinating Indian corn plants and, months later, observing the results of her experimental crosses. By simply using her uncanny ability to find patterns in the color of corn kernels, she developed a theory that called into question the fundamental nature of genes and eventually revolutionized our understanding of genetics.

Unlike Mendel's audience, McClintock's was impressed—but not positively. In short, some thought a brilliant geneticist who was a member of the prestigious National Academy of Sciences had gotten a bit confused. What she proposed was heresy to the existing view of genes and chromosomes, and so they concluded that she was out of touch with reality, not that their model was wrong.

Often scientists forget that the models they develop are supposed to be useful tools for understanding and interpreting the workings of the system under study. Instead they become wedded to the model and are unable to see and hear information that does not support it. Their model acts like a filter, letting expected information pass through to their brains but blocking the information they don't expect.

This surely was the case during the 30 years McClintock tried to explain to her colleagues what she saw in those corn kernels. Like Mendel, she saw patterns where everyone else saw disarray. She open-mindedly asked what the patterns she saw revealed about the behavior of genes and chromosomes. She did not aim for a specific answer to support existing theory, Instead she sought true and deep understanding.

Other scientists were blind to the patterns and deaf to her reasoning. When they couldn't understand her message, they chose to ignore it because it did not fit with their preconceived notions of what was "supposed" to happen. Eventually, after transposons were discovered in bacteria, yeasts, and fruit flies, her fellow scientists could finally hear what she had to say. In 1981, Dr. McClintock received the MacArthur Laureate Award and the Lasker Award in Basic Medical Research; in 1982, Columbia University's Horwitz Prize; and in 1983, the Nobel Prize for her work on transposons.

---

amino acid sequence in a protein. Proteins are made by other proteins (enzymes) that use the gene as a manufacturing guide. Proteins cannot manufacture proteins without this guide, but the guide alone is not sufficient for protein production.

Genes are on chromosomes. Genes are not "on" chromosomes. A chromosome is composed of a very long molecule of DNA that is bound to proteins named histones (see chapter 4). A gene is a given length of DNA that is translated into a chain of amino acids. The genes plus the bound proteins are the chromosome.

Genes replicate themselves. Genes do not replicate themselves. Proteins (enzymes) replicate genes using the gene as a guide.

DNA is like a blueprint. This is a very useful metaphor, up to a point, Then it, like all metaphors, becomes more constraining than useful.

DNA is like a blueprint because it contains information for building proteins just as architectural blueprints contain information for building a structure. But DNA is much more than that. DNA not only has information for building the structure of an organism but it also contains information for making enzymes, hormones, receptor molecules, transport proteins, and antibodies.

In other words, to be like DNA, an architectural blueprint would have to contain instructions for

- *making* the tools and machines used to erect the building
- *fabricating* the electrical wires, polyvinylchloride pipes, dry wall, and insulation
- *securing* the money to fund the construction
- *designing* the precise details of heating and cooling units, security system and telecommunications networks

3. Genes, Genetics, and Geneticists • 63

- *creating* the workers, supervisors, attorneys, bankers, and real estate agents
- *populating* the building and directing the tenants' activities, and even
- *dismantling* the building and constructing another

and much, much more! That would be some blueprint!

Genes determine who we are. Genes do not determine who we are. Genes *influence* who we are.

Envision all of the factors that could influence the process of constructing a building from the architectural blueprint to the appearance of the final product. A few immediately come to mind: choice of materials, worker competence, available funds. Because of these variables, nonidentical buildings could easily be constructed from the same blueprint. Now recall that much more information is included in a DNA "blueprint," providing many more opportunities for extraneous factors to shape the final product.

As we discussed in chapter 2, ignoring the complex realities of genetics—genes are only one factor that contribute to a trait (phenotype), many genes influence a single trait, and any one gene affects many traits—may lead to misunderstanding and misusing genetic information.

Understanding Evolution

In addition to being the repositories of information for synthesizing proteins, genes are the units of evolutionary change. So a complete understanding of genes and genetics comes not only from learning about the molecular structure and function of genes but also from understanding and appreciating the role they play in evolution. After all, the study of evolution is the study of genetic variation over time and space.

Defining evolution

Evolution is simply the change in the frequencies of certain genes in a species' gene pool. More specifically, it is the change in the relative proportion of alternative forms of a gene (alleles) in a population.

Evolution depends on two interrelated processes:

- the creation of genetic variation
- the selection of certain of these variants at the expense of others

The agents that create genetic variation are mutation and **recombination.** The primary agent that selects

certain variants is called **natural selection.** Because the creation of genetic variation and the selection of certain variants are essential to evolution, we must look at each process in more detail if we are to understand evolution.

Creating genetic variation

As you might expect, because genetic variation drives evolution, nature has many mechanisms for generating genetic variation. All of these mechanisms are commonly grouped into two broad categories: mutation and recombination.

Types of Mutations

Biologists use the term "mutation" to refer to the changes in the genetic information a cell carries. Two classes of mutations, based on the amount of genetic information (DNA) that is changed by the mutation, are usually recognized:

1. Changes in large segments of DNA molecules are termed **macromutations, macrolesions,** or **chromosomal aberrations.** Examples of macromutations include
   - changes in the total *amount* of genetic information because of the loss (**deletion**) or addition (**duplication**) of a gene(s) or even entire chromosomes
   - changes in the *positions* of the genes relative to one another without any change in the total amount of genetic information (**inversions** and **translocations**) (Figure 3.6).

Figure 3.6 Diagrams of the four types of chromosomal aberrations, or macromutations. The letters denote genetic loci.

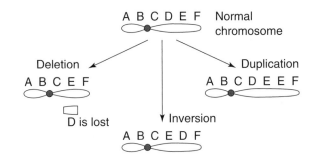

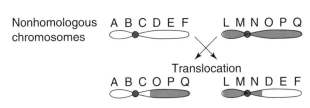

2. Changes in the sequence of nucleotides of single genes are termed **micromutations, micro- lesions,** or **point mutations,** because a relatively small amount of DNA is changed.

Effects of Mutations

What is the effect of a mutation on an organism? It depends on what has changed. Some mutations are beneficial; others are devastating; many are neutral. Whether it is harmless or devastating does not *necessarily* depend on whether or not it's a macro- or micromutation. Sometimes changing a single nucleotide has horrible effects on the phenotype. And yet, a complete doubling of genetic information has been very important in the evolution of many plants. In general, however, as you would expect, the larger the amount of genetic information that is changed, the more likely it is that the mutation will be harmful.

More information on the effects of mutations on protein structure and function and resulting phenotypes will be provided in chapter 4. Because the topic at hand is evolution, we will focus here on mutations as sources of a genetic variation that is then fed into the evolutionary process.

From the point of view of the evolutionary biologist, the two types of mutation vary greatly because they affect the evolutionary process in different ways. Micromutations, or point mutations, create new forms of genes, or, as an evolutionary biologist would say, they can add new alleles to the existing gene pools if they are not harmful.

The effects of macromutations on the evolutionary process cannot be described in a single sentence because they vary in both type and significance according to the exact change that has occurred.

Deletions and duplications. When the total amount of genetic information changes, the evolutionary result may be profound, as in the immediate speciation that has sometimes followed chromosome number doubling, or it may be totally insignificant. For example, when a chromosome is lost or gained, the individual is often sterile, as in the case with XO and XXY humans. While this condition is devastating to the individual, the mutation is irrelevant from an evolutionary perspective because the genetic change does not become incorporated into the gene pool.

Another type of macromutation, gene duplication, can be both directly and indirectly important in evolution. If the gene duplication results in increased production of a protein that contributes to the survival of the individual, it will be favored by natural selection. The indirect benefits of gene duplication result from a different scenario. If a single gene is represented by a number of copies, some of these duplicate genes can mutate into new allelic forms without the loss of the original gene's function.

Inversions and translocations. Macromutations that involve changing the positions of genetic loci relative to one another also vary in the effect they have on the evolutionary process. When a section of a chromosome breaks, rotates 180°, and then reinserts itself into the same chromosome (an inversion) or when nonhomologous chromosomes break and exchange pieces of chromosomes (translocation), the result is usually partial sterility at best. Most gametes produced by individuals heterozygous for translocations do not contain a full complement of the genes involved in the translocation (Figure 3.7). Consequently, macromutations that rearrange the position of genes on chromosomes are not a common source of genetic variation "usable" for driving evolutionary change. However, in certain cases, translocations and inversions have become established components of a species' gene pool. Because they deter appropriate pairing and crossing over in meiosis (see below), the genes in the inverted or translocated segment become tightly linked and are transmitted as a single unit. Sometimes these linked sets of genes are important evolutionarily because they represent specific gene combinations that are highly adaptive and confer a selective advantage.

Transposable Elements

A special form of mutation that occurs in both prokaryotes and eukaryotes involves transposable elements, or "jumping genes." These transposable elements, which are also called transposons, are short segments in the DNA that include a gene for a protein called a transposase. This protein causes the short piece of DNA to "jump" from its original location into a new, often random location somewhere else in the genetic material of the cell. Sometimes the jumping involves forming a new copy of the transposon, sometimes not (Figure 3.8).

The insertion of a transposon at a new location can pose a problem: the transposon might land in the middle of an important gene, alter the nucleotide sequence within the gene, and prevent the synthesis of the protein. In 1992, scientists linked the occurrence of a genetic disease to a transposition event. Neither parent of the child with the genetic defect was a carrier of the gene for the disease. A detailed examination of the child's defective gene showed that a transposon had inserted into it, causing a mutation. To

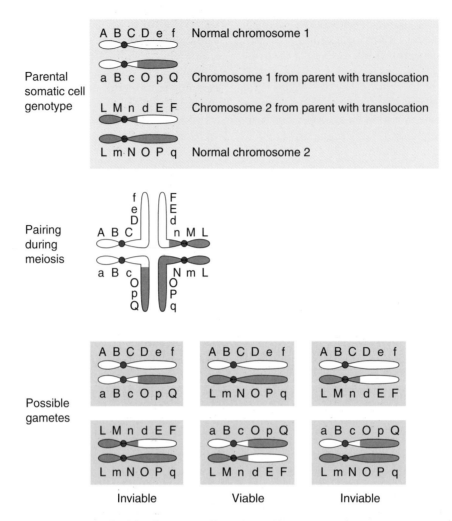

Normal chromosome 1

Chromosome 1 from parent with translocation

Chromosome 2 from parent with translocation

Normal chromosome 2

Parental somatic cell genotype

Pairing during meiosis

Possible gametes

Inviable   Viable   Inviable

Figure 3.7   Diagrams of the disruptive effect of a translocation on chromosome pairing in meiosis and of the resulting gametes. The parental genotype possesses one normal version of each of the two chromosomes involved in the translocation event. The letters indicate genetic loci. Lowercase and uppercase letters denote different alleles for the same trait. Four of the six possible gametes derived from the parental genotype are inviable because they do not contain a complete complement of genes.

date, scientists have found transposons in plants, animals, yeasts, and bacteria.

Recombination

The second mechanism for generating genetic variation is recombination, the joining of genetic information from two sources to produce new genetic combinations. Recombination differs from mutation in that specific genes are not changed but simply reassorted. Exactly how the genetic material from two sources gets combined varies with the mode of reproduction of the organism.

Sexual reproduction and recombination. For almost all organisms that reproduce sexually, reproduction involves two processes:

- the production of gametes that differ genetically from the individual producing them
- the fusion of these gametes to create an individual that differs genetically from both parents

Recombination occurs during *both* processes. Let's start with the most straightforward example of recombination: fertilization.

Fertilization is the fusion of two gametes (eggs and sperm or pollen) into a single zygote. When the gametes, which contain genetic information, fuse, the resulting zygote's genetic information is a new combination of genetic information from the two parents. Thus, the offspring that develops from the zygote contains genetic information from two sources and is

A. Transposons that move by **replicative transposition** are duplicated in the process of jumping, a copy-and-paste mechanism.

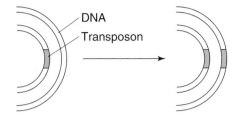

B. Nonreplicative transposons move in a cut-and-paste manner. No copy is made.

Figure 3.8 Replicative and nonreplicative transposons.

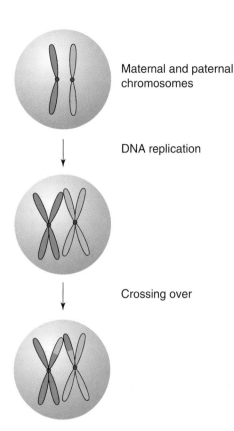

Figure 3.9 Crossing over between homologous chromosomes occurs during gamete production. As a result, the gametes will differ from each other and the parental genotype.

genetically novel: it is not genetically identical to either parent. All offspring of sexual reproduction, including you, are genetic recombinants.

Recombination also occurs during the production of these gametes through meiosis. Increased genetic variation in populations is the result of two associated meiotic events: the independent assortment and segregation of nonhomologous chromosomes and the crossing over that occurs between homologous chromosomes.

As a result of independent assortment and segregation of nonhomologous chromosomes in meiosis, a gamete differs genetically from the cell that gave rise to it by having one-half as much genetic material and a random assortment of the maternal and paternal genetic material. So each gamete contains genetic material from two sources (maternal and paternal) in new combinations. The number of possible combinations of nonhomologous chromosomes that could be generated in the production of each human gamete is 2^{23}, or over 8,000,000.

Independent assortment and segregation alone are a rich source of genetically variable gametes, but another recombination event occurs in meiosis. During the pairing of homologous chromosomes, genetic material from the paternal chromosome is exchanged with genetic material from the maternal chromosome through the process of crossing over (Figure 3.9). The

resulting chromosome is a recombinant: a hybrid containing genes of both maternal and paternal origins, or a novel combination of genetic material from two sources. A graphical summary of the relationship between sexual reproduction and recombination is provided in Figure 3.10.

Asexual reproduction and recombination. Many organisms do not reproduce sexually. In single-parent or asexual reproduction, a copy of the genome of the parent is passed along in its entirety to the offspring (Figure 3.11). The offspring is therefore genetically identical to the parent organism and is often called a clone of the parent. The group of organisms resulting from repeated reproduction is called a clonal population, because every member of the population is genetically identical to every other member.

A typical example of asexual reproduction is the division of a bacterium to produce two bacterial cells. These two cells then divide, the four progeny divide, and so on until a large clonal population is generated. Bacteria are not the only organisms that reproduce asexually. Yeasts and some other fungi, protozoans,

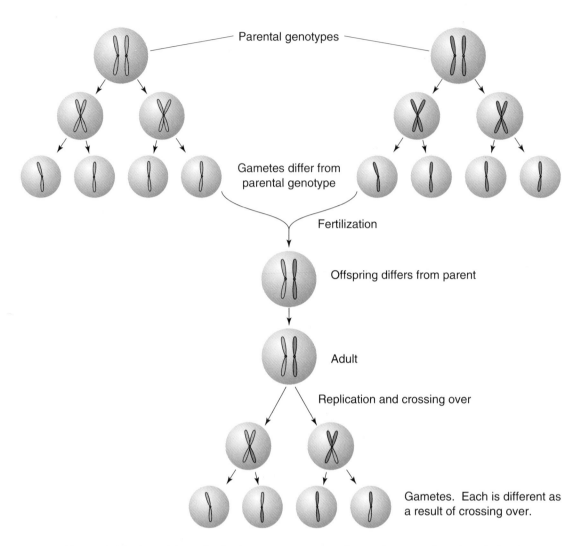

Parental genotypes

Gametes differ from parental genotype

Fertilization

Offspring differs from parent

Adult

Replication and crossing over

Gametes. Each is different as a result of crossing over.

Figure 3.10 Sexual reproduction, genetic variation, and recombination. Genetic variation is created during three stages of sexual reproduction. When gametes are produced, they differ genetically from the parental genotype because they have half the amount of DNA and a random assortment of paternal and maternal chromosomes. During fertilization, genetic material from two sources is combined, creating an offspring that differs genetically from both parents. Finally, during gamete production, crossing over between maternal and paternal (homologous) chromosomes occurs, creating "within-chromosome" genetic variation.

and algae can reproduce by simple fission or by sexual reproduction. A few species of animals reproduce asexually through parthenogenesis (in which females produce offspring without being fertilized by males). Many plants can propagate themselves asexually; humans take advantage of this to duplicate desirable plants by rooting leaves and cuttings.

You can see from comparing Figures 3.10 and 3.11 that populations of asexually reproducing organisms would be far less genetically diverse than sexually reproducing populations if they had no other mechanism for generating genetically variable individuals. In fact, though, bacteria have three natural mechanisms by which genetic materials from two sources can be

combined: conjugation, transformation, and transduction. Each of these processes is not only important in providing genetic variation for the evolution of bacteria but also plays an important role in human interactions with microbes. Later in this book you will learn the importance of these three natural processes to research and technology and also some of their medically important consequences.

Conjugation (Figure 3.12) is a process by which one bacterium transmits a copy of some of its DNA directly to another bacterium. Conjugation is often called bacterial sex, which in a strict biological sense is the mixing of genetic information. In conjugation, the genetic mixing occurs in the recipient cell and

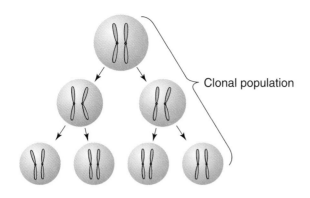

Figure 3.11 Asexual reproduction. Offspring are genetically identical to the parent. Genetic variation is created through mutation.

does not, in and of itself, result in the production of offspring. So bacterial sex is not the same thing as sexual reproduction. After genetic exchange, a bacterium may undergo asexual reproduction as usual, and its new genetic information would be passed to its offspring. (See the activity *Conjugative Transfer of Antibiotic Resistance in* Escherichia coli.)

In transformation (Figure 3.13), a new combination of genetic material is created when cells take up DNA from the medium around them. In nature, some types of bacteria are easily transformed, while others are not. Scientists have learned to manipulate many kinds of cells to render them more susceptible to transformation in the laboratory. (See the activity *Transformation of* Escherichia coli.)

In transduction (Figure 3.14), a virus intermediary carries DNA from one bacterial cell to a second cell. During a typical virus infection, viral nucleic acid is replicated in the host cell, and viral coat proteins are made. Near the end of the infection cycle, the viral nucleic acid is packaged into the new virus coats, and new virus particles are released. For transduction to occur, some of the bacterial DNA must be packaged into a virus particle by mistake. This bacterial DNA is then injected into another bacterium and combines with the resident DNA. No infection occurs, since no virus genome was injected. (See the activity *Transduction of an Antibiotic Resistance Gene.*)

Selecting certain genetic variants

After genetically variable individuals are created by any of the methods just described, these individuals are exposed to the indifferent force of natural selection. Natural selection acts on individuals, but

Figure 3.12 Bacterial conjugation. Genetic material is exchanged between F⁺ and F⁻ cells.

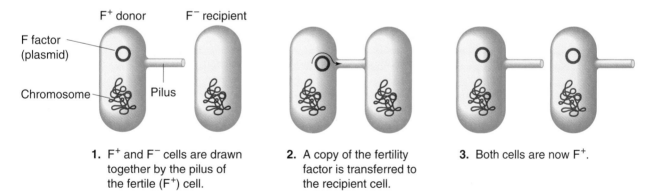

1. F⁺ and F⁻ cells are drawn together by the pilus of the fertile (F⁺) cell.

2. A copy of the fertility factor is transferred to the recipient cell.

3. Both cells are now F⁺.

Figure 3.13 Transformation. A cell takes up free DNA from its environment, integrates it into its chromosome, and expresses the encoded products.

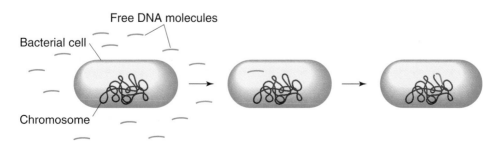

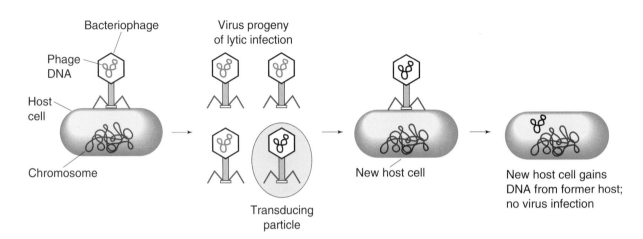

Figure 3.14 Transduction. A virus serves as a conveyer of genetic material from one organism to another. In the example here, the organism is a bacterium, and the virus is a bacteriophage.

individuals do not evolve. Populations evolve. How does this happen?

What Is "Survival of the Fittest"?

Natural selection is a measure of reproductive success but not necessarily of survival. Populations contain individuals that vary in their abilities to both survive and reproduce. This variation provides the raw material for evolutionary change. The fittest organisms are those that produce the most offspring that survive to reproductive maturity. If, however, an organism excels at skills required for survival but leaves relatively few offspring, it is not "fit."

On the other hand, traits that have questionable or even negative survival value persist in a population if they are beneficial in obtaining mates. Think of the peacock's tail. It is difficult to imagine how a 5-foot-long tail of brightly colored, iridescent feathers could have positive survival value. But the peahens think those tails are gorgeous, so males with long, brightly colored tails are the "fittest."

Natural selection favors (selects for) those organisms that produce the most offspring and rejects (selects against) those that produce fewer offspring. The genes of the organisms that natural selection has favored are passed to the next generation in greater numbers than the genes of organisms selected against. In this way, gene frequencies change over time, and this change is the process of biological evolution.

Observable Variation and Natural Selection

While populations are characterized by variation, only some of the individual variation that exists has a genetic basis. Some of the variation we observe in populations is exclusively phenotypic variation caused by environmental factors. Such variation may contribute to differences in survival and reproductive success and thus be subject to natural selection, but it is *not* the raw material for evolution. Traits that contribute to an increase in fitness will evolve only if they have a genetic basis.

So only some of the variation that exists in nature is significant in an evolutionary sense. Some isn't significant now and may or may not be significant in the future, depending on its degree of heritability and the selective pressures that act on that variation.

Natural Selection as a Honing Device

Some people have a somewhat sacred view of evolution. They have faith that through natural selection, evolution has honed all traits in all organisms so that there is a perfect fit between an organism and its environment. That opinion is appealing but naive. Most traits are present simply because they're not harmful—at that time in that environment. Given another environment or another time, they might be.

Even though evolution is a change in gene frequencies, natural selection acts on phenotypic traits. At any one time, only some of the traits of an organism are the focus of natural selection. Others are irrelevant to selection and just happen to get passed on with the favored genes. At that time, operating in that environment, natural selection is not "interested" in those traits. As a result, it is not molding a fit between those traits and the environment. Next year, however, those neutral traits might become the focus of natural selection, or a useful attribute one year may be a handicap, or linked to a handicap, the next.

For example, imagine that a mutation produces a color pattern that provides a lizard with better camouflage. If having that trait means the lizard lives longer and has more offspring, that gene will be passed to the next generation in greater numbers. It will drag with it other neutral genes that may have nothing to do with leaving more offspring.

On the other hand, the new gene and the resulting color pattern might make the lizard more susceptible to predation. In that case, the gene would be selected against, as would the lizard's neutral genes and even other genes with positive survival value.

Now imagine a situation in which the color pattern makes male lizards more susceptible to predation but is also the color pattern that female lizards prefer when choosing mates. In nature, we often have selective pressures working in opposition to each other.

In addition, as described above, some of an organism's traits may not depend entirely, if at all, on its genetic makeup. Only traits that have a genetic basis are subject to natural selection. Therefore, if a trait is caused largely by environmental factors, it could not have been "honed" to perfection by natural selection.

Many people fail to realize that disorder, uncertainty, and, therefore, flexibility are common in biological systems, particularly at and above the level of the organism. That understanding may seem irrelevant to biotechnology, but, in fact, it is critical for accurately analyzing the environmental issues of biotechnology.

Variation, chance, and change

The pivotal concepts of evolutionary biology are contained in the abbreviated description of evolution: variation, chance, and change. A true understanding of genes, biology, and even biotechnology is possible only if you comprehend and appreciate the role these factors play in driving evolution and shaping life on earth. Therefore, they deserve a little more attention.

Variation

Genetic variation is the grist of the evolutionary mill. The great biologist Theodosius Dobzhansky, who was instrumental in elucidating the genetic basis of evolution, said that "nothing in biology makes sense except in the light of evolution." Therefore, understanding the nature of genetic variation is essential to understanding all of biology, not just evolution.

Evolutionary biologists have always studied genetic variation, even if they did not realize it. Charles Darwin, who did not know about genes and hereditary

mechanisms, focused on phenotypic variation between species in an attempt to explain the diversity of species on earth. He studied polymorphisms (different types) visible to the naked eye: birds' beaks, iguanas' feeding behavior, and the shape of tortoises' shells. Later evolutionary biologists, understanding the relationship between genes and proteins, used electrophoresis to look at protein variation as a reflection of genetic variation. Today's evolutionary biologists are able to study genetic variation by looking directly at DNA variation through restriction fragment length polymorphism (RFLP) and nucleotide sequence analysis. Irrespective of the method used, when they ask questions about the amount and type of genetic variation, biologists are constantly changing focal lengths:

- How much variation is there within a population?
- How much variation is there between populations (within a species)?
- How much variation is there between species?

They also ask the flip sides of these questions: How much genetic similarity is there at each of these levels?

In studying biology, always keep in mind that both similarity and dissimilarity are informative and therefore important. All species are both genetically alike and genetically different from one another. Both genetic unity and genetic diversity characterize all of the earth's species. Within a species, the same is true: all individuals of a species (or a population) are genetically alike and genetically diverse.

Because a continuum of genetic variation and similarities exists within a species and extends outside of a species to related species, determining where one species stops and another begins can be difficult and, at times, arbitrary. Where one chooses to demarcate a species boundary is flexible and, more often than you might expect, subject to lively debate. A large and dynamic branch of biology, taxonomy, is devoted to the question: "Where should we draw the line?"

Because of that same genetic continuum, no one gene makes a fish a fish and an oak tree an oak tree any more than one gene that I have and you don't makes me more human than you.

Chance

Many people envision nature prior to human intervention as ideal and ordered: organisms perfectly adapted to unadulterated environments living in balance with each other through time. These people operate on the tacit assumption that Mother Nature would keep a tidy house with a "place for everything and everything in its place" if humans would stop introducing mess and disorder. While this idyllic view of nature may be appealing, it bears no resemblance to reality. The planet and its species are constantly changing independent of any human presence.

Life on earth is like a game of chance governed by constantly changing rules. An organism's opponent, natural selection, is also the game's unpredictable referee, rewriting the rule book on every play.

For biological organisms, every round of reproduction is like a new hand of cards. Chance determines which genes are combined. These new genotypes give rise to new phenotypes. The path from genotype to phenotype is susceptible to environmental influences, many of which are random occurrences.

These new phenotypes then enter the game uncertain of how they will fare, because some of the rules have changed since the last round. Certain phenotypes might be improvements—in that environment at that time. These phenotypes will be selected for, causing the proportion of certain cards in nature's genetic deck to shift before the next hand is drawn. That same phenotype in a different environment at another time might be the kiss of death.

Given such an uncertain scenario, what type of a team would you put together? You would probably choose to hedge your bets and construct a team of diverse talents. Evolution has encouraged biological organisms to do just that. A species persists from one year to the next by producing genetically variable offspring, some of which manage to survive for another round.

And yet the greater the diversity within a species, the less perfectly adapted to its environment that species is. So organisms attempt to balance diversity—a measure of adaptability—and adaptedness. For any species, a constant and very dynamic tension exists between these two properties.

Change

The idea that environmental factors lead to evolutionary change by exerting selective pressure on organisms is familiar. These environmental forces acting on species are constantly changing, and a species' survival depends upon its ability to respond with evolutionary changes. When you think of environmental factors as selective agents, you probably envision aspects of the physical environment, like drought. And yet, other living organisms, the biological environment, make up the most important component of a

species' environment. So if one species responds to physical environmental changes with evolutionary changes, its evolutionary change becomes an environmental change for a second species.

Also keep in mind that the relationship between organisms and their environment is not linear but circular. The activities of organisms continually alter the earth's physical environment, often to the detriment of other species. Changes in the physical environment may then place new selective pressures on other organisms. Organism-driven environmental changes exert effects on individuals in the same species as well. Individuals in the same species compete for access to resources. Some win at others' expense.

Constant flux characterizes the natural world. Species are continually evolving—changing genetically—in response to the physical environment and to each other. It really is a jungle out there.

Understanding the details of evolutionary change may seem unrelated to biotechnology, but it is not. You will see that our underlying assumptions about evolution and the resulting state of the natural world drive some of the concerns people have about certain applications of biotechnology. These same assumptions encourage people to assign more power to biotechnology than it is due and to assume the tasks in front of us are easier than they really are.

Selected Readings

Allen, Garland E. 1978. *Thomas Hunt Morgan: the Man and His Science*. Princeton University Press, Princeton, N.J.

Crick, Francis. 1988. *What Mad Pursuit: a Personal View of Scientific Discovery*. Basic Books, New York, N.Y.

Dobzhansky, Theodosius. 1970. *Genetics of the Evolutionary Process*. Columbia University Press, New York, N.Y.

Dobzhansky, Theodosius, Francisco Ayala, G. Ledyard Stebbins, and James W. Valentine. 1977. *Evolution*. W. H. Freeman & Co., San Francisco, Calif.

Futuyma, Douglas. 1992. *Evolutionary Biology*. Sinauer, Sunderland, Mass.

Iltis, Hugo. 1932. *The Life of Mendel*. Norton, New York, N.Y.

Judson, Horace F. 1979. *The Eighth Day of Creation: Makers of the Revolution in Biology*. Simon & Schuster, New York, N.Y.

Keller, Evelyn Fox. 1983. *A Feeling for the Organism: the Life and Work of Barbara McClintock*. W. H. Freeman & Co., New York, N.Y.

Lagerkvist, Ulf. 1998. *DNA Pioneers and Their Legacy*. Yale University Press, New Haven, Conn.

Mayr, Ernst. 1969. *Populations, Species and Evolution*. Harvard University Press, Cambridge, Mass.

Morgan, T. H., A. H. Sturtevant, H. J. Muller, et al. 1915. *The Mechanism of Mendelian Heredity.* Holt, New York, N.Y.

Stubbe, Hans. 1973. *History of Genetics.* MIT Press, Cambridge, Mass. (Translated by T. Waters.)

Watson, James. 1968. *The Double Helix.* Atheneum, New York, N.Y.

Watson, James, and Francis Crick. 1953. Molecular structure of nucleic acids: a structure for deoxyribonucleic acid. *Nature* 171:737.

Watson, James, and Francis Crick. 1953. Genetical implications of the structure of deoxyribonucleic acid. *Nature* 171:964.

Watson, James, and John Tooze. 1981. *The DNA Story: a Documentary History of Gene Cloning.* W. H. Freeman & Co., San Francisco, Calif.

An Overview of Molecular Biology

Gene Structure and Function

A hallmark of living systems is that they reproduce themselves. For many years, one of the greatest mysteries of science was the puzzle of how the tiniest seed or fertilized egg could contain all the information needed for the development of an entire organism. Classical geneticists deduced that individual traits are determined by invisible information units they called genes. They presumed that each cell duplicated its genes before dividing so that each daughter cell could receive a complete set. They also presumed that the genes present in sperm and egg cells would carry the genetic information to the next generation. Then, somehow, those genes would direct the development of a new individual.

It was clear that the genetic material must be capable of two extremely important functions. First, it must be in a form that can be copied extremely accurately so that correct information is transmitted from cell to cell and generation to generation. Second, its information must somehow be translated into a living organism. What molecule could possibly fulfill these two complex and critical requirements? Scientists naturally assumed that the genetic material must be a very complicated molecule or molecules.

Today we know that DNA is the genetic material of all life on earth. However, at first, that discovery was so astonishing that many scientists refused to believe it. Why? Because DNA is such a simple molecule. How could such a simple molecule be responsible for the extraordinary diversity of life forms on this planet? The determination of the structure of DNA and the cracking of its genetic code are among the most important scientific achievements of the 20th century. The elegant simplicity of DNA structure and the beautiful efficiency with which that structure provides for the two essential functions of the molecule can inspire both the poet and the engineer within each of us.

Structure of DNA

The genetic material of all life on this planet is made of only six components. These components are a sugar molecule (deoxyribose), a phosphate group, and four different nitrogen-containing bases: adenine, guanine, cytosine, and thymine. The essential building block of the DNA molecule is called a nucleotide or, more precisely, a **deoxynucleotide.** A deoxynucleotide consists of a deoxyribose molecule with a phosphate attached at one place and one of the four bases attached at another (Figure 4.1). The carbon atoms of the deoxyribose sugar portion of a nucleotide are always numbered in the same way. The base is always attached to carbon 1, and the phosphate group is always attached to carbon 5.

In a DNA molecule, thousands or millions of these nucleotides are strung together in a chain by connecting the phosphate group on the number 5 carbon of one deoxyribose molecule to the number 3 carbon of a second deoxyribose molecule (a free water molecule is created in this process). Figure 4.2 shows an example in which three nucleotides are connected. The linkages formed between the deoxyribose molecules via the phosphate bridge are called **phosphodiester bonds.** Because the nucleotides are held together by bonds between their sugar and phosphate entities, DNA is often said to have a sugar-phosphate backbone. Notice that the ends of the molecule in Figure 4.2 are labeled 5′ and 3′ for the carbon atoms that would form the next links in the chain at either end.

The sugar-phosphate backbone of DNA is an important structural element, but all of the information is contained in the four bases. The key to the transmission of genetic information lies in a characteristic of these bases: that adenine and thymine together form a stable chemical pair, and cytosine and guanine form a second stable pair. The pairs are formed through weak chemical interactions called **hydrogen bonds.** These two pairs, adenine-thymine and cytosine-guanine, are called *complementary base pairs* (Figure 4.3). In a

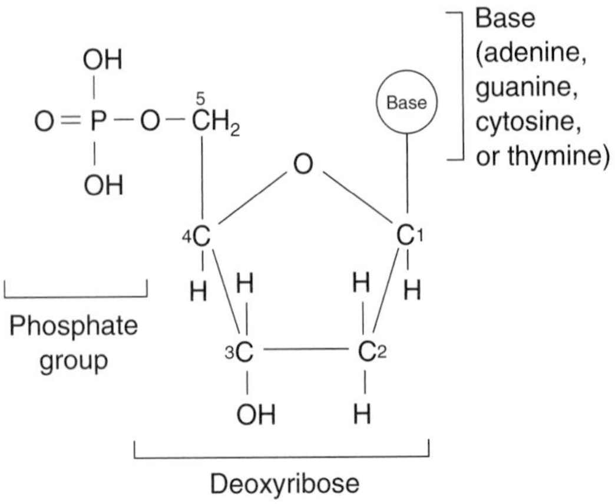

Figure 4.1 The nucleotide. Carbon atoms of the deoxyribose sugar portion are numbered according to chemical convention.

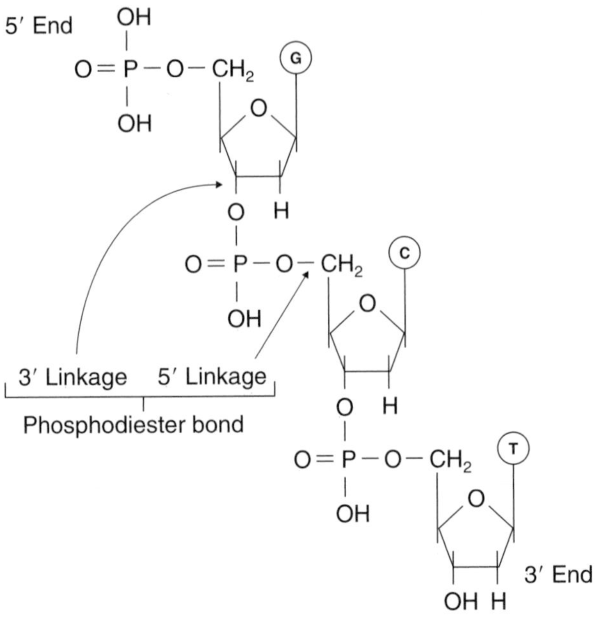

Figure 4.2 A trinucleotide.

DNA molecule, two sugar-phosphate backbones lie side by side, one arranged from the 5′ end to the 3′ end, and the opposite strand arranged from the 3′ end to the 5′ end. The bases attached to one strand are paired with their partner bases attached to the opposite strand (Figure 4.3). Thus, the order of the specific bases on one strand is perfectly reflected in the order of the complementary bases on the other strand. Knowing the sequence of bases on one strand allows us to deduce the base sequence on the complementary strand. (See the activity *Constructing a Paper Helix.*)

Figure 4.3 Complementary base pairs in DNA.

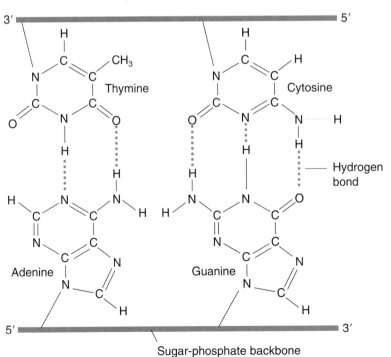

DNA is often drawn as a flat molecule because that shape is easy both to draw and to look at. In reality, each of the two sugar-phosphate backbones is wrapped around the other in a conformation called a double helix. The base pairs are on the inside of the helix, like the rungs of a ladder. For ease of representation and viewing, DNA can be presented by using a model in which each backbone is represented as a thin ribbon and the base pairs are represented schematically (Figure 4.4). In fact, since the nucleotide sequence of one strand specifies the nucleotide sequence of the other strand, a DNA molecule or region of a molecule is very often represented by the sequence of only one strand, always written in the 5′ to 3′ direction.

DNA function: faithful replication

The structure of DNA immediately suggests how DNA carries out the first critical function of genetic material: faithful replication. You can see that either of the two strands of DNA can be used as a template, or pattern, to reproduce the opposite strand by using the rules of complementary base pairing. When a cell is ready to replicate its genetic material, the two opposite strands are gradually "unzipped" to expose the individual bases. Each exposed strand is used as a template to form two new strands (Figure 4.5A). The result is two daughter DNA molecules, each composed of one parental strand and one newly synthesized strand, and each identical to the parent (Figure 4.5B).

How does DNA replication occur in a cell? DNA is duplicated by enzymes, the protein workhorses of cells. Special cellular enzymes work together to unzip the DNA double helix, capture free nucleotides, pair the correct new nucleotide with the template base, and make the new bonds of the growing sugar-phosphate backbone. The central player of this protein team is called **DNA polymerase.** It is the enzyme that actually makes the correct base pairs and forms the new phosphodiester bonds. Finally, some of the DNA replication enzymes "proofread" the new DNA strand, checking for errors in base-pairing and correcting any errors they find. These diligent enzymes ensure that very few errors occur during DNA replication so that genetic information is transmitted correctly. (See the activity *DNA Replication*.)

DNA function: information transmission

Although the structure of DNA immediately suggests how the molecule can fulfill the requirement for faithful transmission through duplication, it is not so obvious how such a simple molecule can determine the development of creatures as complex and varied as a blue whale or a rose. To understand how the structure of DNA elegantly fulfills this requirement, it is necessary to think about what makes a whale a whale or a rose a rose.

What does make a whale a whale? The answer is, its proteins. Just as proteins carry out the complicated task of duplicating the whale's DNA, other proteins carry out nearly every other function necessary to whale life. Structural proteins form the bricks and mortar of its skin, muscles, organs, and tissues; other proteins synthesize additional structural components like bones and lipids. Transport proteins carry oxygen, nutrients, hormones, and other important molecules throughout its body and around its cells. Protein receptors imbedded in cell surfaces bind with great specificity to the whale's hormones, enabling the whale to grow and develop properly. Proteins of the whale's immune system defend it from infection. Protein catalysts (enzymes) digest the whale's foods, synthesize fats for its blubber, replicate its DNA for transmission to baby whales, and carry out all the other metabolic tasks necessary for whale cell life. Thus, proteins provide the structure and carry out the functions of the whale. The same is true for a rose, a fruit fly, a bacterium, a human being, and every other life form on earth.

Figure 4.4 Ribbon model of DNA.

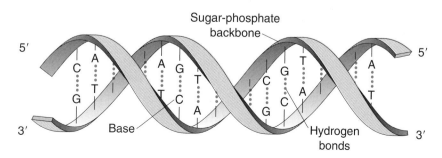

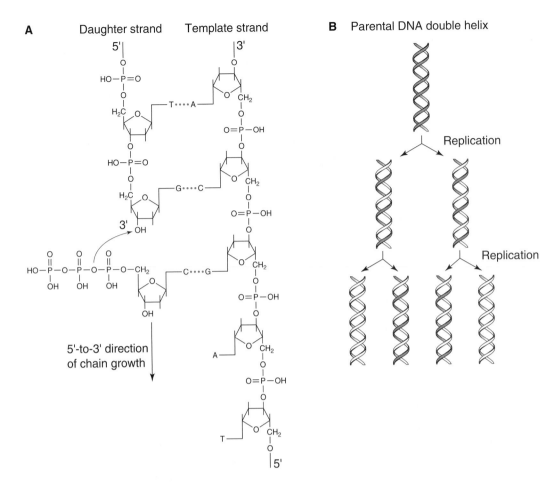

A Daughter strand   Template strand   **B**   Parental DNA double helix

5'-to-3' direction of chain growth

Replication

Replication

Figure 4.5   DNA replication. (A) Base pairing between an incoming nucleotide and the template strand of DNA guides the formation of a new daughter strand with a complementary base sequence. (B) In each round of DNA replication, each of the two DNA strands is used as a template for the synthesis of a new complementary strand, resulting in two daughter molecules, each with one "new" and one "old" strand.

It's not simply the nature of an organism's proteins that give the organism a particular identity. After all, the proteins in different animals can be pretty similar—muscle proteins, hemoglobin, enzymes to replicate DNA, and so on. The proteins of closely related animals can be very much alike (for an example of how evolutionary biologists use comparisons of similar proteins in different animals, see *Evolutionary Studies* in chapter 5 and the activities in the section *Molecular Biology and Genetics*). Even organisms as different as animals and plants have some types of proteins in common.

Much of the diversity in nature is due to the organization of proteins within an organism, particularly structural proteins and those that synthesize additional structural body components. In some ways, the organization of similar proteins into different body plans is analogous to using a pile of bricks in con-

struction. You could assemble them into the wall of a building or into a barbecue pit, or even a sidewalk. That analogy is an oversimplification, of course, since proteins do differ between organisms, and the more distantly related the organisms, the more they differ. The organization of protein synthesis during the development of an organism leads to its unique form, whether it is a whale, a rose, a chrysanthemum, or a human. An organism does what it does and looks like what it is because of the nature of its proteins and how they are organized during development.

Back to questions. What determines the organization of protein synthesis during the development of an organism? Can you guess the answer? More proteins. When these proteins fail to function properly, the results can be dramatic, such as flies that grow legs where they should have antennae, or fly larvae with two tail ends and no head. But the bottom line is that, in the end, proteins organize the body, compose or synthesize its structural components, and carry out its metabolic activities.

So for DNA to dictate the development of organisms, the information in DNA must somehow be converted into proteins, the "stuff" of organisms. How does this conversion occur? To answer this question, we must know a bit more about the makeup of proteins.

Proteins

So what exactly is a protein? A protein is a chain of amino acids. Amino acids are small organic molecules made mostly of carbon, hydrogen, oxygen, and nitrogen. Proteins are made from a pool of 20 different amino acids by connecting a few or thousands of these amino acids in various orders to form chains. For a simplistic analogy, think of making a chain of pop-beads from an assortment of colored beads. But a protein is more than a straight chain of beads. You must also imagine that the chain is folded and coiled into a specific three-dimensional shape (Figure 4.6).

What enables a protein to perform its function? A protein's function is made possible by its unique three-dimensional structure. For an example, let's consider a single receptor protein embedded in the outer membrane of a cell. Our imaginary receptor protein looks like an irregularly shaped glob with interesting nooks and crannies. The shapes of the nooks and crannies are absolutely critical to the protein's function: one cranny on the outside of the cell membrane is the place where a growth hormone molecule must fit exactly to signal the cell to grow. Other nooks and crannies on the inside are sites that fit precisely to other molecules for communication with the rest of the cell. Through interaction at these sites, the receptor protein can tell the cell that the hormone signal has arrived. (For an example of a receptor protein and what happens when it fails to function, see *An Adventure in Dog Hair, Part II: Yellow Labs*.)

Our imaginary receptor illustrates a crucial point about protein function. *A protein depends on its ability to fit to or bind to other molecules (sometimes other proteins) to carry out its functions. That ability is determined by its three-dimensional structure.*

What determines the three-dimensional structure of a protein? When amino acids are assembled into a protein chain, that chain immediately folds back upon itself to assume the most "comfortable" or energetically stable shape. The most energetically stable shape is determined by the interactions of the individual amino acids that make up the protein. Therefore, the identities of the component amino acids and the order in which they occur in the chain govern the final three-dimensional structure of the protein. The order

Figure 4.6   A protein is a chain of amino acids (represented by beads) that folds into a specific three-dimensional shape. The three-letter abbreviations on the beads are standard for specific amino acids (see Figure 4.23).

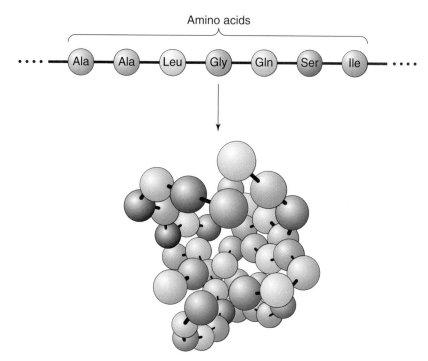

of the amino acids in the chain is thus extremely important to the function of the protein. As you can imagine, the possibilities for constructing different and unique protein chains are almost limitless. (Imagine how many unique chains of pop-beads you could make using 20 different colors of beads.) This variety is fortunate, considering the many and varied functions that proteins must perform. In fact, the forms and functions of proteins are so central to molecular biology that the second portion of this chapter is devoted to them.

But for now let us return to the question of how DNA controls the development of an organism. We have seen that an organism is the sum of its proteins. We have also seen that a protein's function depends upon its three-dimensional structure. Furthermore, its structure depends upon the sequence of amino acids in the protein chain. *DNA determines the characteristics of an organism because it determines the amino acid sequences of all the proteins in that organism.*

How does DNA determine an amino acid sequence? DNA contains a **genetic code** for amino acids in which each amino acid is represented by a sequence of three DNA bases (see Table 4.1). These triplets of bases are called **codons**. The order of the codons in a DNA sequence is reflected in the order of the amino acids assembled in a protein chain (Figure 4.7). The complete stretch of DNA needed to determine the amino acid sequence of a single protein is a **gene,** the unit of heredity defined by the classical geneticists. The complete set of genes in an organism is called its **genome**.

Protein synthesis

The process by which proteins are produced from the genetic code has several steps. DNA is essentially a passive repository of information, rather like a blueprint. The "action" of making a protein occurs at special sites in the cell called **ribosomes**. Therefore, the first step in protein synthesis is to relay the information from the DNA to the ribosomes. To accomplish this step, cellular enzymes synthesize a working copy of a gene to carry its genetic code to the ribosomes. This working copy is called **messenger RNA (mRNA)** (RNA is a close molecular relative of DNA). mRNA carries the genetic code for a protein to the ribosomes. In the second step of protein synthesis, the codons in the mRNA must be matched to the correct amino acids. This step is carried out by a second type of RNA called **transfer RNA (tRNA)**. Finally, the amino acids must be linked together to make a protein chain. The ribosome (which is made of proteins and RNA) performs this function. When the protein chain is complete, a genetic "stop sign" tells the ribosome to release the new protein into the cell.

RNA
Protein synthesis therefore requires a second type of nucleic acid molecule: RNA. Like DNA, RNA is made up of nucleotides composed of a sugar, a phosphate, and one of four different organic bases (see Figure 4.1). However, there are three important differences between DNA and RNA, two of them chemical and one of them structural. The chemical differences are that (1) instead of the sugar deoxyribose, RNA contains the sugar **ribose** (hence the name *ribo*nucleic acid), and (2) instead of the base thymine, RNA contains the base **uracil** (Figure 4.8). The sugar-phosphate

Table 4.1   The genetic code

First base in DNA triplet	A				G				C				T			
Second base in DNA triplet	A	G	C	T	A	G	C	T	A	G	C	T	A	G	C	T
Choices for third base in DNA triplet	A G / C T	A G / C T	A C G T	A G C / T	A G / C T	A C G T	A C G T	A C G T	A G / C T	A C G T	A C G T	A C G T	A G / C T	A G / C T	A C G T	A C / G T
Amino acid encoded	Lysine / Asparagine	Arginine / Serine	Threonine	Isoleucine / Methionine / Isoleucine	Glutamic acid / Aspartic acid	Glycine	Alanine	Valine	Glutamine / Histidine	Arginine	Proline	Leucine	Stop / Tyrosine / Stop	Tryptophan / Cysteine	Serine	Leucine / Phenylalanine

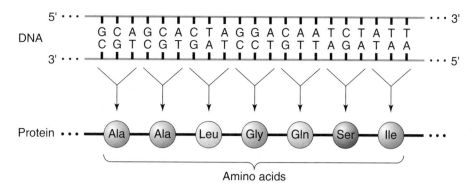

Figure 4.7 The base sequence of DNA determines the amino acid sequence of proteins.

backbone of RNA is linked together like the DNA backbone, and the bases are attached to the number 1 carbon, as in DNA. The important structural difference is that, although RNA bases can also form complementary pairs, RNA is usually composed of only a single strand of sugar-phosphate backbone and bases. It does not have the base-paired double helix structure of DNA (Figure 4.9), although it is capable of pairing with other single strands of DNA or RNA. As we shall see below, the single-stranded structure of RNA is ideally suited to its task of transferring information.

Synthesis of mRNA

The first step of protein synthesis is to make mRNA. This process resembles DNA replication in many ways. First, the DNA double helix must be unzipped

to reveal the information-containing bases. Then, complementary nucleotides (that contain the sugar ribose, since we are making RNA) are paired with the exposed bases. During the synthesis of RNA, the base uracil substitutes for thymine and pairs with adenine. Phosphodiester bonds are made between the nucleotides, and the new mRNA contains a base sequence that is exactly complementary to the template DNA strand. The process of using a DNA template to create a complementary mRNA molecule is called **transcription** (Figure 4.10). tRNA and ribosomal RNA molecules are also encoded in DNA and synthesized by transcription, but unlike mRNA, they are not translated into protein.

There are two major differences between transcription and DNA replication (compare Figure 4.5 and

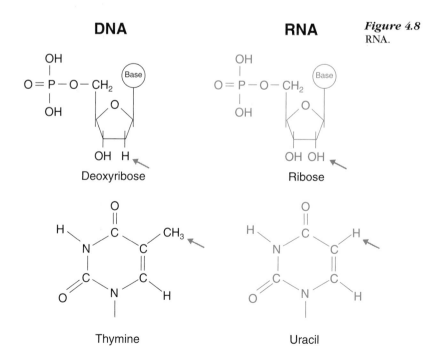

Figure 4.8 Chemical differences between DNA and RNA.

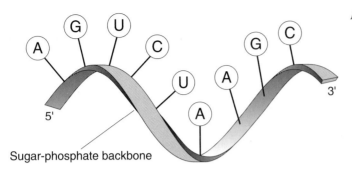

Figure 4.9 A single-stranded RNA molecule.

Sugar-phosphate backbone

Figure 4.10). In DNA replication, both strands are used as templates to generate two new strands for two new helices. In RNA synthesis, only one DNA strand is used as a template, and only a single RNA strand is made. The second difference is that the new mRNA molecule is released from the DNA template as it is made. The DNA double helix "zips back up" as the mRNA is released. Newly synthesized DNA remains part of a new DNA helix, paired with its parent strand.

How is mRNA actually synthesized in the cell? By now you know at least part of the answer: enzymes. The RNA-synthesizing enzyme (**RNA polymerase**) has an interesting task. Not only must it select the cor-

rect complementary nucleotides and link them together (as does its counterpart, DNA polymerase), but it must also decide where a gene is. A DNA helix can contain thousands or millions of base pairs. The RNA polymerase must determine exactly where to start and stop synthesizing RNA so that it will transcribe a complete gene.

How does RNA polymerase know where to start making mRNA? The answer is that special genetic "traffic signals" are built into the DNA base sequence. One very important traffic signal is called a **promoter**. A promoter is a special sequence of DNA bases that tells the RNA polymerase to start synthesizing RNA. As you might imagine, other signals tell RNA polymerase

Figure 4.10 Transcription. (A) Base pairing between an incoming ribonucleotide and the DNA template guides the formation of a complementary mRNA molecule. The DNA template closes behind the RNA synthesis site, releasing the new RNA molecule. (B) In transcription, a single DNA strand is used as a template. The RNA transcript is released, leaving the DNA molecule intact.

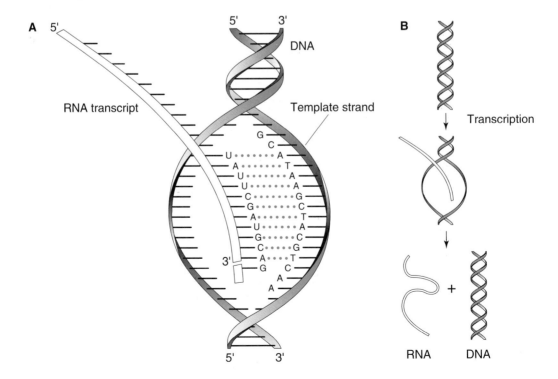

to stop synthesizing RNA and leave the DNA template. These signals are called **terminators**.

Using mRNA To Make a Protein

After transcription is complete, the mRNA moves to the ribosome, the site of protein synthesis. The ribosome recognizes the mRNA and holds it in proper alignment for its codons to be read correctly. Look back at Figure 4.7 and think about this for a minute. The task at hand is to translate the DNA base code in the mRNA to amino acids, as shown in the figure. You can translate a DNA base code to amino acids by looking at the genetic code table (Table 4.1). How could a cell do this with molecules? What if you had a molecule whose one end fit exactly to one and only one codon and whose other end was connected to the correct amino acid? If you had one of these molecules for each of the codons in Table 4.1, you would have a molecular "key" for translating codons to amino acids. This is what cells do—your cells, bacterial cells, lizard cells, and the mold cells on the old sandwich in your refrigerator. The molecules that make up the molecular key are another type of RNA: tRNA molecules.

tRNA molecules are folded in on themselves to resemble cloverleafs. At the tip of one of the lobes is a sequence of three bases called an **anticodon** (Figure 4.11). This anticodon pairs exactly with one of the codons on the single-stranded mRNA, using the rules of complementary base pairing. At the other end of the tRNA molecule is an amino acid. Each different tRNA is connected to the right amino acid for its anticodon. How does that happen? Enzymes, again. Every cell contains a host of exquisitely specific enzymes

(called aminoacyl synthetases) whose job it is to recognize individual tRNAs and attach the correct amino acid to them, so that each type of tRNA has a specific amino acid attached to it. The result is that, when the anticodon on the tRNA pairs with its codon on the mRNA, the correct amino acid is brought to the ribosome.

The ribosome holds the mRNA molecule so the tRNAs pair with their complementary codons one at a time, in order. As the tRNAs bring in the correct amino acids, the ribosomes link the amino acids into a growing protein chain. Once an amino acid has been linked to the chain, the tRNA molecule is separated and released from the mRNA-ribosome complex (Figure 4.12). This process, in which the mRNA base sequence is translated to a protein amino acid sequence, is called (amazingly enough) **translation**.

Protein synthesis is a complicated process involving a variety of interactions between enzymes and RNA molecules. How does a ribosome distinguish an mRNA molecule from other RNA molecules in the cell? The answer is that every mRNA molecule contains more than just the codons needed to make a protein. It also contains traffic signals for the ribosome.

Foremost among these regulatory elements in bacteria is the signal for ribosomal recognition. In bacteria, this element commonly has the sequence 5'-GAGG-3' or 5'-AGGA-3' located 8 to 13 nucleotides upstream of the initiation codon. The ribosome recognizes the element and binds to the mRNA there. How does binding occur? The bacterial ribosomal RNA contains a base sequence complementary to the ribosomal recognition element on the mRNA, and the rRNA and mRNA associate through base pairing at that site.

Figure 4.11 A tRNA molecule. Complementary base pairing between different portions of the tRNA molecule maintains its shape.

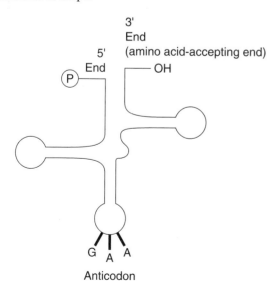

Anticodon

Figure 4.12 Translation. Complementary base pairing between the anticodons of incoming tRNA molecules and the codons of the mRNA guides the formation of the amino acid chain.

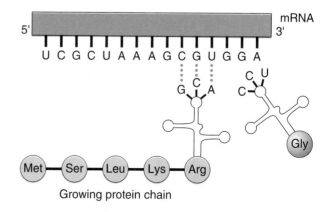

Growing protein chain

The ribosome recognition element is followed by an **initiation codon** (usually AUG, which encodes methionine), where protein synthesis actually begins. At the end of the coding region, there is a stop signal (a **stop codon**; see Table 4.1). There are no tRNAs that pair with stop codons. Instead, proteins called termination factors bind to stop codons and cause protein synthesis to terminate. Since the recognition sequence, initiation codon, and stop codon are present in the base sequence of the mRNA, they must also be encoded in the original DNA template that was transcribed to make the mRNA.

A summary of the major genetic "traffic signals" is presented in Figure 4.13. The DNA molecule is a marvelous storehouse of complicated information. It contains not only the blueprints for the amino acid sequences of every protein an organism makes but also the traffic signals that direct the cell to interpret the information properly. As we shall see below, DNA even contains signals that allow a cell to regulate the synthesis of its proteins to meet the needs of its environment.

Although all cells on earth use the same genetic code and synthesize proteins via mRNA and tRNA, there are important differences between broad categories of cells with respect to the flow of information from genes to proteins. These broad categories of cells are *eukaryotic* cells and *prokaryotic* cells.

Eukaryotic and prokaryotic cells

Eukaryotic cells are the cells present in most organisms with which we are familiar, such as fungi, plants, and animals. These organisms are also called **eukaryotes**. Their genetic material is located in an organized nucleus enclosed in a nuclear membrane. Eukaryotic

cells also contain other membrane-bound organelles such as mitochondria, chloroplasts, lysosomes, and endoplasmic reticula. While most plants, animals, and fungi are multicellular organisms that may contain many specialized tissues and cell types, there are also single-celled eukaryotes. Examples of single-celled eukaryotes include yeasts, protozoans, and green algae.

The cells of **prokaryotes** have a much simpler structure. They lack membrane-bound organelles (including those listed above) and an organized nucleus. In prokaryotic cells, the genetic material is located in the cytoplasm. The two major groups of prokaryotes are eubacteria and archaea. Eubacteria are the familiar bacteria and blue-green algae (sometimes called cyanobacteria). Archaea (also called archaebacteria) are a group of prokaryotic organisms that inhabit extreme environments. They are genetically and biochemically similar to one another but are as different from eubacteria as the eukaryotes are. Archaea are interesting in their own right but are generally not considered in this discussion. When we talk about prokaryotes or bacteria in this book, we are referring to eubacteria. Both eubacteria and archaea are single-celled organisms. There are no true multicellular prokaryotes.

Because bacteria grow rapidly, are easy to grow in large quantities, and have much smaller, more easily manipulated genomes than do eukaryotes, they are a favorite tool of biotechnologists. One of the major achievements of biotechnology has been to use bacteria to produce large quantities of proteins (such as human growth hormone) that are normally made in small quantities in eukaryotic organisms. This achievement is possible because the genetic code used in prokaryotes and eukaryotes is the same.

Differences in the Molecular Biology of Prokaryotes and Eukaryotes

Although their genetic codes are the same, the genetic traffic signals that direct the processing of the coded information differ in prokaryotes and eukaryotes. For example, prokaryotic and eukaryotic promoters and RNA polymerases are different. A typical bacterial promoter consists of the sequence 5'-TTGACA separated by 17 bases from the sequence TATATT-3' (some variation is permitted). Eukaryotic promoters are more variable. One component of a eukaryotic promoter is the sequence TATA located about 30 bases upstream of where transcription begins in yeasts, 60 bases upstream in mammals. This component is often called the TATA box. Usually, two more promoter components are found somewhat further upstream of the TATA box. These are the sequence CCAAT and a GC-rich sequence.

Figure 4.13 Major genetic traffic signals in bacteria. These signals tell RNA polymerase where to begin and end transcription, enable the ribosome to recognize mRNA, and direct the ribosome to start and stop protein synthesis. RRE, ribosome recognition element.

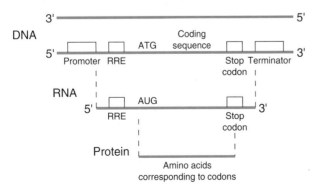

A more important difference is that, in contrast to prokaryotic RNA polymerase, eukaryotic RNA polymerase alone is not efficient at starting transcription from a promoter. Eukaryotic promoters are associated with a variety of additional DNA sequences where various transcription factors bind and stimulate transcription (see *Transcriptional Regulation through Activation,* below).

The recognition of mRNA by ribosomes in eukaryotes is different from the prokaryotic system described above and appears to be more variable. In the simplest model, eukaryotic ribosomes are thought to bind at the 5′ end of the mRNA and to search down the mRNA for the first initiation codon to begin translation. Experiments in some laboratories suggest that this simple model does not explain all situations. In both prokaryotes and eukaryotes, translation begins at the initiation codon AUG. In *Escherichia coli,* initiation of protein synthesis also requires three proteins called **initiation factors**. In eukaryotes, many more protein factors are required. Both prokaryotic and eukaryotic genes contain stop codons as a signal to the ribosome to release the mRNA.

A more fundamental difference between prokaryotic and eukaryotic genes is that prokaryotic genes usually exist as a continuous sequence on a DNA molecule, and several are often transcribed from the same promoter. In contrast, eukaryotic genes are usually transcribed one at a time and are often encoded in many small pieces separated by stretches of noncoding DNA called **introns.**

When eukaryotic cells transcribe one of these split genes, a very long precursor RNA that includes all of the introns and the coding regions (exons) is synthesized first. Next, eukaryotic cells edit the RNA. In a process called **splicing,** the intron sequences are se-

lectively cut out, and the exons are pieced together to make a functional mRNA (Figure 4.14). As you might imagine, the mRNA contains coded signals at the beginning and end of each intron that direct the cell to remove the intron. After the introns have been removed and the exons have been spliced together, the functional mRNA moves to the ribosome for translation. Splicing does not occur in eubacteria, and bacteria lack the enzymes necessary for splicing eukaryotic RNA. (See the activity *From Genes to Proteins.*)

These differences in promoters, ribosome recognition sequences, and splicing not only have fascinated researchers but also have made life interesting for biotechnologists seeking to transfer information from eukaryotes to bacteria. As you might guess, simply transferring a gene from a mammal to a bacterium usually does not result in the production of a functional protein. In addition to the genetic information, biotechnologists must also provide the correct processing signals to the new host cell and "presplice" the gene if it contains introns. Many standard procedures have been developed to simplify these processes (see chapter 5).

Regulating gene expression

Cells must regulate the synthesis of their proteins in order to respond to environmental conditions. For example, most bacteria have genes encoding enzymes capable of breaking down quite a variety of sugars for energy. However, synthesizing these enzymes would be a waste of the cell's energy if the sugars were not available to the cell. So most bacteria synthesize an enzyme that breaks down a particular sugar only if that sugar is present in its environment.

Conversely, most bacteria encode enzymes capable of synthesizing all of the amino acids the bacteria need

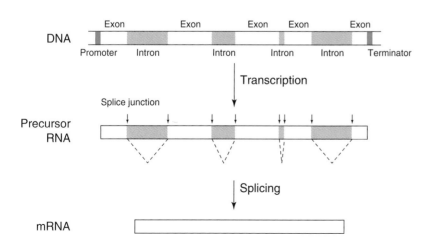

Figure 4.14   Splicing of precursor RNA to create mRNA.

to make their proteins. Yet synthesizing these enzymes would be wasteful if the needed amino acids were already present in the environment, so most bacteria produce the amino acid-synthesizing enzymes only if amino acids are not available to them. Finally, many organisms, even bacteria, undergo some form of change during their life cycles. Plants and animals develop from a single fertilized egg. Some bacteria respond to adverse environments by forming spores. These structural changes also require controlled gene expression.

How is gene regulation achieved? There are many mechanisms. The best-understood ones involve regulation of transcription, the synthesis of mRNA. The rate of degradation of specific mRNAs can also be controlled. The translation rate of mRNA molecules is often regulated. Any step between DNA and protein is a potential regulatory target. However, the most common model for gene regulation involves regulation of transcription.

Transcriptional Regulation through Repression

In addition to promoters, many genes contain sites to which regulatory proteins can bind. These regulatory sites are often very near the promoter. In the most typical scenario in prokaryotes, the binding of the

regulatory protein prevents transcription of the gene either by blocking access to the promoter or by preventing progression of RNA polymerase along the gene. Regulatory proteins that bind to DNA and prevent transcription are called **repressors**.

The sugar lactose can be used by the bacterium *E. coli* (and many other bacteria) as an energy source. *E. coli*'s lactose utilization genes are lined up in a row along its chromosome and are transcribed from a single promoter into one long mRNA. Several proteins are translated from this long message. The collection of lactose utilization genes is called the *lac* operon.

When lactose is present in the environment, *E. coli* synthesizes lactose utilization proteins from the *lac* operon. The lactose utilization proteins allow *E. coli* to derive energy from the sugar. When no lactose is present, these proteins are not synthesized. How does *E. coli* achieve this appropriate regulation of its *lac* genes? The following description is somewhat simplified but gives the basic idea.

E. coli synthesizes a *lac* repressor protein (Figure 4.15). In the absence of the sugar lactose, this protein binds to the *E. coli* chromosome at a special site (the *operator*) near the promoter of the *lac* genes and prevents transcription of the genes. Consequently, the

Figure 4.15 Transcriptional regulation of the *lac* operon. P is the promoter; O is the operator.

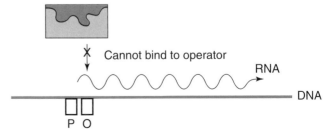

A. No lactose in cells: active repressor prevents transcription.

B. Lactose in cells: inactive lactose-repressor complex allows transcription.

bacterium does not waste energy making lactose-utilizing enzymes when there is no lactose in the cell.

If lactose is present, however, the bacterium needs enzymes for using it. The *lac* regulation system allows the cell to respond beautifully to this new need. When lactose enters the cell, it interacts with a special site on the *lac* repressor protein. This interaction renders the repressor unable to bind to its site on the *E. coli* DNA, presumably by changing the shape of the protein. The repressor releases the DNA, leaving the gene free to be transcribed. The cell then can make the lactose-using enzymes and take advantage of the new energy source (Figure 4.15).

The *lac* genes of *E. coli* are "turned on" when the appropriate sugar is present and are otherwise "turned off." Sometimes, however, it is better for a cell to have genes normally turned on and to turn them off only under special circumstances. An example of this type of regulation is found in the transcription of the tryptophan-synthesizing enzymes of *E. coli*.

Tryptophan is an amino acid that, like other amino acids, is essential for protein synthesis. *E. coli* has genes that encode enzymes for synthesizing trypto-

phan from scratch (the *trp* genes). Like the *lac* genes, the *trp* genes are also lined up on the chromosome and transcribed from one promoter. Since *E. coli* constantly needs tryptophan for making new proteins, the *trp* genes are normally turned on. Occasionally, however, a lucky *E. coli* might find itself in an environment where tryptophan is plentiful. In this case, the bacterium conserves energy by stopping the synthesis of the *trp* proteins. Stopping expression of the *trp* genes in response to environmental changes is also achieved through a repressor protein (the following description is again somewhat simplified).

E. coli synthesizes a *trp* repressor protein, but the shape of the *trp* repressor does *not* allow it to bind to the chromosome (Figure 4.16). In its native state, the *trp* repressor cannot attach to its site near the promoter of the *trp* genes and therefore does not interfere with their transcription.

If the concentration of tryptophan inside the cell rises, however, the excess tryptophan interacts directly with the inactive *trp* repressor. Tryptophan binds to a special site on the protein, and its binding changes the shape of the repressor. The altered shape of the protein allows it to bind to the regulatory site near the promoter of the

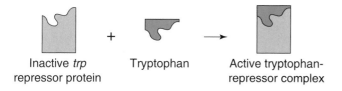

Inactive *trp* repressor protein + Tryptophan → Active tryptophan-repressor complex

Figure 4.16 Transcriptional regulation of the *trp* operon. P is the promoter; O is the operator.

A. Low tryptophan concentration: inactive repressor allows transcription.

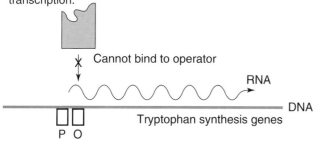

Cannot bind to operator

RNA

DNA

Tryptophan synthesis genes

P O

B. High tryptophan concentration: active repressor complex binds to operator and prevents transcription.

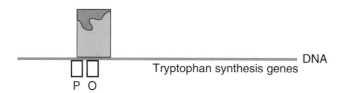

DNA

Tryptophan synthesis genes

P O

trp genes. When the active repressor binds, transcription of the *trp* genes is turned off. Thus, the presence of excess tryptophan in the cell leads to a shutoff of the tryptophan-synthesizing enzymes.

Transcriptional Regulation through Activation

The *lac* and *trp* operons provide excellent examples of transcriptional regulation in prokaryotes. Bacterial cells are easy to culture and manipulate, and transcription and gene regulation were first studied in them. When techniques for studying these processes in eukaryotic cells became available, scientists expected to find similar mechanisms at work. However, many years of work have suggested that the typical regulation mechanism in eukaryotic cells involves activation rather than repression. In these cells, most promoters are apparently not recognized efficiently by RNA polymerase alone. Instead, eukaryotic RNA polymerase requires helper proteins called **transcriptional activators** or *transcription factors* to help it bind to a promoter and begin transcription. Some of these activators associate with the RNA polymerase enzyme and do not bind to DNA themselves. Other activators bind to special base sequences in DNA and then interact with RNA polymerase to help it bind to a promoter (Figure 4.17).

The DNA-binding sites of these activators are often called **enhancers** because their presence enhances transcription from the associated promoter as long as the appropriate activator protein is present. The activity of genes can be regulated by the availability of the necessary transcriptional activators. For example,

Figure 4.17 Activator proteins are needed for transcription in eukaryotic cells.

A. No activators present: RNA polymerase cannot bind to promoter.

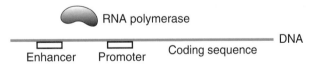

B. Activators present: RNA polymerase can bind to promoter and synthesize RNA.

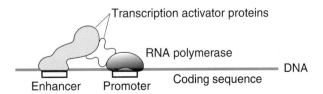

genes encoding the antibody proteins of the immune system have a specific enhancer, the immunoglobulin enhancer, associated with their promoters. This enhancer works only to enhance transcription in B lymphocytes (the antibody-producing cells of the immune system), because only B lymphocytes contain the transcriptional activator protein that binds the immunoglobulin enhancer. As a result, only your B lymphocytes make antibody proteins, even though all the cell nuclei in your body contain the antibody genes.

Another example of gene regulation through enhancers is the response of genes to steroid hormones such as estrogen (not all hormones are steroid hormones; other hormone types exert their effects on cells through different mechanisms). Estrogen and other steroid hormones pass through the cell membrane and bind to specific receptor proteins inside the cell. When the hormone binds, the receptor changes its shape in a way that allows it to move to the nucleus, where the receptor-hormone complex acts as a transcriptional activator by binding specific enhancers. Thus, specific genes are turned on in response to the presence of the hormone.

Alteration of Promoter Recognition

Alteration of promoter recognition is a form of transcriptional regulation that occurs when a drastic change in gene expression is needed—a new set of genes must be transcribed and/or a currently transcribed set must be turned off. For example, when the bacterial virus T4 invades a host *E. coli* cell, the *E. coli* RNA polymerase begins to transcribe a few of the T4 genes that have *E. coli*-style promoters. One of these genes encodes a protein that binds to the host RNA polymerase. The modified RNA polymerase can subsequently recognize only the virus's other promoters, which have a base sequence different from that of the normal host promoters. In this way, the virus switches transcription from the host genes to its genes.

Another example of this type of regulation is found in the bacterium *Bacillus subtilis*. This organism forms durable, dormant spores in response to adverse environmental conditions. Spore formation requires the expression of a number of genes that are not active during the normal life cycle of the bacterium. In addition, normal gene activity all but ceases during the spore stage. To achieve this gross change in gene expression, *B. subtilis* synthesizes a special protein that binds to its RNA polymerase and causes it to recognize only the special promoters controlling sporulation genes.

Repression of Translation

Regulation of gene expression can be exerted through control of the rate of translation of an mRNA.

A good example of translational repression can be found in the synthesis of the ribosomal proteins of *E. coli.* The ribosomes of *E. coli* are made up of large ribosomal RNA molecules and several proteins that bind to specific regions of the RNA. The genes for the ribosomal proteins are lined up in operons. A single mRNA is transcribed from each operon and translated into several proteins. As the proteins are translated, they find free ribosomal RNA in the cytoplasm and bind to their recognition sites. When all available ribosomal RNA is complexed with protein, one of the proteins from that operon instead binds to the translation initiation region of the operon for the ribosomal proteins' mRNA. The binding of that protein to this mRNA prevents further translation. For each one of the ribosomal protein operons, one of the encoded proteins acts as a repressor of translation. Through this mechanism, a balance between the amount of available ribosomal RNA and ribosomal proteins is achieved.

These examples of simple gene regulation mechanisms are typical. We want to point out that repression occurs in eukaryotic cells and activation occurs in prokaryotes, too. Many steps in protein synthesis other than transcription can also be regulated. However, a central theme of gene regulation is that it involves interactions between proteins and other molecules: additional proteins, small molecules, RNA, and DNA. These interactions are dependent (as are all protein functions) upon the three-dimensional structure of the proteins. At the end of the second part of this chapter (*Protein Structure and Function*), we will look closely at the structures of two DNA-binding regulatory proteins and how they interact with DNA.

Genomic organization

The genes encoded in DNA must be accessible to enzymes at the proper times for replication and transcription. Considering the extreme length of the DNA molecule in most organisms, this is a staggering requirement. Storage of DNA in cells is therefore not a haphazard affair, with the DNA "just lying there." DNA storage is a highly organized, complex phenomenon that is not well understood. Many scientists today are working to understand how cells manage their DNA information libraries.

Chromosomes

The physical packing of DNA inside cells presents a problem because of the extreme length of the DNA molecule relative to the size of the cell itself. If the DNA of *E. coli* were stretched out, it would be 1,000 times longer than the *E. coli* cell (see Figure 4.18). The problem faced by eukaryotic cells is even more amazing: the DNA of a single human cell would stretch

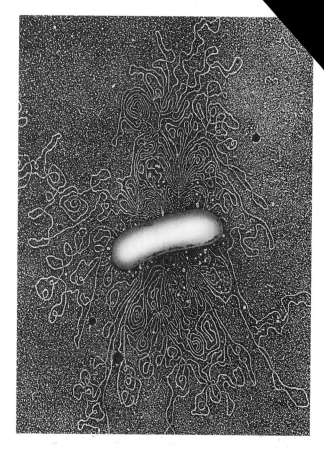

Figure 4.18 *Escherichia coli* osmotically shocked to release DNA. (Photograph courtesy of VU/©K.G. Murti.)

2 m, yet the cell itself has a diameter of only 1/50 of a millimeter. It is clear that cellular DNA must be very highly folded. In cells, DNA is folded and packed in association with proteins. The DNA-protein material is called **chromatin,** and the packed structure is called a **chromosome**.

In eukaryotes, the genome is usually divided among several different linear chromosomes located in an organized nucleus. The ends of linear chromosomes are capped by structures called telomeres, which are segments of DNA consisting of short repeated sequences assembled into an unusual formation that includes a loop at the very end of the chromosome. Telomeres are not replicated by DNA polymerase. Instead, they are maintained by an enzyme called telomerase, which adds nucleotides to the ends of chromosomes, synthesizing the repeated sequence.

Telomeres are essential for protecting the ends of linear chromosomes and may be linked to cell aging and death. Cells taken from higher eukaryotes and grown in culture have limited life spans, undergoing an average number of cell divisions before they die. The number of divisions they undergo is correlated with the

...anism from which the cells were ...f the individual organism. As the ...ted divisions, the average length ...decreases. It is thought that this ...ction for the ends of the chromo-...y lead to cell death. Interestingly, ...cer cells, which are immortal in ...k over time. It appears that the perpetuation of telomeres is linked to immortality, at least for cancer cells.

Chromosomal DNA is packed in an extremely ordered manner with multiple layers of folding and coiling of the DNA molecules around the chromosomal proteins. Yet even though the DNA is tightly packed, all of the DNA encoding essential proteins is available for transcription at the proper time. How the cell organizes the complex problem of storage and accessibility is not understood. Some scientists think that the large amount of noncoding DNA present in higher organisms (see below) may play an important role in DNA organization.

In most bacteria, DNA is present as one long circular molecule that is associated with special proteins in a single chromosome. These proteins are thought to hold the DNA in a condensed form so that it fits neatly into the cell. The circular chromosome is present in the cytoplasm, attached to the cell membrane.

Plasmids

Some cells, in particular bacteria, contain small rings of DNA outside of their chromosomes. These small circular pieces of DNA are called **plasmids**. Plasmids are replicated by the cell's enzymes and inherited by progeny bacteria. They typically contain a few thousand base pairs of DNA and encode a few proteins. None of these proteins are normally essential to the survival of the bacterium. Plasmids often contain genes for drug resistance or for poisons to kill rival bacteria and, in disease-causing bacteria, for proteins that contribute to the disease process. Because they are small and convenient to work with, plasmids are a favorite tool of biotechnologists. (See the activity *Recombinant Paper Plasmids* and the section *Transfer of Genetic Information*.)

Mitochondrial and Chloroplast DNA

Mitochondria and chloroplasts each contain their own circular DNA molecules, which are also referred to as genomes. These organelles carry out the essential functions of ATP synthesis and photosynthesis, respectively. Their circular genomes encode some, but not all, of the proteins needed for their essential functions. They also encode ribosomal and tRNAs, which are used to carry out translation inside the organelles. However, in both cases many of the proteins needed for the organelle's

function are encoded in nuclear DNA, synthesized in the cytoplasm, and transported into the organelle.

Inheritance of traits encoded by mitochondrial or chloroplast genes is different from the inheritance of other traits in eukaryotes. When egg and sperm unite to form the zygote, essentially all the cytoplasm, and thus the mitochondria and chloroplasts, are provided by the egg. Therefore, with possible rare exceptions, mitochondrial and chloroplast genes are inherited solely from the mother. Variegated leaves are a trait encoded in chloroplast DNA. Defects in human mitochondrial DNA are associated with several different genetic diseases. The pattern of maternal inheritance of chloroplast and mitochondrial genomes also gives researchers a tool for following family trees (see the Reading *Mitochondrial DNA in DNA Typing*).

Virus Genomes

Viruses are not cells. They consist simply of genetic material enclosed in a capsule generally made of protein. Viral genetic material can be either DNA or RNA. Viruses require a host cell to make copies of themselves. When a virus infects a cell, it introduces its genetic material into that cell. The cell's enzymes and ribosomes transcribe and translate the viral genetic material and eventually reproduce virus particles. Outside of a host cell, viruses are biochemically dormant and are not even considered living.

It is easy to imagine a DNA virus genome substituting for the cell's own genes, but how does an RNA genome work? An RNA virus introduces its RNA genome into its host cell. When the viral RNA enters the cell, the host cell translates it as it would any other mRNA. Some RNA viruses encode in their RNA a message for a special enzyme that uses RNA as a template to synthesize new RNA, in rather the same way that DNA polymerase replicates DNA. While the viral enzyme is synthesizing more viral RNA, the host cell translates the RNAs to make viral proteins. Other RNA viruses encode an enzyme that uses RNA as a template to synthesize DNA. This enzyme is called **reverse transcriptase**. Once the viral RNA genome has been copied into DNA, the cell machinery of the hapless host transcribes it for the virus.

Viruses are extremely specific about the host cells they can infect. Not only do viruses usually infect only one type of organism (such as humans or cats), but they are limited to certain cell types within the organism. For example, a "cold virus" infects only the cells of the upper respiratory tract, while a "stomach virus" infects only the cells of the digestive tract. This specificity explains why a pet dog does not get its master's cold. However, a few viruses, such as rabies, normally infect several different host species. And once in a while, a virus acquires the ability (possibly through mutation) to infect a new host.

In addition to viruses that infect humans or other animals, there are plant viruses specific to certain plant tissues and bacterial viruses specific to certain bacterial strains. Bacterial viruses have played an important role in the development of molecular biology and have been given a special name: **bacteriophages**. Electron micrographs of several viruses are shown in Figure 4.19.

Figure 4.19 Electron micrographs of various viruses. (A) Bacteriophage lambda (×275,000; photograph courtesy of VU/©K. G. Murti); (B) purified bacteriophage T4 (photograph courtesy of F. P. Booy; reprinted from J. D. Karam et al., ed., *Molecular Biology of Bacteriophage T4,* ASM Press, Washington, D.C., 1994); (C) tobacco mosaic virus (×144,000; photograph courtesy of VU/©K. G. Murti); (D) vesicular stomatitis virus (rabies group) (×100,000; photograph courtesy of VU/©K. G. Murti).

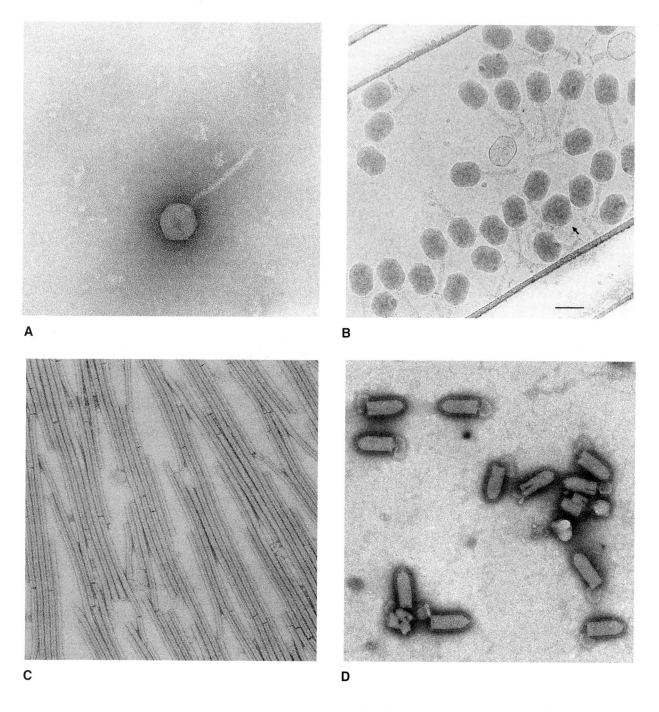

A

B

C

D

a virus infects a plant cell, animal cell, or ...cell, the mechanism by which it recognizes ...host is similar. Proteins that make up the ...sule of the virus recognize and bind to a specific molecule, often a protein, on the surface of the host cell. This specific molecular recognition is what determines the extremely limited range of host cells a virus can infect.

Viruses as Tools in Biotechnology

Viruses have proved to be useful tools for biotechnologists because of some of the properties mentioned above. Viruses are essentially containers of genetic material. They deliver that material into very specific cells. Scientists have developed ways of packaging new genetic material into several different types of viruses. The virus particles then inject that material into their host cells. Delivery of genetic material via viruses is an important method of gene transfer. (See the activity *Transduction of an Antibiotic Resistance Gene.*)

Noncoding DNA

How much DNA do organisms have? Naturally, it depends on the organism. The smallest viruses may have only a few thousand base pairs. Since the "average" protein requires 1,200 bases of coding sequence, these viruses encode only a few proteins. A bacterium such as *E. coli* contains about 4 million base pairs in its genome, while the human genome is composed of about 3 billion base pairs. Lest we grow smug about the complexity of the human genome, it bears noting that the genome of the mud puppy (an amphibian similar to newts and salamanders) is estimated to contain about 50 billion base pairs, and that of the lily contains 250 billion.

Does all of this DNA encode proteins? In bacteria and viruses, most of it appears to. Eukaryotes are a different story. In fact, only a small fraction of the DNA of multicellular eukaryotes (such as amphibians, mammals, and plants) is coding sequence.

What is all that extra DNA in eukaryotes? A large portion of it is repeated sequences that are present throughout the genome, sometimes in large clusters. Some of the repeated sequences appear to be transposons or parts of transposons. No one knows what the purpose of any of this DNA is or even if it has a purpose. This DNA has sometimes been called "junk DNA," though it seems presumptuous to label it junk before we understand it. A more neutral name for it is **noncoding DNA.** When you imagine the genomes of bacteria and viruses, you may imagine compact arrays of genes along the DNA. However, when you imagine your own genome or the genomes of plants or animals, you must imagine vast stretches of repeated DNA sequences, occasionally broken up by genes. Puzzling, isn't it?

Mutations

Any change in a DNA sequence is called a mutation. It is a fact of life that mutations happen. They result from normal cellular processes and unavoidable environmental hazards. During DNA replication, the DNA polymerase enzyme occasionally makes errors that escape the proofreading that occurs during DNA replication. Environmental factors such as ultraviolet light or mutagenic chemicals damage DNA regularly. Most of this damage is corrected by **DNA repair enzymes,** which are proteins that recognize and repair abnormalities in DNA, but occasionally the repair enzymes miss something. When DNA polymerase attempts to use a damaged base as a template during DNA replication, it often cannot read the base properly and so inserts an incorrect base into the new strand, thus creating a mutation. Genetic events such as transposition (see *Transposable Elements,* chapter 3) result in insertion or loss of segments of DNA, also changing the original sequence. Errors in cell division or recombination can lead to a rearrangement of the segments of chromosomes or even a change in chromosome number.

What is the effect of a mutation? It depends. It depends on where the mutation is in the DNA, exactly what it is, and, often, what environment the organism inhabits. For example, a sequence change in one of the many noncoding regions would probably not have any effect on the organism. Similarly, a mutation in an intron sequence in a eukaryote might not have any effect unless it involved a processing signal for the splicing enzymes. Even changes in coding sequences may not have any effect on a protein. Many amino acids are encoded by more than one codon. For example, the codons TTT and TTC each encode the amino acid phenylalanine (Table 4.1). A mutation that changed TTT to TTC would not have any effect on the protein. Mutations with no effect on a protein are often called "silent" mutations.

In addition, many amino acid changes may not alter the function of a given protein in a significant way. Recalling the example of the receptor protein in the cell membrane, imagine that a mutation occurred in a region of the protein apart from the specific hormone recognition area. As long as the change did not distort the overall shape of the protein or impair its interaction with the cell, it might not affect the function of the protein.

Although many mutations are harmless, others can be devastating. Mutations in genetic traffic signals such as promoters and ribosome recognition sequences can completely shut down the synthesis of a protein. Sometimes a single base change in a coding region can result in an amino acid substitution that severely harms or destroys the protein's ability to perform its function. Recall again our example of the receptor protein embedded in the cell membrane. If anything distorted the shape of the receptor protein, the hormone might no longer fit and/or the receptor might no longer be able to communicate with the rest of the cell. The cell would lose its ability to respond to the body's signals for growth. Thus a change in the protein's shape could have disastrous consequences for the organism. Similar examples could be given for proteins of the immune system that must recognize specific invaders, for transport proteins that carry specific nutrients, and so on.

Although most mutations are harmful or neutral, they can also be beneficial. Changes in the amino acid sequence of a protein might make it more resistant to heat, which could be an advantage if the organism's environment is becoming warmer. An alteration in the shape of another protein might allow it to bind to and break down a different type of sugar, which could be an advantage if the organism's environment contained that sugar. If a change in a protein's function is not immediately fatal, it might actually help the organism under the right environmental conditions.

One consequence of mutations and recombination is that the chromosomes of all sexually reproducing individuals differ in many ways. Although all humans have similar chromosomes and produce the same sets of essential proteins, the exact DNA sequences in those chromosomes varies. One obvious source of variation is the same variation that causes us to look different: one individual has genes encoding blue eyes, and another has genes encoding brown eyes. A less obvious source of variation is the accumulation of changes in noncoding regions of DNA. Great variety can be present in these regions, particularly in the number of repeated sequences present, with no outwardly observable effects. In fact, it is extremely unlikely that any two individuals (except identical twins) would share the same sequences in all of their noncoding DNA. It is this variety that makes possible the new procedures of DNA fingerprinting. (See chapter 5 and *Analyzing Genetic Variation: DNA Typing*.)

At the end of this chapter we will look at specific examples of a harmful and a helpful mutation: the nucleotide change, the amino acid changes, the effects on protein structure, the effects on the proteins' functions, and the effects of the changes in protein behavior on the phenotype of the organism. To put all of these things together, we first need to look at the link between genes and phenotype: proteins.

Protein Structure and Function

Genes are important because they supply information that directs the synthesis of proteins. It is the proteins that confer a phenotype on the cell: its biochemical capabilities, its shape, its communication channels, and so on. The function of a protein is determined by its three-dimensional shape, which is determined by the nature and sequence of its amino acids. The relationship between amino acid sequence and three-dimensional structure has been called the "second half of the genetic code," because it is the three-dimensional structure that leads to function and phenotype. Unfortunately, the relationship between amino acid sequence and three-dimensional structure is not simple. It would be wonderful if we could predict the three-dimensional structure of a protein from its amino acid sequence, but we can't. At least not yet.

Over the past several years, the three-dimensional structures of hundreds of proteins have been painstakingly determined using the techniques of X-ray crystallography and nuclear magnetic resonance (see chapter 5). These structures have greatly increased our understanding of how amino acid chains fold up to be energetically stable proteins. Here's a summary of the basic principles.

Amino acids and peptide bonds

There are 20 amino acids that normally make up proteins. Amino acids have the general chemical structure shown in Figure 4.20. The R in the figure can be any of 20 different so-called side chains. These different side chains give the 20 amino acids their separate identities.

When amino acids are joined to make a protein chain, the OH group on one end of the amino acid reacts with the NH_2 group of another amino acid. A water molecule (H_2O) is lost, forming what is called a **peptide bond** (Figure 4.21A). A protein consists of many amino acids joined together via peptide bonds. Like a

Figure 4.20 General structure of an amino acid. R signifies one of the 20 different side chains shown in Figure 4.23.

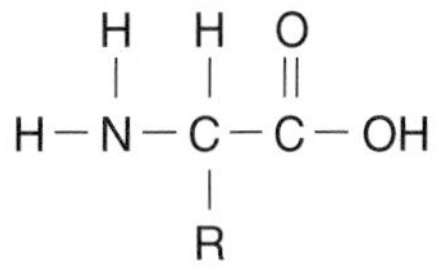

Figure 4.21 To form proteins, amino acids are joined by peptide bonds.

A. Peptide bonds are formed between the NH₂ group of one amino acid and the COOH group of another, with the formation and loss of a water molecule. The peptide bond is shown in the shaded area.

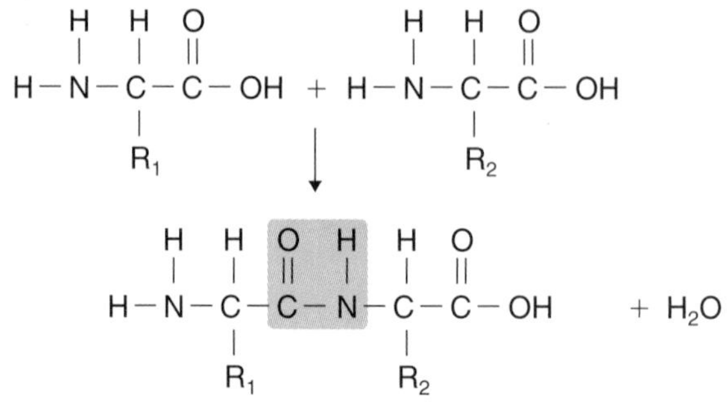

B. A protein is a polypeptide backbone with various amino acid side chains (Rₙ).

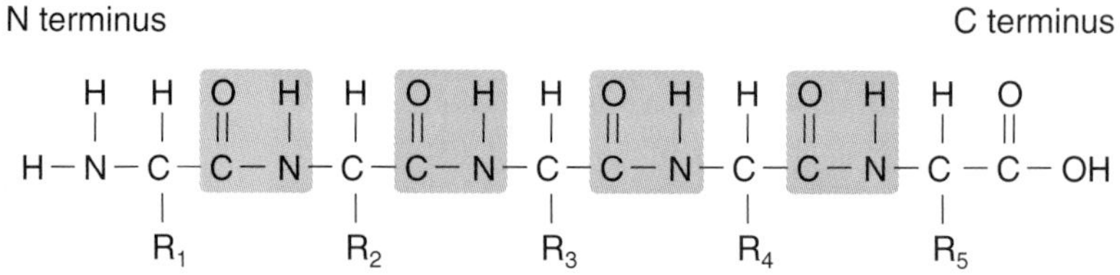

DNA strand, a protein backbone has a direction, too. One end has a free NH₂ group, and the other has a free COOH group. These ends are called the *N terminus* and the *C terminus*, respectively. The overall effect is that a protein has a uniform peptide backbone with various amino acid side chains (Figure 4.21B). The identity and order of the side chains in a protein are called the **primary structure** of the protein. Primary structure is a direct consequence of the DNA base sequence in the gene encoding that protein.

You have probably already figured out that if all proteins have a uniform peptide backbone, then the nature of the amino acid side chains must be the factor that governs how an individual protein folds into a three-dimensional structure. That is true, and fortunately, one particular property of side chains is the most important for influencing three-dimensional structure. This property has to do with how well the individual side chains interact with water molecules. To understand this property, let's start with a quick review of covalent bonds.

Polar and nonpolar covalent bonds

The atoms in a protein molecule are held together by **covalent** chemical bonds, which consist of electrons shared by two atomic nuclei. If the electrons are shared equally by the nuclei, then their negative charge is distributed evenly over the area of the bond and balanced by the positive charges in the nuclei. However, nuclei of certain elements attract electrons more strongly than other nuclei do (the ability to attract electrons is called **electronegativity**). If a strongly electronegative element forms a covalent bond with a less electronegative element, the electrons tend to be found near the strongly electronegative nucleus. You can think of the electronegative nucleus as pulling the bond electrons away from the less electronegative nucleus. This uneven distribution of electrons creates a partial negative charge around the electronegative nucleus and a partial positive charge at the other nucleus. Chemical bonds with this type of uneven charge distribution are called *polar* bonds because they have positive and negative poles (Figure 4.22).

Amino acids (and proteins) are made primarily of carbon, hydrogen, oxygen, and nitrogen. Of these four elements, oxygen and nitrogen are strongly electronegative, and carbon and hydrogen are not. Thus, bonds between oxygen and hydrogen, oxygen and carbon, nitrogen and hydrogen, and nitrogen and carbon are polar. However, many of the bonds in a protein molecule are between carbon and hydrogen and are not polar. Whether a larger chemical group like the side chain of an amino acid is polar or not depends on its constituent chemical bonds.

The chemical structures of the side chains of all 20 normal amino acids are shown in Figure 4.23. The

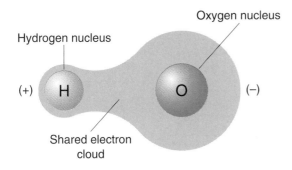

Oxygen nucleus

Hydrogen nucleus

(+)   H   O   (−)

Shared electron
cloud

Figure 4.22   A polar covalent bond. Although the oxygen and hydrogen nuclei are sharing two electrons, the highly electronegative oxygen nucleus tends to draw them away from the weakly electronegative hydrogen nucleus. As a result, the oxygen end of the bond acquires a partial negative charge, while the hydrogen end is partially positive.

structures are grouped according to whether the side chains are fully charged (ionized) at physiological pH, are polar (partially charged), or are nonpolar. A fourth group, the aromatic amino acids, are called that because, like other aromatic compounds, they have a ring structure in their side chains. Of this group, phenylalanine is nonpolar, and the other two are somewhat polar.

Polarity and Stability

The polarity of the amino acid side chains is fundamentally important to protein structure, because polarity determines how stably a side chain interacts with other elements in the protein and in the cellular environment. A compound that has an overall polar character or is actually electrostatically charged is energetically stable when it associates with other compounds with complementary charges or partial charges. The complementary charges neutralize each other. Nonpolar molecules do not associate with charged or polar molecules in an energetically favorable manner. Instead, they associate comfortably with other nonpolar molecules.

We stated above that the single property of amino acids that most determines three-dimensional protein structure is their ability to interact stably with water. Water is so important because the intracellular environment is water based, as are other body fluids. How stably an amino acid side chain interacts with water therefore determines how "comfortable" it is when exposed to the intracellular fluid.

Here is the bottom line about water. Water consists of two polar oxygen-hydrogen bonds and is a very polar molecule (Figure 4.24). It thus associates comfortably with other polar or charged molecules. For this reason, molecules that are electrostatically charged or

that are polar are called **hydrophilic** (water loving). Since nonpolar molecules do not associate comfortably with water, they are called **hydrophobic**. Hydrophobic amino acid side chains (the nonpolar ones in Figure 4.23) do *not* associate stably with the intracellular fluid. Hydrophilic amino acid side chains (the charged and polar ones in Figure 4.23) *do* associate stably because their charges or partial charges can be neutralized by complementary partial charges of polar water molecules.

Hydrogen Bonds

One type of neutralization that is particularly important in considering protein structure is the hydrogen bond, the same kind of bond found between base pairs in DNA (see Figure 4.3). Hydrogen bonds are not covalent bonds. They are much weaker and form when two highly electronegative nuclei "share" a hydrogen atom that is formally bonded to only one of them. The partial positive charge on the bonded hydrogen neutralizes the partial negative charge on the second electronegative nucleus (Figure 4.25). Hydrogen bonds form only between the three most electronegative elements: oxygen, nitrogen, and fluorine. Of these three elements, only oxygen and nitrogen are common in biological systems, so you can forget about fluorine when thinking about protein structure. In proteins, the most important groups involved in hydrogen bonding are N, NH, O, OH, and CO groups. Water can form hydrogen bonds with all of them (Figure 4.26). Look at the hydrogen bonds in DNA in Figure 4.3, and you will see the same groups.

The fundamental consideration of protein structure

Now that we have looked at the features of covalent bonds that are important for understanding protein structure, let's see what it all boils down to. A protein molecule is a long peptide backbone with a mixture of charged, polar, and nonpolar amino acid side chains. Cytoplasm is a watery environment, so the charged and polar side chains will be stabilized through interactions with water molecules there. However, the hydrophobic amino acid side chains do not associate stably with water; they are more stable when clustered together away from water.

It appears that the basic rule underlying protein structure is, as much as possible, to fold up hydrophobic amino acid side chains together in the interior of the protein, creating a water-free hydrophobic environment. Hydrophilic side chains, meanwhile, are stable when exposed to the cytoplasm on the surface of the protein molecule. This is not to say that you would never find a hydrophilic

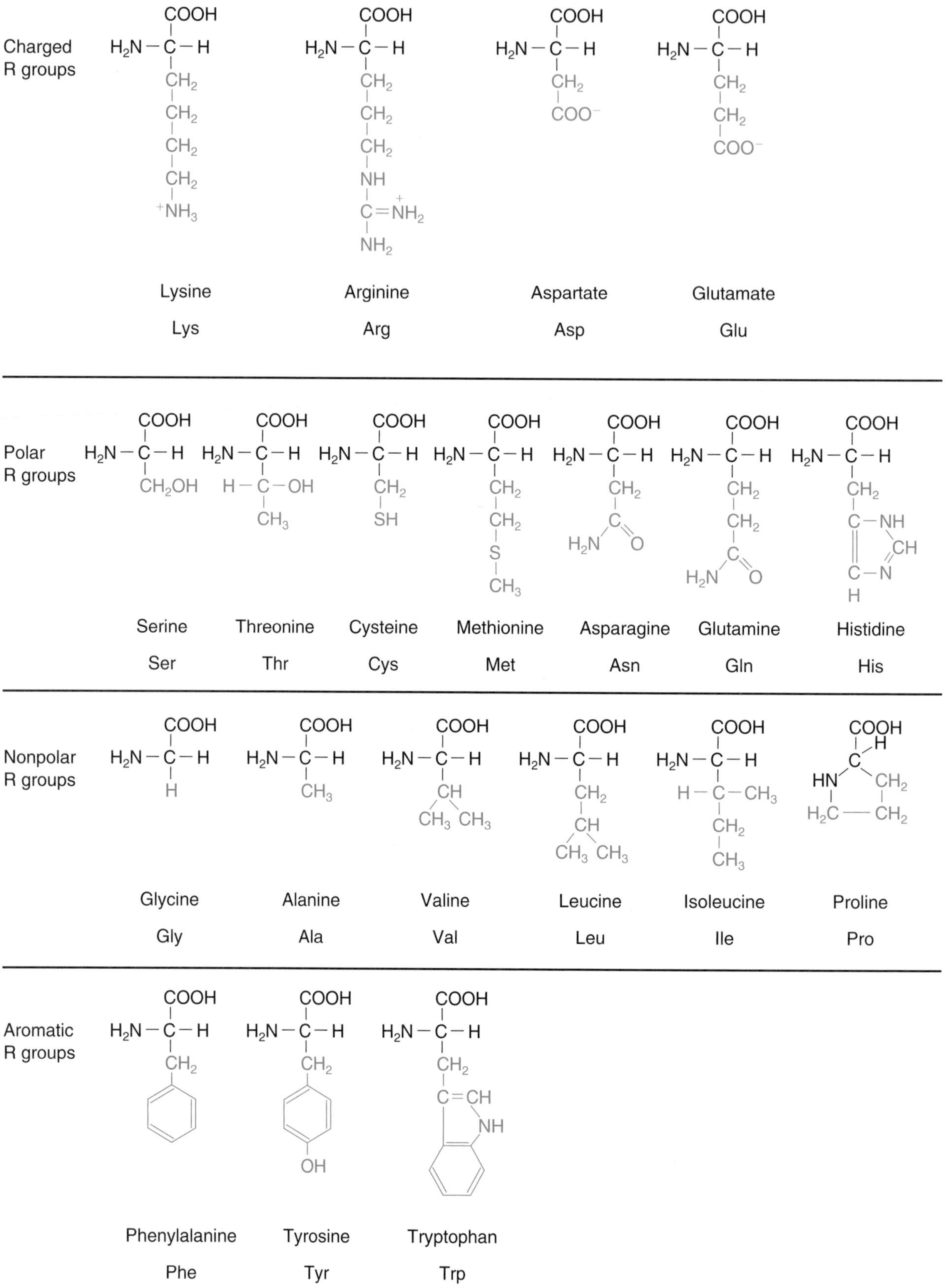

Figure 4.23 The amino acids commonly found in proteins. The three-letter abbreviation for each is shown beneath its full name.

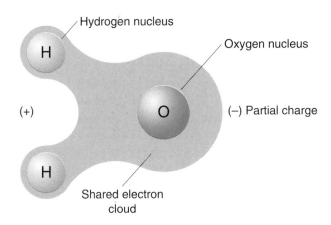

Figure 4.24 Water is a very polar molecule. The strongly electronegative oxygen nucleus hogs the electrons it shares with the hydrogen nuclei.

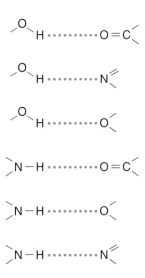

Figure 4.26 Common hydrogen bonds in biological systems.

amino acid on the interior of a protein or a hydrophobic one on the surface, but in general, the rule holds good. A protein is therefore said to have a hydrophobic core. The three-dimensional structure of each individual protein can be thought of as a solution to the problem of creating a stable hydrophobic core, given that protein's primary structure. However, every protein structure must solve one common problem. As you might guess, solving that problem forms another theme in protein structure.

Taking Care of the Hydrophilic Backbone

There is one major problem in folding a protein to create a hydrophobic core: the backbone. Look at Figure 4.21. The peptide backbone is full of NH and CO bonds, and both kinds of bonds are highly polar. On the surface of a protein, these partially charged bonds can be readily neutralized through hydrogen bonding with water. However, for a protein structure to be stable, the partial charges of the peptide backbone must also be neutralized inside the protein core, where there is no water. The solution to this problem is a major factor in determining protein structure.

The fundamental solution to the problem of the peptide backbone in the hydrophobic interior is for the

Figure 4.25 A hydrogen bond (dotted line) is a weak electrostatic attraction between opposite partial charges.

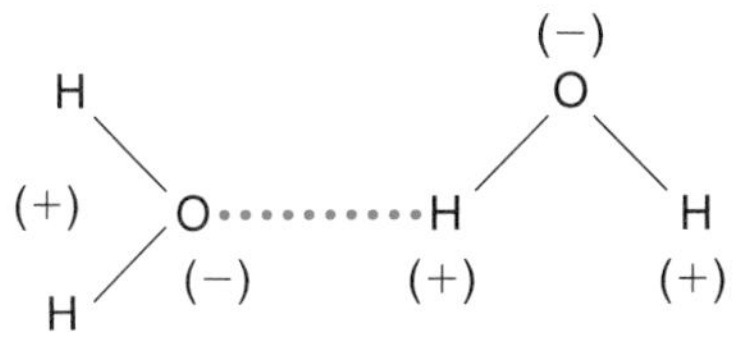

backbone to neutralize its own partial charges. The NH groups can form hydrogen bonds with the CO groups (Figure 4.26), neutralizing both. Since every amino acid contributes one NH group and one CO group to the backbone, this solution is very convenient. However, because of geometric constraints, the NH and CO groups from the same amino acid are not in position to form a hydrogen bond with one another. Instead, the peptide backbone must be carefully arranged so that the NH and CO groups along it are in position to form hydrogen bonds with complementary groups elsewhere along the backbone. Two basic arrangements work well, and these two arrangements form major components of protein structure.

The first self-neutralization arrangement for the peptide backbone is for the backbone to form a helical coil, as if it were winding around a pole. The amino acid side chains point outward, away from the imaginary pole. The NH and CO groups along the backbone form hydrogen bonds with complementary groups above or below them on the pole, as shown in Figure 4.27. This arrangement is called an **alpha helix**.

In the second self-neutralization arrangement, stretches of the peptide backbone lie side by side, so that a CO group on one backbone can form a hydrogen bond with an NH group on the adjacent backbone (Figure 4.28). The amino acid side chains point alternately above and below the plane of the backbone. This arrangement is called a **beta sheet,** and the individual stretches of backbone involved in the sheet are called beta strands. Beta sheets are usually not flat, but twisted.

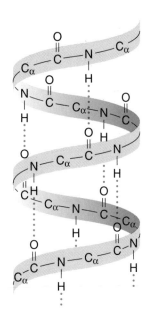

Figure 4.27 The alpha helix. C$_\alpha$ indicates the carbon atoms with side chains, which are not shown.

Secondary structure

Within a protein molecule, particular stretches of the amino acid chain may assume an alpha-helix or beta-sheet conformation. Certain amino acid sequences favor the formation of each one, although we cannot predict these structures from the primary structure with perfect accuracy. The regions of alpha helix and beta sheets within a protein are referred to as **secondary structure**.

So, within the hydrophobic core of a protein, some segments of the backbone may be found in the alpha-helix conformation while other segments may be arranged as beta sheets. (Some proteins are formed entirely from alpha helices; others, entirely from beta strands.) These secondary structures are often connected to one another via stretches of amino acids on the surface of the protein, where the partially charged backbone does not need to assume a particular secondary structure because it is neutralized by water in the cellular environment. The active sites of enzymes often involve these unorganized loops of amino acids, probably because the loops are freer to change conformation to bind a substrate.

Drawing Proteins
If you look at a picture of a protein in which all the atoms are shown, you can't tell much (Figure 4.29). The picture contains too much information, too many atoms to tell what the big picture is. For purposes of

Figure 4.28 A beta sheet. C$_\alpha$ indicates the carbon atoms with side chains, which are not shown.

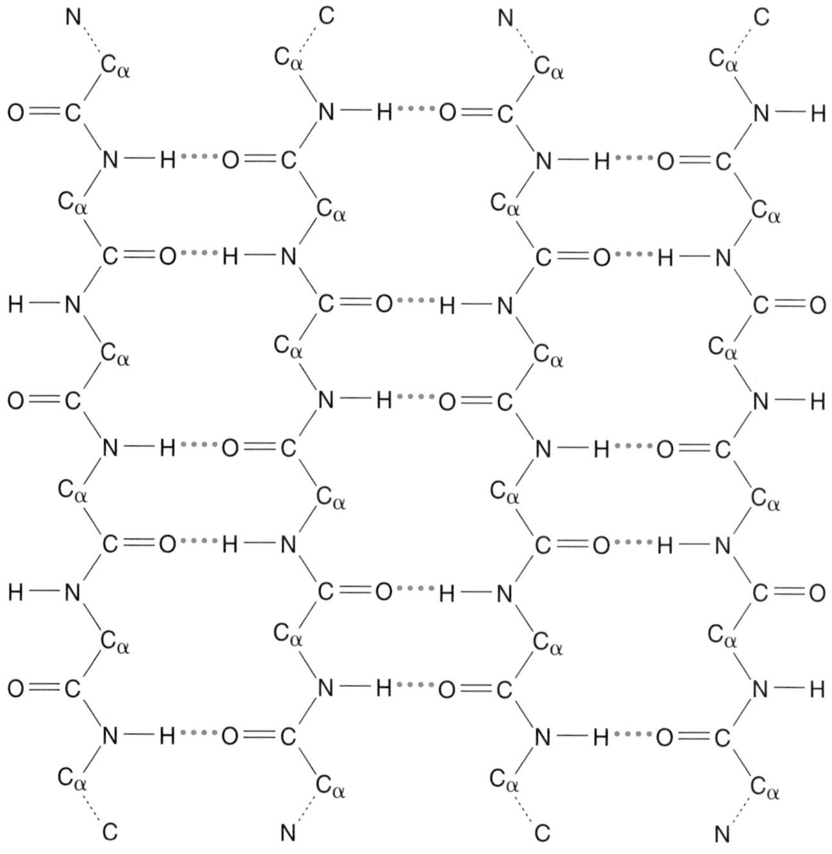

Figure 4.29   In this drawing of the replication termination protein of *E. coli,* each sphere represents an atom. Even though this protein is small, its structure is complex. (Drawing courtesy of Stephen White, in whose laboratory the structure was determined.)

understanding the overall structural plan, it is most helpful to look at just the configuration of the backbone. One popular way of showing this configuration is to draw the backbone as a ribbon, with alpha helices coiled and beta strands indicated with arrowheads. The arrows point toward the C terminus of the amino acid chain. Free loops are uncoiled regions of ribbon without arrowheads. Examples of some protein structures drawn in this manner are shown in Figure 4.30.

Look closely at the drawings in Figure 4.30. Plastocyanin (panel A) is composed of beta strands connected by loops. The center part of flavodoxin (panel B) is a twisted beta sheet. The strands of the beta sheet are connected by regions of alpha helix. The center of triose phosphate isomerase (panel C) also consists of beta strands, but they are arranged somewhat differently. Try using your finger to follow the peptide backbones of all three structures from N to C. You can see that adjacent strands in a beta sheet do not have to come from contiguous stretches of the backbone but may be segments that are widely removed in the primary structure.

You can also see from the flavodoxin and triose phosphate isomerase structures that alpha helices are apparently at the surfaces of these proteins. This arrangement is fairly common because of a handy property of alpha helices. Since the amino acid side chains stick out from the center of the imaginary barber pole, one side of the pole can have mostly hydrophilic side chains while the other has mostly hydrophobic ones. With such an arrangement, the helix can sit comfortably at the protein surface, its hydrophobic side buried

in the protein core and its hydrophilic side exposed to the cellular fluid. Neat, isn't it?

Structural Motifs

Simple combinations of a few secondary-structure elements occur frequently in proteins. Protein structure scientists call them *motifs.* You can think of a motif as a little module of protein structure; many proteins are put together by assembling combinations of these modular motifs. Of course, that statement makes it sound as if the structure is independent of the primary amino acid sequence, which it is not. But if you simply compare lots of protein structures, it does seem as though various structural motifs form a sort of "tool kit" for assembling larger proteins. A couple of examples of structural motifs are the helix-loop-helix, the beta turn, the beta-alpha-beta, and the beta barrel. These motifs are shown in Figure 4.31. There are many others.

Look at Figure 4.30. You can find three of the structural motifs in Figure 4.31 within the proteins of Figure 4.30. Plastocyanin has many beta-turn motifs. Both flavodoxin and triose phosphate isomerase contain many beta-alpha-beta motifs (trace the backbones; you will discover there is a helix between each pair of beta strands). The center of triose phosphate isomerase is a beta barrel. Some structural motifs are associated with functional activities, though most are not. For example, one variety of helix-loop-helix (also called helix-turn-helix) is a DNA-binding motif that occurs in a variety of DNA-binding proteins. When these proteins bind their target DNA sequences, the helices of this motif sit next to the DNA, and the side chains on one of the helices reach into the major

A. Plastocyanin

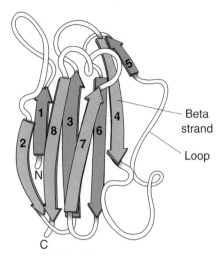

Beta strand

Loop

B. Flavodoxin

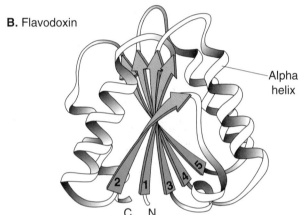

Alpha helix

C. Triose phosphate isomerase

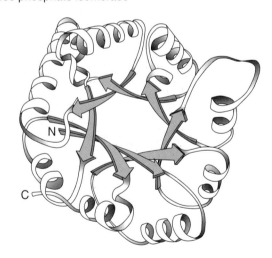

Figure 4.30   Ribbon drawings of protein structures. The beta strands in panels A and B are numbered in order from the N terminus to the C terminus of the amino acid chain. (Drawings courtesy of Jane Richardson.)

A. Helix-loop-helix

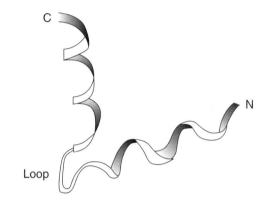

Loop

B. Beta turn

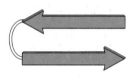

C. Beta-alpha-beta

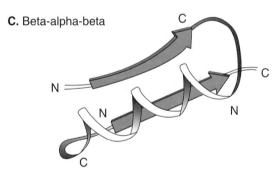

D. Beta barrel

Figure 4.31   Some protein structure motifs. (Panels A and C are from C. Branden and J. Tooze, *Introduction to Protein Structure,* Garland Publishing, Inc., New York, N.Y., 1991; panels B and D are from A. Lehninger, D. Nelson, and M. Cox, *Principles of Biochemistry,* 2nd ed., Worth Publishers, Inc., New York, N.Y., 1993.)

groove and make hydrogen bonds with the edges of specific bases there. (For an example, see *Structure and function*, below.)

Domains

Several secondary-structure motifs usually combine to form a stable, compact three-dimensional structure called a domain. Small proteins may consist of a single domain (such as those pictured in Figure 4.30); larger proteins may fold into several separate domains. The structures of individual domains within a protein and the way multiple domains fit together are called the **tertiary structure** of the protein.

Domains appear to be fundamental units of protein structure and function. Domains are usually formed from continuous stretches of amino acids and therefore are translated from continuous regions of mRNA. In multifunctional proteins, it is not uncommon to find that the protein folds into several domains and that each domain is associated with one function. Sometimes it is possible to separate the domains of a protein (through brief enzymatic digestion), and we sometimes find that the separate domains retain their individual functions.

An example of a bifunctional protein with two domains is the repressor protein of the bacterial virus lambda. The lambda repressor folds into two domains; the first 92 amino acids fold into the N-terminal domain, and amino acids 132 through 236 fold into the C-terminal domain. The other 40 amino acids connect the two domains (Figure 4.32A). Each domain of the lambda repressor has a separate function.

The two functions of lambda repressor are to bind to the correct operator DNA and to form *dimers* by binding to a second molecule of itself. A dimeric protein is a stable association of two copies of the same polypeptide chain (see *Quaternary structure*, below). Each of the chains is called a *monomer*. In lambda repressor, the folded C-terminal domain of one monomer binds to the C-terminal domain of a second repressor monomer to create the dimer. The N-terminal domains bind to DNA via a helix-turn-helix motif (Figure 4.32B). The N- and C-terminal domains can be separated by digesting the 40-amino-acid linker region, which is more vulnerable to digestion because it is not folded tightly into a tertiary structure. After separation, the N-terminal domain still binds its DNA recognition sequence but cannot dimerize, and the C-terminal domain can dimerize but cannot bind DNA. Not all domains retain a function when they are separated, but many do.

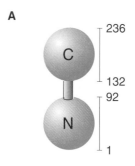

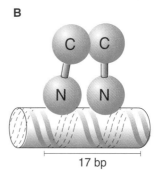

Figure 4.32   Domain structure of the bacteriophage lambda repressor protein. (A) The N-terminal domain consists of amino acids 1 through 92, and the C-terminal domain consists of residues 132 through 236. (B) The repressor forms dimers through the interaction of the C-terminal domains. The N-terminal domains bind to a specific DNA sequence. (From M. Ptashne, *A Genetic Switch,* 2nd ed., Blackwell Scientific Publications and Cell Press, Cambridge, Mass., 1992.)

Quaternary structure

Many functional proteins consist of a single amino acid chain, but many contain more than one polypeptide chain. These chains can be multiple copies of the same chain, as in the dimeric lambda repressor described above. They can also be assemblies of different polypeptides; *E. coli* RNA polymerase contains five different chains encoded by five different genes. The identity and number of the polypeptide chains and how they fit together in the final protein are called the protein's **quaternary structure**.

Do polypeptide chains automatically fold into the correct secondary and tertiary structures as they are synthesized and then associate with the correct additional polypeptides into quaternary structures? Many of them do. However, the folding of some proteins is assisted by other proteins that have been named *chaperone proteins*. Some molecular chaperones seem to act by increasing the rate of final folding; others actually guide the folding itself as well as the assembly of multiple polypeptide chains into complex quaternary structures.

Before we go on with our consideration of protein structure, stop and think about what protein folding might mean to a biotechnologist who wishes to move the gene(s) for a protein into a completely different type of cell and then obtain a functional protein. Examples of this kind of operation would include moving a eukaryotic gene or genes into a prokaryotic system or moving an animal gene or genes into a plant system. Substituting the correct DNA traffic signals so that the new cell can transcribe and translate the DNA may be only a small part of the battle. Getting the amino acid chain(s) to fold correctly may be much more challenging. Moving a gene from one system to another does not guarantee the production of an active protein.

Stability of protein structure

Three-dimensional protein structures are held together largely through the relatively weak chemical interactions of hydrogen bonds and the favorable interactions of hydrophobic side chains in the interior. Anything that disrupts these weak interactions—heat, extremes of pH, organic solvents, detergents—can alter the folding of the protein. The most extreme form of alteration is the complete unfolding of the amino acid chain, a process called **denaturation.** Denaturation of a protein is often irreversible. You can observe denaturation by frying an egg. The egg white protein albumin is soluble in its native state. As you heat it, the albumin denatures and coagulates, forming a white solid. Cooling the cooked egg does not reverse the process.

Proteins vary in their stability. One way to quantify stability is to measure the temperature at which a given protein denatures. This is sometimes called the melting temperature (T_m) of the protein. Some proteins require much higher temperatures to unfold than others do. For example, the enzymes of organisms that inhabit hot springs and ocean thermal vents are stable at very high temperatures. No one single thing makes these enzymes more thermostable. It appears that many different aspects of their primary and tertiary structures contribute to their heat resistance.

One feature of protein structure, however, makes a significant contribution to stability. This feature exploits the special properties of one amino acid, cysteine. In an oxidative environment, two properly positioned cysteine residues can react with each other to make a *disulfide bridge* (Figure 4.33A). The intracellular environment is not oxidative, so disulfide bridges are rare inside cells, but many extracellular proteins contain them. Disulfide bridges anchor regions of the protein in a specific configuration, stabilizing the structure (Figure 4.33B) Disulfide bridges can form between distant portions of the same domain or between different polypeptide chains within a quaternary structure.

Similar domains are found in different proteins

We said previously that domains appear to be fundamental units of protein structure and function. Scientists believe this statement because similar domains appear in different proteins, sometimes many different proteins. Some of the domains appear to be mostly structural; others are connected with specific functions. For example, a DNA-binding domain called the homeodomain (rhymes with Romeo-domain) is found in a large number of transcriptional activator proteins that interact with a specific type of enhancer sequence. (See *Transcriptional Regulation through Activation*, above.) These proteins activate different sets of genes and are found in a diverse set of organisms, including worms, fruit flies, and humans. Even so, the proteins all bind to DNA via their homeodomain. The amino acid sequences of the homeodomains are very similar in these proteins.

Another example of a functional domain found in many different proteins is the domain that binds the enzyme cofactor nicotinamide adenine dinucleotide (NAD). A number of enzymes use NAD as a cofactor in oxidation-reduction reactions. Each of these enzymes has a domain that binds NAD, and the structure of that domain is practically identical from enzyme to enzyme. In fact, many of these enzymes have two domains, the common NAD-binding domain and a second unique domain containing the active site at which the substrate binds. Surprisingly, the amino acid sequences of NAD-binding domains vary, but the structure is almost perfectly conserved from enzyme to enzyme.

A repeated domain with no known biochemical activity is the kringle. It consists of about 85 amino acid residues folded into a shape that reminded some scientists of a certain Danish pastry, the kringle; thus, it was named. Kringle domains are found in a variety of proteins.

Modular Proteins

You may be wondering now if it is possible to put together proteins by connecting domains like modules or tinker toys. The answer is yes. Although not all proteins are made this way, many are, with perhaps a few

A. Two cysteine side chains can form a disulfide bridge.

Peptide
backbone

$$-CH_2-SH + HS-CH_2- \longrightarrow -CH_2-S-S-CH_2-$$

Figure 4.33 Disulfide bridges stabilize protein structure. (Panel B is from C. Branden and J. Tooze, *Introduction to Protein Structure,* Garland Publishing, Inc., New York, N.Y., 1991.)

B. A disulfide bridge stabilizes the structure of this domain of an immunoglobulin protein (antibody).

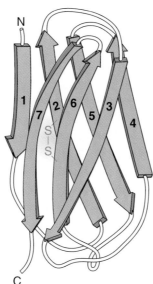

unique regions thrown in. Figure 4.34 shows some modular protein structures. These modular structures suggest that many genes did not evolve from scratch. Rather, it looks as if many genes were patched together from pieces or copies of preexisting genes, resulting in proteins that contain domains common to many other proteins. It would be esthetically pleasing if protein domains were encoded by exons, with introns providing the bridges between them. For some domains, such as the epidermal growth factor domain in Figure 4.34, this possibility seems to be true. The domain is encoded by a single exon. Unfortunately, the gene segments encoding more domains are distributed among several exons, with no apparent pattern to the exon-intron structure. So the modular structure of proteins is not always related in a logically obvious (to us at this time) way to the modules of coding sequence in eukaryotic genes.

The modular structure of some proteins has implications for biotechnology. Nature has produced proteins with many different functions by joining similar domains in different ways over the course of evolution. Using recombinant DNA technology (described in chapter 5), we can now swap domains, too, by swap-

ping portions of genes. For example, the portion of the lambda repressor gene encoding the first 92 amino acids (the DNA-binding domain) can be replaced with a similar domain from a different repressor protein. The hybrid protein works as a repressor, and it recognizes the DNA-binding site of the second protein.

Structure and function

Protein structure is an interesting topic in and of itself, but it is so important because protein function depends on it. Let's look at a few specific examples: a structural protein, two DNA-binding proteins, and an enzyme.

Keratin: A Structural Protein

The **keratins** are a family of similar proteins that make up hair, wool, feathers, nails, claws, scales, hooves, and horns, and are part of the skin. Keratin fibers also form part of the cytoskeleton. To fulfill their function, these proteins must be very strong. In addition, they must not be soluble in water (it would be quite unhandy if your hooves dissolved), even though most proteins are. Let's look at how the structure of keratin makes strong, water-insoluble fibers possible.

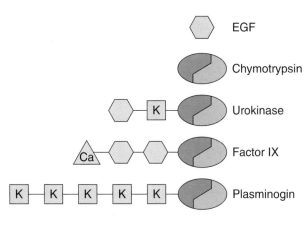

EGF

Chymotrypsin

Urokinase

Factor IX

Plasminogin

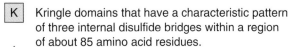

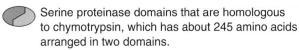

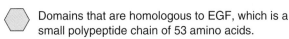

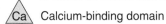

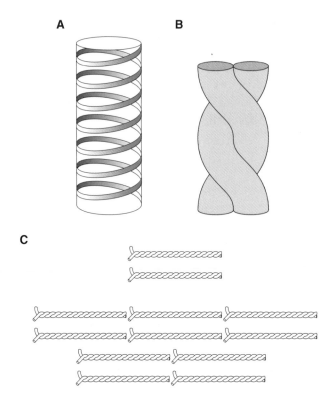

Domains that are homologous to EGF, which is a small polypeptide chain of 53 amino acids.

Serine proteinase domains that are homologous to chymotrypsin, which has about 245 amino acids arranged in two domains.

K Kringle domains that have a characteristic pattern of three internal disulfide bridges within a region of about 85 amino acid residues.

Ca Calcium-binding domain

Figure 4.34 Domain structures of some modular proteins. Epidermal growth factor (EGF) is a protein that signals several cell types to divide. The other four proteins are protein-cleaving enzymes with a variety of physiological roles. (From C. Branden and J. Tooze, *Introduction to Protein Structure,* Garland Publishing, Inc., New York, N.Y., 1991.)

Figure 4.35 Keratin, a structural protein. (A) A single keratin molecule forms a long alpha helix. (B) Two keratin alpha helices then wrap around each other. (C) Two-chain coils lie end to end and side by side, forming fibers. (Panel B is from J. D. Watson et al., *Molecular Biology of the Gene,* vol. 1, 4th ed., Addison Wesley Longman Publishers, Inc., 1987. Panel C is from A. Lehninger, D. Nelson, and M. Cox, *Principles of Biochemistry,* 2nd ed., Worth Publishers, Inc., New York, N.Y., 1993.)

The amino acid chain of keratin folds into one long alpha helix. Almost all of its amino acids—alanine, isoleucine, valine, methionine, and phenylalanine—are hydrophobic, and they extend outward from the helical backbone. The presence of all these hydrophobic side chains everywhere violates the general rule that hydrophobic side chains must be buried in the protein's core. As you probably expect, this rule violation is important: it means that keratin is not energetically comfortable surrounded by water molecules and therefore is not soluble in water. Imagine, the reason your fingernails don't dissolve when you wash them is all those hydrophobic side chains.

Instead of associating with the watery environment inside cells, keratin molecules associate with each other in large groups. Their structure, a single long helix, lends itself to forming fibers. First, two alpha helices of keratin wind around each other. Their hydrophobic surface side chains interact favorably in this conformation. These two-chain coils lie end to end and side by side with one to many other coils, forming fibers (Figure 4.35).

These fibers are not only held together by favorable hydrophobic interactions between side chains but are also stabilized by disulfide bridges between the coils. The bridges make the fibers strong and rigid. Different keratin proteins have different amounts of cysteine to make the bridges: the harder the final structure (hooves versus hair, for example), the more disulfide bridges are present. In the toughest keratins, such as tortoise shells, up to 18% of the amino acids are cysteines involved in disulfide bridges.

If you have ever had a permanent wave in your hair, you have manipulated the disulfide bridges of your keratin hair fibers. Recall that the disulfide bridges form only in the right kind of chemical environment (an oxidizing environment; see above). With the right chemicals, an oxidizing environment can be changed to its opposite (a reducing environment), causing the disulfide bridges to break apart.

When you get a permanent, the hair stylist first wraps your hair around small rods. Next, a smelly solution is applied to your hair. The smelly solution contains a *reducing agent*, a chemical that changes the oxidizing environment in your hair and breaks the disulfide bridges holding the keratin fibers side by side (the reducing agent is the component of the solution with the strong odor, too). While this is going on, the hair stylist has also arranged for your hair to be warm, either by placing you under a hair dryer or by putting a plastic bag over your hair. The moist heat breaks some hydrogen bonds that keep the keratin alpha helices stiff, allowing them to relax a little. The net effect is that the keratin helices move a little with respect to one another while they are wound around the rods.

After your hair has incubated sufficiently in the warm reducing environment, the stylist rinses out the reducing agent and applies a neutralizing solution. This solution restores the oxidizing environment, allowing disulfide bridges to re-form. But here's the catch. Your hair has been relaxed around the curling rods, and many of the cysteine SH groups will form disulfide bridges with new cysteine SH groups. These brand-new SH bonds hold the hair fibers in the conformation they were in around the curling rods: a permanent wave (Figure 4.36). Rinsing and cooling your hair allow the keratin helix hydrogen bonds to reestablish themselves, returning your hair to normal except that new disulfide bridges now hold it in a wavy shape.

Lambda and *trp* Repressors: DNA-Binding Proteins

The bacteriophage lambda repressor protein binds to a specific operator DNA sequence in its bacteriophage genome and prevents RNA polymerase from transcribing certain genes, as do the bacterial *lac* and *trp* repressor proteins. As described above, a single lambda repressor polypeptide chain folds into two domains (Figure 4.32). The C-terminal domains of two polypeptides bind to one another, creating a dimeric protein.

The N-terminal domains of the lambda repressor bind to operator DNA sequence via a helix-turn-helix motif. The lambda operator DNA sequence is symmetrical; each N-terminal domain of repressor interacts with identical bases. Figure 4.37A shows how the helix-turn-helix motifs of the two N-terminal domains sit on the operator DNA. The amino acid side chains of helix 3 (part of the helix-turn-helix motif) contact specific bases within the DNA binding sequence (Figure 4.37B). When the protein binds to DNA, helix 3 sits along the DNA molecule so that these specific contacts can occur. Figure 4.37C shows one of these amino acid-base contacts in detail.

From this example you can see that both the overall structure of the lambda repressor and its specific amino acid sequence are important to its function. The helix-turn-helix motif is oriented within the protein so that helix 3 can sit alongside the DNA molecule. The specific amino acids within helix 3 must contact specific bases for the binding to work, so the protein only binds to its recognition sequence. (For further information, see the works by Ptashne in *Selected Readings*.)

Now let's revisit a protein we met earlier in this chapter, the repressor of the tryptophan operon, to see

Figure 4.36 The biochemistry of a permanent hair wave.

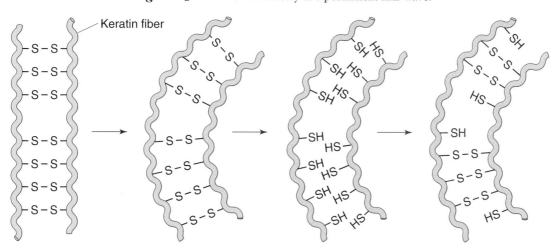

1. Normal hair
2. Hair is wrapped around a curling rod.
3. Reducing agent breaks disulfide bonds.
4. Neutralization allows disulfide bridges to re-form; new disulfide bonds hold hair in a wavy conformation.

A. The orientation of helix-turn-helix DNA-binding regions on operator DNA

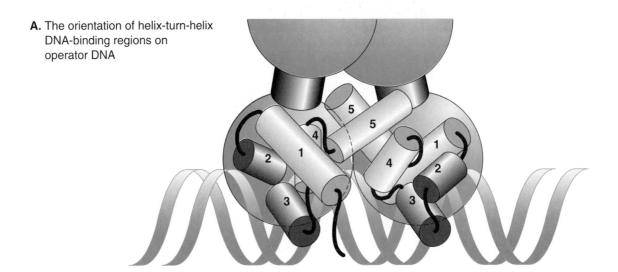

B. The amino acid sequence of helix 3. The specific bases in the operator sequence that are contacted by amino acid side chains are indicated by arrows.

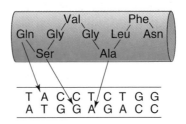

C. The contact between the first glutamine side chain in helix 3 and its target base

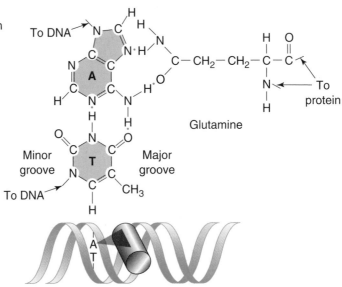

Figure 4.37 Binding of lambda repressor protein to DNA. (From M. Ptashne, *A Genetic Switch,* 2nd ed., Blackwell Scientific Publications and Cell Press, Cambridge, Mass., 1992.)

how a DNA-binding protein itself can be regulated by a second molecule. Recall that the *trp* repressor protein can bind both to the amino acid tryptophan and to a specific DNA base sequence, the *trp* operator sequence, but that the repressor *only* binds to the operator DNA when it is also binding to tryptophan (Figure 4.16).

Like the lambda repressor, the *trp* repressor is a dimeric protein that binds to DNA via a helix-turn-helix motif. The amino acid sequence of the DNA-binding helix is different, so it binds a different sequence of bases. Unlike the lambda repressor, the *trp* repressor's structure doesn't position the DNA-binding helix correctly for interacting with DNA.

Each of the *trp* repressor monomers has the helix-turn-helix motif; however, the two monomer chains interact in such a way that the DNA-binding helices are folded in toward the main body of the protein and aren't positioned to fit alongside the DNA molecule (rather like the folded-in claws of a crab). Here's where tryptophan comes in. The repressor protein interacts with two molecules of tryptophan (one per monomer). The tryptophan molecules fit into the structure like wedges between the DNA-binding helices and the body of the protein, forcing the DNA-binding helices to swing out (imagine the crab's claws unfolding outward from its body; Figure 4.38). Now the helices are positioned to bind their DNA recognition sequences. The tryptophan-repressor complex binds the *trp* operator and blocks further transcription of the *trp* operon.

Chymotrypsin: An Enzyme

Since enzymes are proteins that cause a chemical reaction, any discussion of their structure and function has to include the chemical reaction they promote. Chymotrypsin is a digestive enzyme that breaks down other proteins; it is called a *proteinase* (or protease, pronounced PRO-tee-ace). Proteinases cleave the peptide bonds in protein backbones.

Chymotrypsin belongs to the family of proteinases called serine proteinases. They are called serine proteinases because a serine side chain participates in the actual cleavage of the substrate molecule (serine is pictured in Figure 4.23). The digestive enzyme trypsin is also a serine proteinase; the domain structures of some other serine proteinases are shown in Figure 4.34.

Chymotrypsin cleaves one peptide bond at a time, cutting a single polypeptide chain into two shorter ones (Figure 4.39A; see Figure 4.21 to review peptide bonds and polypeptides). The cleavage reaction occurs in two steps (Figure 4.39B). First, the peptide bond is cut, freeing one half of the original molecule but leaving the other half covalently bonded to the catalytic serine side chain. In the second step of the reaction, the bound substrate chain is released, restoring the enzyme to its original form. Both of these steps require a basic amino acid side chain in the proper position to hold onto a hydrogen. In chymotrypsin, the basic side chain is a histidine (Figure 4.23). A close-up of the first step of the reaction is shown in Figure 4.39C.

Figure 4.38 Binding of the amino acid tryptophan to the *trp* repressor protein changes the conformation of the repressor so that it can bind to DNA. (Reprinted by permission from *Nature* 327:591–597, 1987, Macmillan Magazines Ltd.)

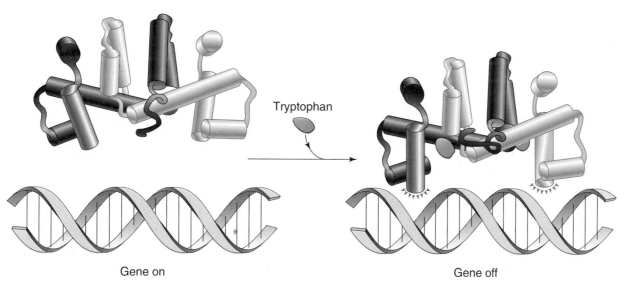

Tryptophan

Gene on

Gene off

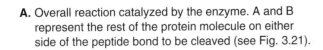

A. Overall reaction catalyzed by the enzyme. A and B represent the rest of the protein molecule on either side of the peptide bond to be cleaved (see Fig. 3.21).

$$A-\underset{\underset{O}{\parallel}}{C}-\underset{\overset{H}{|}}{N}-B \ + \ H_2O \ \longrightarrow \ A-\underset{\underset{O}{\parallel}}{C}-OH \ + \ H-\underset{\overset{H}{|}}{N}-B$$

B. The overall reaction proceeds in two steps. E—OH represents the chymotrypsin enzyme with—OH on the catalytic serine side chain (see Fig. 3.23).

Step 1. $\quad E-OH \ + \ A-\underset{\underset{O}{\parallel}}{C}-\underset{\overset{H}{|}}{N}-B \ \longrightarrow \ E-O-\underset{\underset{O}{\parallel}}{C}-A \ + \ H-\underset{\overset{H}{|}}{N}-B$

Step 2. $\quad E-O-\underset{\underset{O}{\parallel}}{C}-A \ + \ H_2O \ \longrightarrow \ E-OH \ + \ HO-\underset{\underset{O}{\parallel}}{C}-A$

C. Close-up of step 1, showing the role of the catalytic serine side chain and its histidine helper.

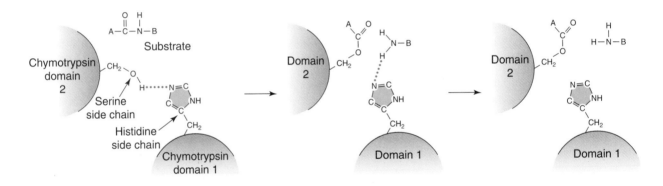

D. The domain structure of chymotrypsin, showing the positions of the serine (S195) and histidine (H57) side chains.

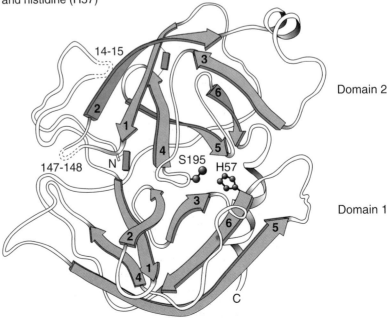

108

The structure of the serine proteinases positions the catalytic serine and the basic side chain so that the cleavage reaction can occur. A schematic of the structure of chymotrypsin is shown in Figure 4.39D. The protein has two domains, and the serine and histidine are each in a different one. They are located within loop regions where the domains fit together. Other amino acids in the loop help bind the polypeptide substrate in the proper position for cleavage, and some specific ones, which vary from proteinase to proteinase, confer substrate specificity. Chymotrypsin, for example, prefers to cleave peptide bonds adjacent to aromatic amino acid side chains, while trypsin prefers positively charged side chains. All these features of serine proteinase activity are understood in detail. We decided it was beyond the scope of this book to describe them all and instead refer you to the excellent book by Branden and Tooze that is listed in *Selected Readings*, where you will find a more complete discussion.

Effects of mutations on protein structure and function

Earlier in this chapter, we defined a mutation as any change in a DNA sequence and stated that the effects of a mutation depended on what, if any, effect it has on the production or function of proteins. In this discussion, we focus on some specific effects of amino acid changes on proteins.

From the examples given previously, you have probably realized that the function of a protein really depends both on its three-dimensional structure and, often, on specific amino acids within that structure. For example, there are many helix-turn-helix DNA-binding proteins. For them to bind DNA, the helix-turn-helix structure has to be maintained. However, the base sequence these proteins bind to depends on the identities of the amino acids within the DNA-binding helix. In addition, many proteins have multiple jobs to do, and these jobs are usually carried out by different regions of the protein, often different domains (recall the lambda repressor protein described above: one domain binds DNA, and the other forms dimers). So the effect of any amino acid change depends on what, if anything, that change does to critical protein structures, whether or not that particular amino acid is specifically involved in a function, and how that function or structure relates to the rest of what the protein does.

The accumulating data about DNA and protein sequences indicate that protein structures and functions can often be maintained through many amino acid changes. The keratin proteins are a good illustration. Their specific amino acid sequences vary; their cysteine content varies; yet they are all recognizable, though different, versions of keratin. The genetic mutations that resulted in different versions of the protein did not destroy structure or function. Another example of this kind is hemoglobin, which has more than 300 known genetic variants in the human population alone. Most of these variants are single amino acid changes with only minor structural and functional effects. Protein structure, it seems, is reasonably robust in the face of many amino acid changes.

A Harmful Mutation

A specific example of a harmful mutation is found in the disease sickle-cell anemia. This disease is the result of one of the more than 300 variations in hemoglobin. In sickle-cell anemia, a single A-to-T mutation changes the sixth codon from GAG to GTG. This mutation changes the sixth amino acid in the 146-amino-acid protein from glutamic acid to valine. In the three-dimensional structure of hemoglobin, the sixth amino acid sits on the surface of the protein. Glutamic acid (the normal amino acid) is hydrophilic, so its side chain is stable when exposed to the intracellular fluid. Valine, however, is hydrophobic (Figure 4.23). Its side chain is more energetically stable when interacting with other hydrophobic molecules.

The problem with the valine side chain is not simply that it is hydrophobic. It doesn't really change the three-dimensional structure of hemoglobin for it to be on the surface, and the mutant hemoglobin can still bind oxygen. The problem is how the hydrophobic surface valine affects the way hemoglobin molecules interact with one another.

As it happens, the hydrophobic valine side chain on the surface just fits into a hydrophobic pocket that is exposed on the hemoglobin molecule when it is not bound to oxygen. The surface valine is not positioned to fit into its own pocket, but it can fit into the pocket on a second molecule. When the mutant hemoglobin molecules give up oxygen in the capillaries, the deoxygenated molecules fit together in a lock-and-key fashion. The surface valine fits into the pocket of another molecule, which itself has a surface valine to fit into another molecule, and so on (Figure 4.40). The

---

Figure 4.39 Mechanism of action of the proteinase chymotrypsin, an example of a serine protease. (Panel D is from C. Branden and J. Tooze, *Introduction to Protein Structure*, Garland Publishing, Inc., New York, N.Y., 1991.)

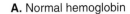

A. Normal hemoglobin

B. Sickle cell hemoglobin

Val-6

Figure 4.40 Representation of sickle-cell hemoglobin aggregation. (A) Normal hemoglobin molecules do not stick together. (B) The hydrophobic patch on the surface of sickle-cell hemoglobin caused by the glutamate-to-valine substitution at position 6 (Val-6) fits neatly into a hydrophobic pocket on a second molecule. Thus, sickle-cell hemoglobin molecules can polymerize in a head-to-tail fashion.

mutant hemoglobin molecules aggregate into long fibers, changing the red blood cells' shape to a sickle form and interfering with circulation of the cells through the capillaries. The impaired circulation gives rise to a number of deadly problems in the afflicted individual.

Thus, the fatal disease sickle-cell anemia is a consequence of hydrophobic interactions and altered protein structure. If the hydrophobic valine did not happen to fit into the hydrophobic pocket on the deoxygenated hemoglobin molecule, you would not get polymerization of the mutant hemoglobin, sickling of cells, and impaired circulation.

A Beneficial Mutation

An example of a beneficial mutation is the inherited condition benign erythrocytosis. Individuals with this condition have highly elevated red blood cell levels. Far from being ill, these individuals have greatly enhanced stamina. One such person, the Finnish athlete Eero Mäntyranta, won three gold medals for cross-country skiing in the 1964 Winter Olympics. Scientists recently determined the molecular basis of benign erythrocytosis.

Red blood cells arise from progenitor cells called stem cells that are found in bone marrow. Stem cells

are stimulated to mature into red blood cells by a hormone. The hormone communicates to the stem cell through a 550-amino-acid receptor protein imbedded in the stem cell's outer membrane. The N-terminal portion of the protein lies outside the cell, forming a docking site for the hormone. When the hormone binds, the C-terminal portion of the receptor, positioned inside the cell, transmits the maturation signal. The C-terminal end of the receptor also contains a docking site for a cellular protein that prevents transmission of the maturation signal. Docking of this cellular protein thus acts as a molecular brake on red blood cell production (Figure 4.41A). The Finnish athlete and other members of his family carry a mutant version of the receptor gene. A G-to-A mutation changes codon 481 from TGG, for tryptophan, to TAG, a stop codon. This single base change causes the athlete's ribosomes to stop synthesis of the receptor protein 70 amino acids early.

Losing the 70 C-terminal amino acids does not disrupt the extracellular hormone-binding domain of the protein, its cell membrane-spanning domain, or the intracellular region responsible for transmitting the maturation signal. However, loss of those 70 amino acids removes the docking site for the braking protein. Thus, Eero Mäntyranta's red blood cell production has no molecular brakes, and he and other mutation-bearing family members have higher-than-normal levels of red blood cells (Figure 4.41B). Their blood can carry more oxygen than normal, so they have enhanced stamina. It is possible that their elevated red blood cell levels would be detrimental in some circumstances. Some reports state that "afflicted" family members have normal life spans, while other reports claim that their average life spans are reduced.

A Disease of Protein Structure?

In the early 1990s an outbreak of so-called mad cow disease struck the British cattle industry. Symptoms of mad cow disease include erratic behavior and resemble symptoms of the sheep disease scrapie. In fact, the British mad cow disease appeared to have been transmitted to cattle through feed made in part from the remains of scrapie-infected sheep. Scrapie and mad cow disease are two of a group of diseases that affect the brain, including the human diseases Creutzfeldt-Jacob disease and kuru.

Using animal models, scientists were able to isolate an infectious agent that transmits scrapie. The agent passed through filters that would exclude bacteria, so it was assumed to be a virus. However, no one could detect either DNA or RNA in the scrapie agent, and it was unaffected by enzymes that destroy nucleic acids. However, the agent was neutralized by enzymes

A. Normal receptor protein

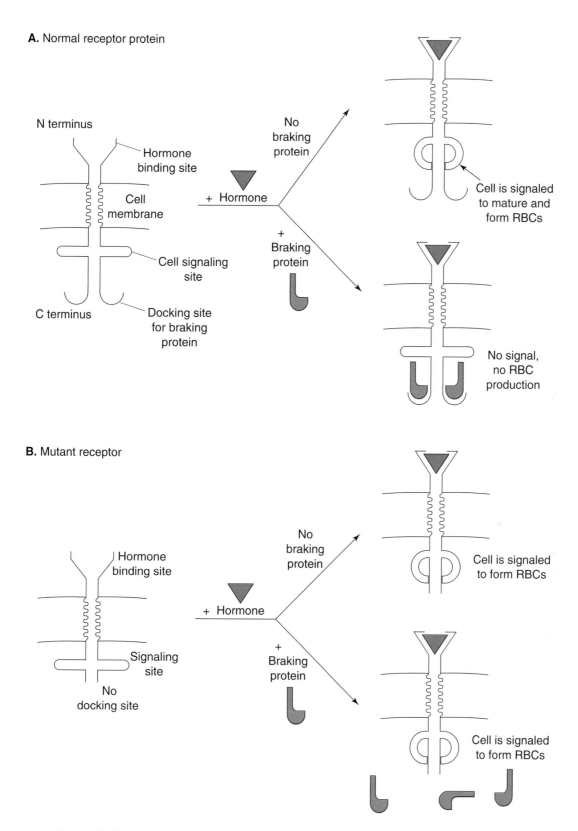

N terminus

Hormone binding site

Cell membrane

Cell signaling site

C terminus

Docking site for braking protein

+ Hormone

No braking protein

+ Braking protein

Cell is signaled to mature and form RBCs

No signal, no RBC production

B. Mutant receptor

Hormone binding site

Signaling site

No docking site

+ Hormone

No braking protein

+ Braking protein

Cell is signaled to form RBCs

Cell is signaled to form RBCs

Figure 4.41   Schematic representation of how the loss of 70 C-terminal amino acids from a receptor protein results in increased red blood cell (RBC) production.

that degrade proteins. Researchers named the mysterious agent a *prion*. How could an agent that apparently contained no nucleic acid transmit a disease? Many scientists assumed the scrapie researchers must be in error.

After 20 years of research, however, the story is becoming clearer. It appears that prion diseases are diseases of protein structure. Normal individuals manufacture the prion protein, and it assumes its normal configuration. Under some circumstances, however, the prion protein can assume an alternative configuration. Once it is in this alternative configuration, it does not return to normal. And when the altered form of the protein comes into contact with the normal form of the protein, the normal form changes to the altered form. As the altered form becomes more and more prevalent, disease begins. Thus, the aberrant form of the protein can act as a disease agent.

Predicting three-dimensional protein structure

Predicting the three-dimensional structure of a protein from its amino acid sequence is a major unsolved problem in structural biology. Structural biologists have searched and are searching for clues in solved protein structures, looking for amino acid patterns that correlate with specific structures. Structure prediction computer programs based upon these statistical studies have been developed, and they are useful but not perfect. If you are trying to predict a three-dimensional structure for a protein whose gene you have recently discovered, you are most likely to be successful if someone else has already discovered a similar protein and determined its structure.

DNA and protein sequence information is stored in large public databases, and computer software is available to compare these sequences (see chapter 30). Proteins with homologous amino acid sequences (sequences that match fairly well) have similar three-dimensional structures and generally have similar functions. So the first step in determining the three-dimensional structure of a protein is to compare its amino acid sequence with those of all proteins in the public databases to see if the structure of a homologous protein has already been solved. If you are lucky and find one whose structure has been determined, you have a good model for your protein. This model can then be used as a basis for identifying where the active sites of your protein are and can guide further experiments. If you don't find a homologous protein with a solved structure, the best you can do is to use the imperfect computer algorithms for structure prediction.

Sometimes you can get clues about what a protein does from amino acid sequence comparisons. For example, scientists identified the gene that is defective in cystic fibrosis (CF) patients in 1989. At the time, they did not know exactly what the protein product of the gene did, although they knew from the symptoms of CF that the protein's function was related to the movement of salt across cell membranes. (This movement is defective in CF patients, causing some cellular secretions to contain abnormal levels of salt and water.) When scientists compared the predicted amino acid sequence of the CF gene product to the amino acid sequences of known proteins, they found that part of the CF protein looked exactly like helical protein domains that span cell membranes. So their amino acid sequence comparison strongly suggested that the CF protein sits in the cell membrane, a reasonable position for a protein involved in transporting things in and out of cells.

It is important to note that we are discussing *predicting* protein structure from the amino acid sequence, not *determining* that structure. The *determination* of a protein's structure is a scientific specialty all its own, requiring expensive instrumentation and highly specialized training. Most molecular biologists who study the function of genes and proteins do not have the training or the instruments to determine a protein's structure. Even if they did, the methods available at this time do not work on all proteins (see chapter 5). So predicting protein structures from DNA (and therefore amino acid) sequences is part of trying to figure out what a protein does and how it does it or what part of a protein is doing a specific thing.

Testing Structure-Function Predictions

Molecular biologists have come up with ways to test predictions about what parts of a protein are involved in specific functions even when they don't know what the structure is. How can they do this? In a way, they imitate nature. They introduce mutations into proteins (via manipulating the DNA sequence of the gene) and determine what the changes do to the protein's function.

For example, if you are studying a newly discovered DNA-binding protein and find that part of its amino acid sequence is consistent with a helix-turn-helix motif, you might suspect that the putative helix-turn-helix is the portion of the protein that binds DNA. Being a scientist, you would like to test your hypothesis.

First, you would have to clone the gene for your protein so that you could work with it (see chapter 5). Then you could introduce specific base changes into the gene to cause specific amino acid changes in the protein. To test your hypothesis, you could change specific amino acids in the region you think might be a DNA-binding helix. If you found that your mutations caused the protein to bind to *different* specific DNA sequences, you would have good evidence that the region of the protein you altered was the DNA-binding region. Would you have proved that the structure was actually a helix-turn-helix? No, but your evidence that that particular area of the protein is directly involved in DNA binding would support the structure prediction.

Molecular biologists use approaches like the mutation experiment just described, often guided by predictions about a protein's structure, to try to learn which regions of proteins are involved in performing specific functions. In the end, though, the only way to be completely certain of a protein's structure is to use one of the instrumental methods described in chapter 5 for determining it, a long and exacting process.

Molecular biology, recombinant DNA technology, and scientific knowledge

Here at the end of this chapter is one final thought. You have just waded through a lot of information about basic molecular processes involving DNA and proteins. We started with the structure of DNA and genes and moved to protein synthesis, then to protein structure and function, and finally back to connections between changes in DNA sequence and changes in protein function. Along the way, we tried to emphasize that all the cellular operations involving DNA are performed by enzymes. The next chapter is a description of how scientists have learned to use some of these enzymes to manipulate DNA and, thereby, proteins. The whole business of using cellular enzymes to manipulate DNA is often called recombinant DNA technology.

Recombinant DNA technology as such seems to make the news in connection with controversial biotechnology products, gene therapy, or scary science fiction, but thousands of scientists around the world are using recombinant DNA technology every day simply to learn more about how life works. Think about the preceding discussion on predicting protein structure and probing protein function. These efforts involve computerized comparisons of DNA sequences from hundreds of genes. They involve identifying and cloning the gene for the hypothetical DNA-binding protein. They involve altering the sequence of the cloned gene and producing protein from the altered DNA sequence to see what the effects of the amino acid changes were. In real life, accomplishing these steps would all involve recombinant DNA technology.

One of the wonderful things (at least for a scientist) about the field of molecular biology is that the technologies it has spawned have made it possible to learn even more basic biology. It seems that the more we learn, the more neat technologies we develop. The new technologies let us design new kinds of experiments for learning even more—almost a chain reaction of knowledge. As you read chapter 5 and beyond, notice how the molecular technologies are applied to many scientific questions as well as to medical and other applications in day-to-day human life.

Summary

All of earth's living creatures use DNA as their genetic material. The elegant structure of this wonderful molecule enables it to perform its two functions: replication and transmission of information. DNA controls the form of an organism by specifying the amino acid sequences of all the proteins in that organism. The proteins, in turn, form many of the organism's building blocks and carry out nearly all of its metabolic processes.

Protein synthesis in all known organisms uses the same genetic code and the same general process. The universality of this most basic life process demonstrates the interrelatedness of all life on the planet. The cells of all creatures synthesize mRNA. The codons of all their mRNAs carry the same meaning. The ribosomes of the simplest bacterium are capable of reading the genetic code and synthesizing the protein from a human brain cell.

All forms of earthly life arose as a result of genetic variation followed by natural selection. Mutation and recombination can result in the formation or addition of new genes, the destruction or deletion of old ones, major and minor changes in protein structures, and changes in gene regulation.

Now that scientists understand the inner workings of heredity and genetic variation, they have begun to experiment with those processes. People are beginning to harness the natural mechanisms of genetic variation and protein synthesis to manipulate the genetic content of organisms and direct their gene expression. Are these experiments fundamentally different from natural evolution? No, except in the sense that humans are at the controls. Are they occurring on a time scale that is immeasurably faster than nature's time scale? Yes.

The realization that humans now have the power to manipulate the genes of living organisms is exhilarating, awe inspiring, and potentially disturbing. It offers the hope of treatment for previously incurable diseases. It holds out promise for an increased and better food supply. At the same time, our newfound power raises a multitude of hard questions. It has now become the responsibility of all of us to understand this power of genetic manipulation. Only through understanding can we hope to use and regulate this power wisely.

Selected Readings

Bayley, H. 1997. Building doors into cells. *Scientific American* 277:62. Examples of protein engineering projects.

Branden, C., and J. Tooze. 1991. *Introduction to Protein Structure*. Garland Publishing, Inc., New York, N.Y. Well-illustrated, readable book about protein structure.

Cuerces-Amabile, C., and M. Chicurel. 1993. Horizontal gene transfer. *American Scientist* 81:332–341. Transfer of genetic information outside of parent to offspring.

Doolittle, R., and P. Bork. 1993. Evolutionarily mobile modules in proteins. *Scientific American* 269:50–56. Protein domain structure.

Gerstein, M., and M. Levitt. 1998. Simulating water and the molecules of life. *Scientific American* 279:100.

Hall, S. 1995. Protein images update natural history. *Science* 267:620–624. How protein structure determinations and new computer models of protein structure are changing the science of biology.

Holtzman, D. 1991. A "jumping gene" caught in the act. *Science* 254:1728–1729. Two cases of human hemophilia A apparently caused by the movement of a transposon.

McGinnis, W., and M. Kuziora. 1994. The molecular architects of body design. *Scientific American* 270:58–66. Genes, proteins, and development.

Moxon, E. R., and C. Wills. 1999. DNA microsatellites: agents of evolution? *Scientific American* 280:94.

Prusiner, S. 1995. The prion diseases. *Scientific American* 272:48.

Ptashne, M. 1989. How gene activators work. *Scientific American* 260:40–47.

Ptashne, M. 1992. *A Genetic Switch*, 2nd ed. Blackwell Scientific Publications and Cell Press, Cambridge, Mass. Excellent short book describing in molecular detail how bacteriophage lambda switches from lysogenic to lytic growth. Protein structure, protein-protein interactions, protein-DNA interactions, and gene regulation are combined in one well-understood biological system.

Rennie, J. 1993. DNA's new twists. *Scientific American* 266:122–132. Current thinking about DNA structure.

Rhodes, D., and A. Klug. 1993. Zinc fingers. *Scientific American* 268:56–65. Protein structure and gene regulation; zinc fingers are one structural motif used for DNA binding.

Richards, F. 1991. The protein folding problem. *Scientific American* 264:54–63. An overview of the attempt to understand what determines three-dimensional protein structure.

Roush, W. 1995. An "off switch" for red blood cells. *Science* 268:27–28. Description of the molecular biology of the stamina-enhancing mutation causing benign erythrocytosis.

Tjian, R. 1995. Molecular machines that control genes. *Scientific American* 272:54–61. Description of the complex assembly of proteins needed for transcription of eukaryotic genes

Wallace, D. 1997. Mitochondrial DNA in aging and disease. *Scientific American* 277:40.

Watson, J., M. Gilman, J. Witkowski, and M. Zoller. 1992. *Recombinant DNA*. Scientific American Books, W. H. Freeman & Co., New York, N.Y. Excellent general reference.

Applying Molecular Biology: Recombinant DNA Technology

The Tools of DNA-Based Technologies

The 20th century witnessed a veritable explosion of knowledge about molecular biology. This new knowledge has demonstrated more clearly than ever that all life on earth is related: every living creature uses DNA and RNA to store and transfer genetic information, and every living creature uses the same genetic code for making its proteins. As their understanding of the inner workings of the hereditary machinery increased, scientists began to wonder if the genetic material and the expression process could be manipulated. They asked questions such as these: Is it possible to transfer a gene from one organism to another and have that gene expressed in the new host? Is it possible to combine two genes into one? Is it possible to regulate gene expression? Can we use the uniqueness of every person's DNA for identification? Can DNA changes tell us anything about evolution? Can we look at DNA and detect mutations that will lead to disease? Can we identify microorganisms on the basis of their DNA sequences? Is it possible to change the genetic makeup of an organism in a way that is advantageous for human society?

Before these questions could be answered, ways to manipulate DNA had to be found. This task turned out to be relatively simple, because DNA is constantly manipulated in nature. It is copied, cut, and rejoined over and over again in living cells. Nature's agents of DNA manipulation are enzymes. DNA-based technology employs these enzymes, which scientists have identified and purified for use in the laboratory.

Cellular enzymes

What are the cellular enzymes used by biotechnologists to manipulate DNA and protein expression? We discuss some of the more important ones here.

Restriction Endonucleases

Restriction endonucleases recognize specific base sequences in a DNA molecule and cut the DNA at or near the recognition sequence in a consistent way

(Figure 5.1). The most commonly used restriction enzymes recognize palindromic sequences (sequences in which both strands read the same in the 5′-to-3′ direction). Restriction endonucleases are made by bacteria and are thought to defend their bacterial producers against invading DNA, such as bacteriophage genomes. Their names indicate the organism from which they were purified (*Eco*RI from *Escherichia coli; Hin*dIII from *Haemophilus influenzae;* and so on). (See the activity *DNA Scissors: Introduction to Restriction Enzymes* for more information.)

DNA Polymerases

DNA polymerases are the enzymes that copy DNA. They synthesize a single new strand that is complementary to the template strand of the parent molecule, adding new nucleotides to the 3′ end of the growing strand. To synthesize new DNA, DNA polymerases absolutely require a template strand and a primer. A primer is any piece of DNA that is base paired to the template strand in such a way that the 3′ end of the primer is available to serve as the starting point for the new DNA. The first base of the new DNA is attached via a phosphodiester bond to the 3′ end of the primer and is complementary to the base on the template strand. Synthesis proceeds as more bases are added to the primer (Figure 5.2). DNA polymerases have been purified from a wide variety of organisms, which is not surprising, since all organisms must copy their DNA. (See the activity *DNA Replication* for more information.)

One group of organisms whose DNA polymerase enzymes have become very important in the laboratory is bacteria that live at very high temperatures (such as in hot springs and thermal vents in the ocean floor). Since their natural environments are very hot, their DNA polymerases function well at high temperatures. Scientists now use these heat-tolerant polymerases to "xerox" DNA fragments via the polymerase chain reaction (PCR), which requires repeated heating of the enzyme-DNA mixture. (See *Amplifying DNA: PCR* later in this chapter and the activity *The Polymerase Chain Reaction: Paper PCR* for more information.)

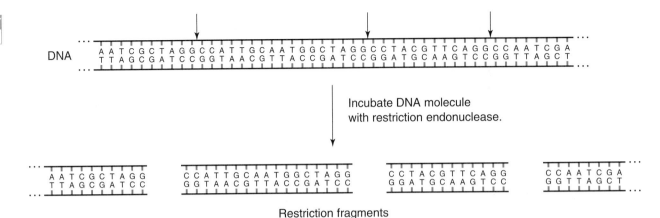

DNA

...
AATCGCTAGGCCATTGCAATGGCTAGGCCTACGTTCAGGCCAATCGA
TTAGCGATCCGGTAACGTTACCGATCCGGATGCAAGTCCGGTTAGCT
...

Incubate DNA molecule
with restriction endonuclease.

...
AATCGCTAGG
TTAGCGATCC

CCATTGCAATGGCTAGG
GGTAACGTTACCGATCC

CCTACGTTCAGG
GGATGCAAGTCC

CCAATCGA ...
GGTTAGCT ...

Restriction fragments

Figure 5.1 Restriction endonucleases recognize and cut specific sites in a DNA molecule. The arrows indicate the cleavage sites of one such endonuclease.

RNA Polymerases

RNA polymerases are the enzymes that read a DNA sequence and synthesize a complementary RNA molecule. RNA polymerases require a special sequence of bases on the DNA template, called a promoter, to signal to them where to begin transcription, but they do not require a primer (Figure 5.3). Like DNA polymerases, RNA polymerases have been purified from many organisms, since all organisms must transcribe their genes.

DNA Ligases

Ligases join pieces of DNA (or RNA) together by forming new phosphodiester bonds between the pieces (Figure 5.4).

Reverse Transcriptases

As their name suggests, reverse transcriptases read an RNA sequence and synthesize a complementary DNA (often abbreviated as cDNA) sequence (Figure 5.5). These enzymes are made by RNA viruses that convert

Figure 5.2 Activity of DNA polymerase.

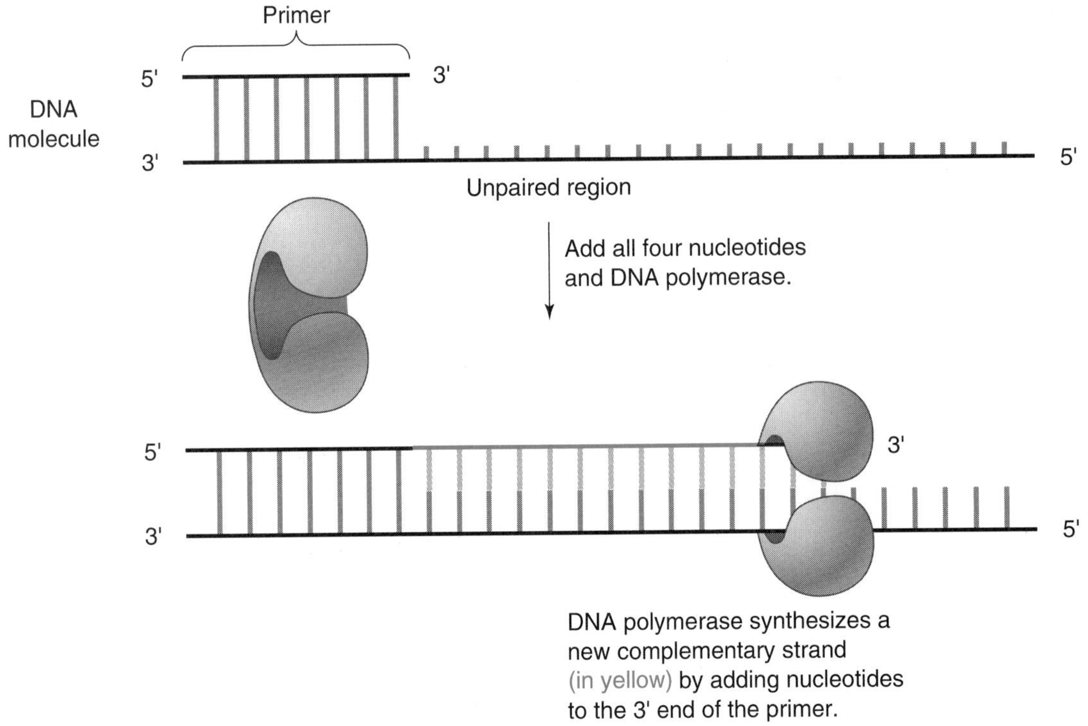

Primer

5'   3'

DNA
molecule

3'   5'

Unpaired region

Add all four nucleotides
and DNA polymerase.

5'   3'

3'   5'

DNA polymerase synthesizes a
new complementary strand
(in yellow) by adding nucleotides
to the 3' end of the primer.

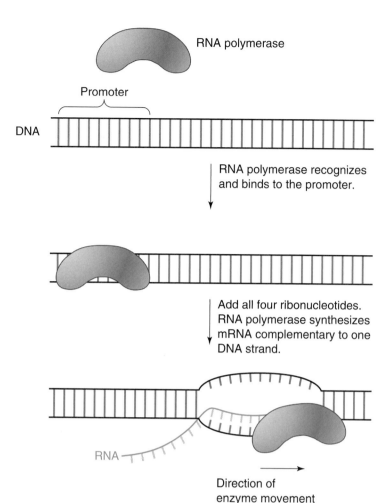

RNA polymerase

Promoter

DNA

RNA polymerase recognizes and binds to the promoter.

Add all four ribonucleotides. RNA polymerase synthesizes mRNA complementary to one DNA strand.

RNA

Direction of enzyme movement

Figure 5.3 Activity of RNA polymerase.

their RNA genomes into DNA when they infect a host cell (see *Virus Genomes* in chapter 4). Reverse transcriptases allow scientists to synthesize a DNA gene from an RNA message. This ability is useful for dealing with eukaryotic genes, since the original genes are often split into many small pieces separated by introns in the chromosome. The messenger RNA (mRNA) from these genes has undergone splicing in the eukaryotic cell, and the introns are gone, leaving only the coding sequences (see *Differences in the Molecular Biology of Prokaryotes and Eukaryotes* in chapter 4). Reverse transcriptase can convert this mRNA into a "pre-spliced" gene consisting only of protein-coding sequences. Why would someone want a prespliced gene? Since bacteria possess no equipment for splicing, they must be given a prespliced version if they

Figure 5.4 Activity of DNA ligase.

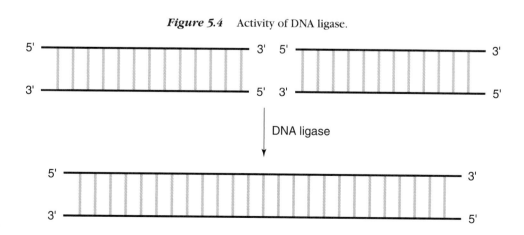

5'   3' 5'   3'

3'   5' 3'   5'

DNA ligase

5'   3'

3'   5'

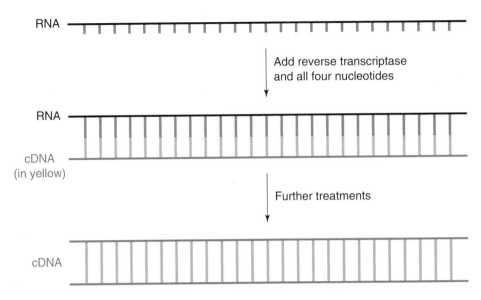

Figure 5.5 Activity of reverse transcriptase.

are to express the correct protein product from a eukaryotic gene. Thus, the process of making cDNA from mRNA is important for expressing eukaryotic genes in prokaryotes.

Translation Extracts

Although they are not enzymes, translation extracts have also been useful to biotechnologists. These extracts are made from eukaryotic cells or bacteria, and they contain ribosomes, transfer RNAs, and other necessary translation components. When mRNA is added to such an extract, translation occurs, and the protein encoded in the RNA message is synthesized.

One of the more time-consuming tasks for the pioneers of biotechnology was purifying enzymes such as those described previously for laboratory use. Not surprisingly, genes for most of these enzymes have now themselves been cloned and engineered for large-scale protein production. Many companies produce and sell these valuable proteins for commercial and research uses. Essentially all of the enzymes used in DNA-based technologies can now be easily purchased.

Natural vectors

In addition to manipulating DNA outside the cell, biotechnologists must be able to introduce the manipulated DNA into the host cell of their choice and have it stably maintained there. Once again, nature provides the tools. The term **vector** is used to describe any vehicle that carries DNA into a host cell. Plasmids and viruses are useful natural vectors both for delivering DNA to a new host cell and for providing for the maintenance of that DNA after the delivery.

Naturally occurring plasmids are used for transfer of DNA by transformation (the uptake of naked DNA by a cell; see *Asexual reproduction and recombination* in chapter 4). In addition to being small (and thus easy to work with), plasmids are replicated by the host cell and distributed to daughter cells during cell division. Therefore, a new gene introduced on a plasmid vector will be stably maintained in progeny cells. The most common form of gene cloning is the insertion of a gene into a plasmid and the subsequent introduction of the plasmid into the desired host. Scientists now have many plasmids tailor-made for cloning procedures. (See the activity *Transformation of Escherichia coli* for more information.)

One particularly useful natural plasmid is the Ti plasmid from the plant pathogen *Agrobacterium tumefaciens*. After the bacterium infects a plant, the Ti plasmid transfers itself from the bacterial cell into the invaded plant and inserts certain of its genes into the plant DNA. Biotechnologists take advantage of the Ti plasmid by adding genes of interest to the plasmid. The altered Ti plasmid is reintroduced into *A. tumefaciens*, the plasmid-containing bacterium is inoculated into a plant, and the altered Ti plasmid then transfers the new genes to the plant's DNA. Unfortunately, only certain plants are susceptible to infection by *A. tumefaciens*. (See the activity Agrobacterium tumefaciens: *Nature's Plant Genetic Engineer* for more information.)

Biotechnologists also employ viruses as agents of DNA transfer. Many viruses normally operate by combining their DNA with that of their host (see the activity *Transduction of an Antibiotic Resistance Gene* for more information). Biotechnologists can insert genes of interest into the genomes of these viruses and then let the virus carry the new DNA into the host cell and recombine it with the host DNA. One advantage of viruses as vectors is their extreme host specificity. Foreign genes can be targeted to specific cell types by the appropriate virus. This characteristic of viruses may make them important players in gene therapy procedures.

Other tools for transferring DNA

Although transformation and transduction are useful means of introducing foreign DNA into cells, they do not work in every case. Especially with eukaryotic cells (both plant and animal), it can be difficult to introduce a foreign gene. Two recent technological methods of getting DNA inside cells are electroporation and the so-called DNA gun. In electroporation, an electric current is used to force DNA across the cell membrane. Although this procedure is also used with bacteria, it is particularly helpful for introducing plasmids into eukaryotic cells.

The other means of DNA introduction is the DNA gun. To use the gun, the desired DNA must first be coated onto microscopic pellets. These pellets are then fired into the target cells. At some frequency, the DNA on the pellets may be maintained and expressed in the target cells. The DNA gun has proved useful for introducing DNA into plant cells that are not susceptible to infection by *A. tumefaciens* and was recently used successfully to introduce DNA into animal cells.

A different approach to the introduction of new genetic material is cell fusion, the actual fusing of two cells into one. Cell fusion is usually achieved by treating the two different cell types with a chemical that affects the cell membranes and promotes the fusion of cells in close contact. For plant cell fusion, the thick outer walls must first be digested away with enzymes, leaving the membrane-bound **protoplasts.** The fused cells do not maintain both sets of chromosomes. Instead, they lose chromosomes until they reestablish the correct number. Chromosomes from either parent can be lost, apparently randomly, resulting in a cell containing some chromosomes from each parent cell. Cell fusion is the basis of monoclonal antibody technology.

Important chemical methods

Biotechnologists and molecular biologists are not the only scientists who study nucleic acids and proteins. Chemists and biochemists have investigated the structures of these molecules for many years, and their efforts have resulted in some extraordinarily useful technologies:

Automated DNA Synthesis

Scientists studying the structure of DNA worked out reaction conditions for chemically synthesizing single-stranded DNA molecules. These methods have been standardized to the point of automation. DNA synthesizers can now make single-stranded DNA molecules over 100 bases long from scratch. Such relatively short single-stranded DNA segments are called **oligonucleotides.** The desired oligonucleotide sequence is typed into a computer that controls the synthesizer, and the synthesizer connects the nucleotides in the proper order. How can we get double-stranded DNA? Synthesize two complementary single strands and allow them to hybridize (form base pairs). Chemically synthesized oligonucleotides are often referred to as synthetic oligonucleotides. Many biotechnology materials companies now provide custom DNA synthesis service. You send them the sequence of the oligonucleotide you want, and they will send you the oligonucleotide.

Automated DNA Sequencing

The enzymatic DNA sequencing reactions described in chapter 17 have also been standardized to the point of automation. Automated DNA sequencers are the workhorses of large sequencing laboratories, such as those involved in the Human Genome Project. They have also become common as shared resources at research universities, where they enable researchers to obtain information about DNA sequences of interest without having to learn the relatively time-consuming and exacting procedures for sequencing DNA manually.

Automated Protein Sequencing

A vital piece of information about any protein is the sequence of its amino acids. Initially laborious chemical methods have now been automated, and the latest protein-sequencing equipment can determine the amino acid sequences of tiny amounts of purified protein. Limitations of protein sequencing are that the protein must be pure and that the accuracy of the sequencing deteriorates as the number of sequential amino acids determined increases. To avoid the problem of inaccuracy, scientists often determine only the first several amino acids on one end of a protein, or they cleave the protein molecule into several smaller fragments that can each be sequenced separately and accurately.

Three-Dimensional Protein Structure Analysis

The function of a protein depends on its three-dimensional structure. Chemists and biochemists now use the sophisticated techniques of X-ray crystallography and nuclear magnetic resonance (NMR) spectroscopy to determine protein structure. These methods were originally used to determine the structures of relatively small molecules. They have been refined and augmented for application to complicated molecules such as proteins. Both methods are difficult and have significant limitations.

X-ray crystallography, the more widely used technique, requires the use of extremely pure protein crystals. Crystals are very regular, packed arrays of molecules, and since proteins can have quite irregular shapes, most are not easy to crystallize, and many cannot be crystallized at all. Even for those that can be crystallized, growing protein crystals large enough for crystallography studies can take months or years. If a suitable crystal is obtained, a beam of X rays is directed into it. The regular array of protein molecules within the crystal diffracts the X rays in a pattern that is recorded on film and used to deduce the arrangement of atoms within the molecules. X-ray crystallography can be used to determine the structure of other classes of molecules, too. For example, X-ray diffraction data obtained from DNA crystals by Rosalind Franklin were used by Watson and Crick to determine the three-dimensional structure of DNA.

Protein NMR uses highly concentrated, pure solutions of protein, avoiding the need for crystals. This technique is limited to very small proteins that remain soluble at the required concentrations. NMR is used in many applications other than determining protein structure. For example, NMR exploits the magnetic properties of certain atomic nuclei (usually ^{1}H in protein NMR) to determine molecular structure. These nuclei have a magnetic moment or spin. When protein molecules are placed in a strong magnetic field, the spins of the hydrogen nuclei align along the field. The nuclei are then excited by applying electromagnetic pulses to the sample. When the nuclei return to their ground state, they emit electromagnetic energy, which can be measured. The frequency of the emitted electromagnetic radiation depends on the precise molecular environment of the individual nucleus. By varying the nature of the applied electromagnetic pulses and measuring the subsequent emissions, we can deduce various molecular properties of the sample.

The structures of the proteins analyzed to date dramatically illustrate the relationship of form and function. Proteins have grooves, pockets, and even pincerlike structures for binding to DNA or other molecules. Some proteins assume a different shape when bound to their targets. Knowing the structure of a protein often allows us to see how the protein performs its function. It also helps molecular biolo-

gists think about how other proteins might work. Knowing the structure of a protein helps biotechnologists design ways to improve the protein.

Fundamental procedures

The tools and technologies described above are employed in several basic procedures that form a core of standard techniques for molecular biologists and biotechnologists. These standard techniques are rather like the basic procedures that a chef learns: making a white sauce, sautéing, beating egg whites, poaching, cooking a sugar syrup. The chef then applies selected techniques to different ingredients to produce a wonderful variety of dishes. Biotechnology is rather like that. Using these basic procedures with all kinds of organisms and experimental systems, biotechnologists and molecular biologists working around the world are producing an amazing array of knowledge, products, and even new basic techniques. What are some of these basic procedures?

Separating DNA Fragments: Gel Electrophoresis

In the course of almost any manipulation of DNA, it is necessary (or at least desirable) to have a look at the various pieces of DNA you are working with. The standard method used to separate DNA fragments is electrophoresis through agarose gels. Gels made of an alternative material are used under some circumstances.

Agarose is a polysaccharide (as are agar and pectin) that dissolves in boiling water and then gels as it cools, like Jell-O. To perform agarose gel electrophoresis, a slab of gelled agarose is prepared, DNA is introduced into small pits in the slab, and then an electric current is applied across the gel. Since DNA is highly negatively charged (because of the phosphate groups), it is attracted to the positive electrode. To get to the positive electrode, however, the DNA must migrate through the agarose gel.

Smaller DNA fragments can migrate through an agarose gel faster than large fragments. In fact, the rate of migration of linear DNA fragments through agarose is inversely proportional to the $\log_{10}$ of their molecular weights. What it boils down to is that if you apply a mixture of DNA fragments to an agarose gel, start current flowing, wait a little, and then look at the fragments, you will find that the fragments are spread out like runners in a race, with the smallest one closest to the positive electrode, the next smallest following it, and so on (Figure 5.6). Because of the mathematical relationship given previously, it is possible to calculate the exact size of a given fragment on the

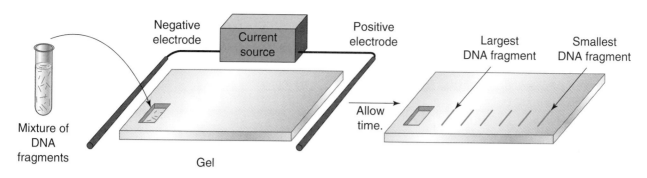

Figure 5.6   Gel electrophoresis of DNA fragments.

basis of its migration rate. (For more information, see *Introduction to Gel Electrophoresis Laboratory Activities* and the activity *DNA Goes to the Races*.)

After gel electrophoresis, the DNA fragments in the gel are usually stained to render them visible. DNA fragments can also be isolated and purified from agarose gels. Electrophoresis through gels made of agarose or other materials is also used to separate proteins (see the activity *Electrophoresis of Amylase Samples*).

Finding Complementary Base Sequences: Hybridization

Hybridization is a natural phenomenon that provides another essential tool for biotechnologists. It occurs as a consequence of the structure of DNA. Hybridization (also called *annealing* or *renaturation*) is the term used to describe the process in which two single DNA strands with complementary base sequences stick together to form a correctly base-paired double-stranded molecule. Hybridization occurs spontaneously: if two complementary single DNA strands are mixed together and left alone, they will hybridize. The time it takes for hybridization to occur is directly related to the length of the DNA sequences involved; as one might expect, short complementary sequences can line up correctly and base pair much faster than long sequences can. (For more information, see the activity *Detection of Specific DNA Sequences: Hybridization Analysis*.)

Why is hybridization an essential biotechnological tool? It provides a way to look for a specific DNA sequence in a mixture or to operate selectively on specific sequences. For example, suppose we want to know whether a specific DNA sequence has been successfully inserted into a bacterium's genome. To find out, we first make a **probe** for that sequence. A probe is a piece of single-stranded DNA that will hybridize only to the DNA of interest (because of its

unique base sequence). Next, we denature the bacterium's DNA (often by heating to 95°C, which causes base pairs to come apart), thus exposing the bases on single strands. Then we add the probe. If the new DNA sequence is present, the probe will hybridize, or "stick," to the sample DNA. A hybridized probe can be detected by a number of methods (Figure 5.7).

Another important use of hybridization is to provide a starting place for additional techniques. A segment of DNA with a hybridized probe is the starting material for a variety of other procedures (see below for examples). Hybridization also works with RNA molecules; complementary RNAs can hybridize to form a double-stranded RNA molecule, and a complementary RNA can hybridize to DNA. As you will see, hybridization is an important step in many of the other basic methods. The ability to synthesize oligonucleotides

Figure 5.7   Hybridization analysis.

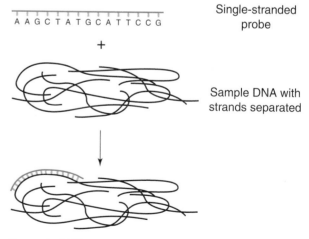

Single-stranded probe

+

Sample DNA with strands separated

If sample DNA contains the base sequence complementary to the probe sequence, the probe will form base pairs with the sample DNA, or hybridize to the sample, and physically stick to it.

with any desired base sequence to use as probes has greatly increased the options for applications of all of them.

Transfer of DNA Fragments: Blotting

Often it is important to know which DNA fragment in a mixture contains a sequence of interest. Gel electrophoresis and staining alone cannot answer this question, because all DNA fragments look alike when stained. By now you know that if we had a specific probe for the sequence of interest, we could perform hybridization on the DNA fragments and determine which one contained our sequence. Unfortunately, hybridization doesn't work well on DNA fragments embedded in an agarose gel. To get around this problem, scientists use a procedure called blotting.

Blotting DNA is analogous to blotting ink writing. If you write something with a fountain pen and cover the writing carefully with a sheet of blotting paper, the pattern of your writing will be exactly transferred to the blotter. DNA blotting works in the same man-

ner. DNA fragments are separated by agarose gel electrophoresis and then the gel itself is covered with a membrane and blotting paper. The DNA fragments in the gel transfer to the membrane in exactly the same arrangement as they were in the gel. Once the DNA fragments are on the membrane, they can be hybridized to probes to test for the presence of specific sequences (Figure 5.8).

This DNA transfer and hybridization technique was first described by a scientist named Southern in 1975. It acquired the name **Southern blotting** for its originator. Later on, other scientists modified the procedure to use with RNA fragments. They named their procedure for transferring RNA from a gel to a membrane **Northern blotting.** Later still, other researchers figured out how to transfer proteins from gels to membranes to do tests on them. The name of this procedure? **Western blotting.** No Eastern blot has yet been invented. (For more information, see the activity *Detection of Specific DNA Sequences: Hybridization Analysis.*)

Figure 5.8 Blotting and hybridization analysis. DNA fragments separated by gel electrophoresis are transferred from the gel to a membrane. The membrane is then exposed to a probe to test for the presence of specific DNA sequences.

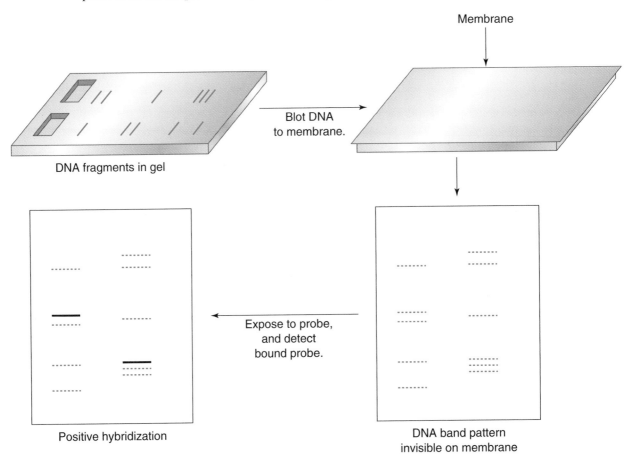

Amplifying DNA: PCR

PCR is a clever procedure that takes advantage of DNA polymerase enzymes and synthetic oligonucleotides to make many copies of a specific segment of DNA. In brief, we synthesize two primers that hybridize to opposite strands of the DNA molecule at the boundaries of the segment to be copied. DNA polymerase then copies both of the strands, starting at the two primers (Figure 5.9). Because the reaction mixture contains primers complementary to both strands of the DNA, the products of the DNA synthesis can themselves be copied with the opposite primer. PCR is actually a chain reaction that results in thousands or millions of copies of a given DNA segment. It is used to amplify DNA for cloning, to detect scarce DNA, and to distinguish different DNA samples (for example, from related viruses). (See the activity *The Polymerase Chain Reaction: Paper PCR* for more information.)

Determining the Base Sequence of a DNA Molecule

The sequence of bases in a gene determines the sequence of amino acids in a protein. If we can read the base sequence of a gene, we can learn about the protein it encodes. Furthermore, if we want to manipulate the gene, it helps to know what we are starting with. The ability to determine the sequences of bases in a DNA molecule is an important tool for research as well as applied science.

The first attempts to determine the sequence of bases in a given piece of DNA were completely chemical approaches. These chemical methods were successful but slow and labor-intensive. The advent of biotechnology has greatly simplified the process. Even so, determining a DNA sequence is still a multistep process that requires precision and care. Sequencing relatively short stretches of DNA (a few hundred bases long) can usually be accomplished in a few hours or days, but sequencing long segments takes much longer. Automated DNA sequencing is much faster.

Current DNA sequencing methods employ DNA polymerase enzymes, synthetic oligonucleotides, and hybridization techniques. The DNA polymerase synthesizes a new DNA strand on a single-stranded template using special nucleotide derivatives. These nucleotide derivatives allow scientists to determine the order in which new bases are added to the growing chain. The

Figure 5.9 PCR makes many copies of a DNA segment lying between and including the sequences at which two single-stranded primers hybridize to the substrate DNA molecule. The primers are usually synthetic oligonucleotides.

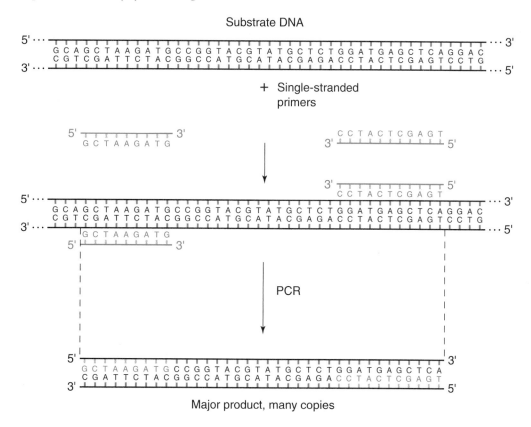

Major product, many copies

order in which bases are added to the new strand reveals the DNA sequence of the template strand through the rules of complementarity. (See the activity *DNA Sequencing* for more information.)

Remember that DNA polymerases must have a primer from which to begin synthesis (see *DNA Polymerases* above). For DNA sequencing, we need a primer that is hybridized to the single-stranded template adjacent to the area that we wish to sequence. If any of the adjacent sequence is known, a complementary oligonucleotide primer can be chemically synthesized. Another option is to clone the region of interest into a vector of known sequence. A primer complementary to adjacent vector sequences can then be used.

RNA can also be sequenced. RNA sequencing is similar to DNA sequencing but uses the enzyme reverse transcriptase. Reverse transcriptase reads the base sequence of single-stranded RNA and synthesizes a complementary DNA molecule (see above). The same strategy of using altered nucleotides to give a detectable sequence is used with reverse transcriptase and an RNA template. Reverse transcriptase also requires a primer. As with DNA sequencing, chemically synthesized oligonucleotides are usually hybridized to the template RNA to provide a starting point for the enzyme.

Cloning DNA

The term **cloning** means the production of identical copies of something through asexual reproduction. When applied to DNA, cloning is usually understood to mean the insertion of a piece of DNA into a cell in such a way that the DNA will be replicated (copied) and maintained. Simply forcing a segment of DNA into a cell does not usually result in the copying and maintenance of that DNA. Cells, whether prokaryotic or eukaryotic, contain a variety of DNA-degrading enzymes that destroy unprotected DNA segments. To clone a segment of DNA (a gene, perhaps), the DNA must be placed into a carrier DNA molecule (a vector) whose structure is immune to the degrading enzymes. The vector protects the cloned segment and provides for the replication and maintenance of the DNA in its new host cell. Natural virus and plasmid vectors were described previously.

The most basic method of cloning DNA begins with cutting the DNA into segments with a restriction enzyme (or a combination of enzymes). The selected vector is cut with the same enzyme(s). The restriction fragments and the cut vector are mixed with the enzyme DNA ligase (see above), which connects the desired fragment with the vector. Plasmids that contain a segment of DNA originally from another source are

often called *recombinant plasmids* (Figure 5.10). (See the activity *Recombinant Paper Plasmids* for more information.)

You are probably wondering what happens to all the other fragments when you ligate vector and insert molecules and whether the vector couldn't simply be reconnected to itself or other fragments. After all, the ligase enzyme is "blind"—it merely connects one DNA end to another. The answer to both questions is yes. Part of the process of cloning is identifying the recombinant molecule you want out of the variety that can be produced. (See the activity *Recombinant Paper Plasmids*.)

Making DNA Libraries

A common application of cloning is to produce what is called a **DNA library** or **genetic library.** To make a library, the entire genome of an organism is digested with restriction enzymes, and the entire batch of products mixed with vector DNA and ligated. The resulting mixture of recombinant plasmids is transformed into a host organism. The pool of transformed cells containing the collection of plasmids is called a DNA library of that organism (Figure 5.11A). The idea is that the resulting recombinant plasmids or viruses will contain fragments covering the whole genome of the organism. A DNA library can be used like a resource. If a scientist knows what gene he wishes to study, he has only to look for it in the library and pull out a cloned version. (See below and the activity *Detection of Specific DNA Sequences: Hybridization Analysis*.)

A special kind of DNA library is often made from eukaryotic organisms. Much of the DNA in eukaryotes is noncoding and would not be of interest to scientists studying the proteins encoded by genes. To avoid cloning vast regions of noncoding DNA, scientists make what is called a **cDNA library.** The first step in making one of these libraries is to isolate mRNA from the organism. Then, reverse transcriptase is used to generate cDNA copies of all the mRNA molecules. Finally, these cDNA copies are inserted into vector DNA molecules en masse, resulting in a cDNA library. The cDNA library contains DNA with the coding sequences of all the proteins the organism was producing when the mRNA was isolated (Figure 5.11B).

Putting the Knowledge and Tools to Work

Given the tools and procedures described above, we are now able to use our understanding of the structure and function of DNA to solve problems and

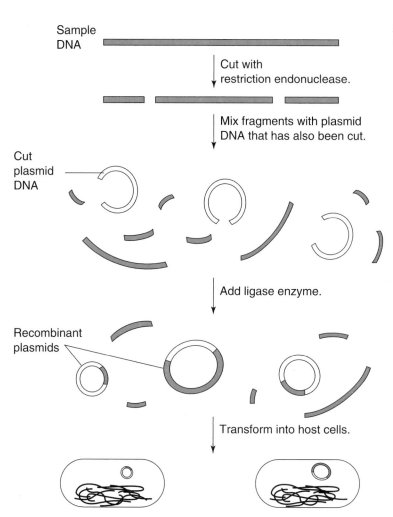

Sample DNA

Cut with restriction endonuclease.

Mix fragments with plasmid DNA that has also been cut.

Cut plasmid DNA

Add ligase enzyme.

Recombinant plasmids

Transform into host cells.

Figure 5.10 Cloning a DNA segment into a plasmid.

make useful products. What kinds of things can we do? The following discussion gives a sample of some of the possibilities.

Analyzing genetic variation

As discussed in chapter 3, genetic variation, both between and within species, is pervasive in the natural world. It provides the raw material on which natural selection acts to create new species and change existing ones. While its importance to evolutionary change is obvious, the ways in which the existence of genetic variation is used in human endeavors is probably less so. We diagnose certain diseases, determine the compatibility of tissue types in organ transplants, produce new and better agricultural crops, find the perpetrators of crimes, and probe the origins of species by investigating genetic variability.

Until recently, determinations such as these were based on phenotypic characteristics, that is, some tangible manifestation of the information contained in the DNA molecule. Phenotypic manifestations, while important, are limited by our ability to observe and measure them. Furthermore, our ability to assess genetic variation by measuring phenotypic characteristics depends on the relationship between genotype and phenotype. Sometimes genetic differences are not detectable at the phenotype level (for example, people with blood group genotypes AA and AO both express the phenotype of blood group A, despite the genetic difference). And adaptation to similar environmental circumstances can cause genetically dissimilar organisms to display similar phenotypic characteristics.

About 30 years ago, more detailed methods of looking at variations in proteins came into widespread use. Investigators could use electrophoresis (see above) to compare sizes and electrophoretic mobilities. Amino acid sequence determinations gave scientists an indirect way of looking at variation in individual genes. Protein sequence comparisons are still used to compare species, particularly in evolutionary science.

A. Genomic DNA library. The insertions in the recombinant
plasmids represent the entire DNA content of the organism.

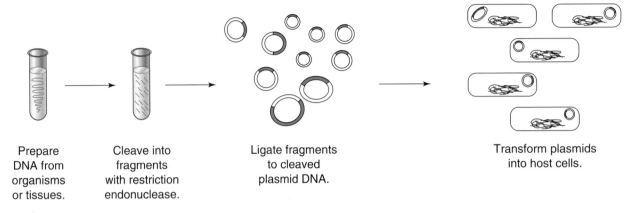

| Prepare DNA from organisms or tissues. | Cleave into fragments with restriction endonuclease. | Ligate fragments to cleaved plasmid DNA. | Transform plasmids into host cells. |

B. cDNA library. The insertions in the recombinant plasmids
represent genes that were being expressed in the sample.

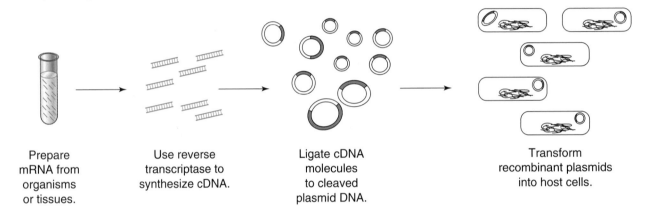

| Prepare mRNA from organisms or tissues. | Use reverse transcriptase to synthesize cDNA. | Ligate cDNA molecules to cleaved plasmid DNA. | Transform recombinant plasmids into host cells. |

Figure 5.11   DNA libraries.

Now we are able to assess genetic variation by looking at the DNA itself. There is much more variation in the DNA of different individuals than we can observe and measure phenotypically, even through amino acid sequence determinations. By looking directly at the DNA, we can make our measures of genetic variation more accurate and precise.

How is genetic variation analyzed with DNA-based technologies? One method is DNA sequence comparisons. This method gives detailed and accurate information but cannot be applied to whole genomes because of their great size. DNA sequence comparisons focus on small, specific regions. To give meaningful data, these target regions have to be areas of significant variation; comparisons of most human genes would show very few if any differences between individuals. For this reason, sequencing studies of individ-

uals of the same or closely related species often focus on mitochondrial DNA. Mitochondria have small circular genomes that undergo mutation at a much higher rate than the nuclear genome; thus, mitochondrial DNA sequences display much more variation between individuals than do most nuclear sequences.

One way to compare whole genomes is through hybridization studies. In this approach, genomic DNAs from two species are prepared then hybridized to each other. The more similar the nucleotide sequences of the two species, the more extensive the regions of base-paired duplex they will form. To measure the extent of the duplex DNA, the hybrid molecules are heated until the strands separate again. The temperature at which the strands separate (the **melting temperature**) is a function of the extent of base pairing. The more perfectly matched the two genomes, the higher the temperature required to separate the strands of the hybrid molecules. The melting temperature thus gives a gross measurement of DNA sequence similarity.

A shortcut to DNA sequence comparisons is to focus only on changes to the short base sequences at which restriction enzymes cut the DNA molecule. Changes in the DNA sequence can create or eliminate these sites. By digesting two DNA molecules with the same restriction enzyme and comparing the lengths of the fragments generated, we can estimate how similar the two molecules are. Two identical molecules will give identical patterns. Two fairly similar molecules will give similar patterns, with perhaps a few differences in fragment size (caused by addition or removal of cut sites through mutations). Scientists estimate the similarity of DNA molecules based on the similarity of their restriction fragment lengths. This approach permits sampling of variation across the entire genome.

Just as an observable phenotypic variation is called a polymorphism (from the Greek for "many forms"), these differences in restriction fragment lengths are also called polymorphisms. Restriction fragment length polymorphisms (RFLPs) are used in DNA fingerprinting, paternity determinations, genetic disease diagnosis, plant breeding, and in evolution and conservation research. (See the activity *Analyzing Genetic Variation: DNA Typing*.)

Evolutionary Studies

Evolution is a unifying theme in biology, and molecular biology has provided powerful new tools for studying it. Evolutionary biologists seek to discover how individual species arose from earlier forms and to understand the mechanisms of the process of evolution itself. Biotechnology has literally revolutionized both endeavors.

Formerly, biologists who sought to describe how modern species arose had to rely on morphological, ecological, and behavioral comparisons between various modern species and between modern and fossil forms to deduce degrees of kinship. DNA and protein sequence comparisons have given these scientists an entirely new set of data to consider. To use molecular data in evolutionary studies, biologists first assemble protein or DNA sequence data from a specific protein or region of the genome in the group of organisms under study. They then measure the degree of difference in the sequences. On the assumption that changes accumulate slowly and relatively steadily, they construct various "trees" that show how the different sequences could have been generated from a common ancestor. An evolutionary tree based on the

amino acid sequence of the protein cytochrome *c* is shown in Figure 5.12. (You will use amino acid sequence data to construct an evolutionary tree in chapter 29.)

Another molecular evolutionary analysis found that the kiwi, a flightless New Zealand bird, is more closely related to the flightless birds of Australia than to the moas, the other major group of New Zealand flightless birds. DNA and protein sequence comparisons as well as hybridization studies found that humans and chimpanzees are more similar to one another than to any other species. These studies suggest that ancestors of modern humans and chimps probably diverged a mere 5 million years ago. Using molecular techniques to probe evolutionary relationships between species is a large and active area of current research.

An interesting twist from the field of molecular evolution is the discovery that we can extract DNA fragments from appropriate ancient samples. DNA is rapidly degraded to small fragments after an organism dies, and only special combinations of circumstances will preserve soft tissue cells with DNA. However, samples preserved in bogs, including 17-million-year-old magnolia leaves, have yielded enough DNA for comparisons with modern species, as have amber-encased insects. It is also possible to recover DNA fragments from bones and teeth, which are more commonly preserved than is soft tissue. PCR is particularly useful in studying ancient DNA because it can amplify the small amounts typically present in specimens.

Early attempts to amplify ancient DNA by PCR were often "learning experiences" for the scientists involved. What they learned was that it is extremely easy to amplify modern DNA contaminants of the ancient sample. In fact, many attempts to amplify ancient DNA resulted in the amplification of the scientists' own DNA. Through these experiences, laboratory personnel learned that extremely stringent protection against contamination is essential in these kinds of experiments.

In one such headline-making study, a group of scientists painstakingly amplified fragments of mitochondrial DNA from a Neanderthal human fossil. They carefully established that the amplified DNA did not come from cells of any of the researchers. An analysis of the base sequence of the amplified DNA revealed it to be very different from the mitochondrial DNA sequences of any known modern human group. So different, in fact, that the researchers concluded that Neanderthals could not be direct ancestors of modern humans.

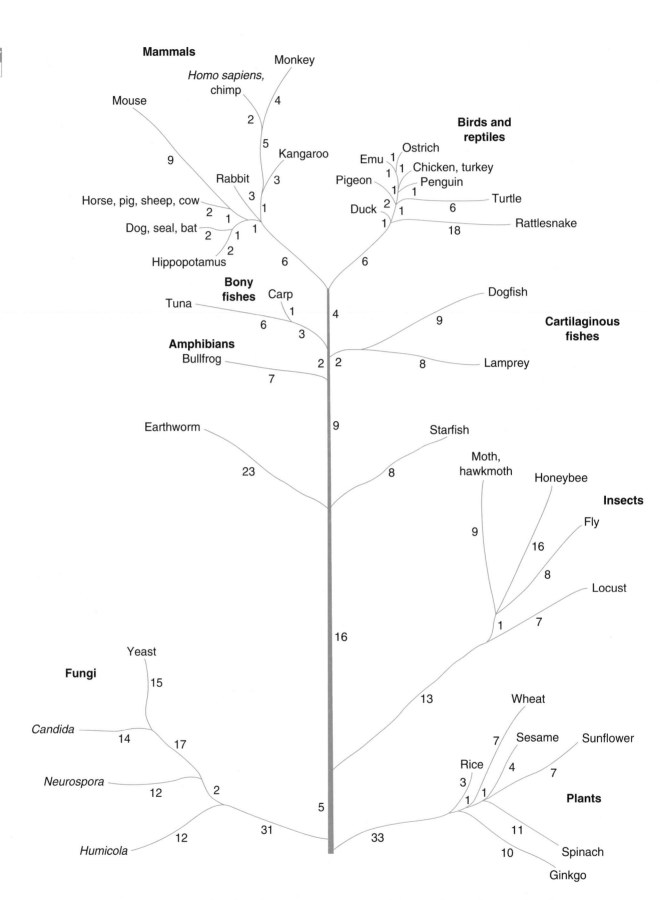

Figure 5.12 An evolutionary tree based on the amino acid sequence of the protein cytochrome *c*. Numbers represent the numbers of amino acid changes between two nodes of the tree. For example, when the common ancestor of cartilaginous fish diverged from the line leading to bony fish, mammals, etc., it evolved and accumulated two amino acid changes in its cytochrome *c* protein before the lines leading to dogfish and lamprey diverged. The dogfish line accumulated nine more amino acid changes in becoming the modern organism; the lamprey accumulated eight changes. Thus, the amino acid sequences of cytochrome *c* in lamprey and dogfish have 17 differences. (From A. Lehninger, D. Nelson, and M. Cox, *Principles of Biochemistry,* 2nd ed., Worth Publishers, Inc., New York, N.Y., 1993.)

We stress that the DNA recovered in these processes is extremely fragmentary and is useful only for comparisons with modern sequences to measure the amount of change. So far, experience with ancient DNA suggests that using it to resurrect extinct species is impossible.

For scientists who seek to understand the mechanisms behind evolution, molecular biology and biotechnology have provided a gold mine of new information. First, studies of genes and genomes from many organisms have underscored how fundamentally similar we all are. For example, studies using hybridization, RFLP analysis, and sequence comparisons suggest only about a 1.6 to 3% difference in the genomes of humans and chimpanzees. From earlier protein and enzyme studies, we already knew that all animals produce about the same complement of enzymes and proteins. Now that we are mapping genes to specific locations on chromosomes, we find that animals' genes are arranged similarly from species to species. Differences in chromosome number and structure seem to have been generated in a process in which chromosomes were cut or broken into large pieces and then put back together in many different ways into different final numbers of chromosomes.

Looking at the DNA sequences of individual genes, we see that nature appears to be quite economical: once a protein function that works has evolved, it is often used over and over. We find similar domains in many different proteins (see chapter 4 for more information). We find proteins that appear to have diverged from two original copies of the same gene. We find similar proteins filling similar roles in widely disparate species (see the activities in Part II, Section D). For example, the proteins that govern body plan in worms, flies, mice, and humans are very similar. The genes (called *homeotic* genes) encoding these proteins are even arranged in the same way in the chromosomes of flies, mice, and humans, and they share similar regulatory sequences. Some scientists are now looking at the regulatory sequences associated with homeotic genes as a way to investigate ancestry. The flood of new information about genome structure is giving evolutionary scientists a wealth of fuel for theories about genome evolution.

DNA Typing

Identifying people who will not or cannot identify themselves has long been a challenge in the courtroom and other arenas. Identification has traditionally been made through phenotype: appearance, voice patterns, blood types, fingerprints, and such. DNA-based technologies now make it possible to "go to the source" of individual identity: a person's DNA.

No two individuals (except identical twins) have identical DNA sequences. It is theoretically possible to identify nearly every individual on earth from his or her DNA sequence, though doing so is not realistic. It *is* possible, however, to make very good predictions about individual identity on the basis of a limited and practical examination of DNA with RFLPs.

DNA-based identification (DNA typing) is based on regions of DNA that vary greatly from individual to individual. When these regions are cut by a restriction enzyme or amplified by PCR, the sizes of the resulting products depend on the sequence of the individual's DNA. When the products are separated by electrophoresis, the patterns constitute a sort of genetic fingerprint of the individual.

To examine an individual's DNA fingerprint by restriction analysis, a sample of his or her DNA is cut with a restriction enzyme, and the resulting fragments are separated by gel electrophoresis. Because the human genome is so large, literally thousands of fragments are generated. To look selectively at the fingerprint regions, the restriction fragments are blotted to a membrane and hybridized to labeled probes for the fingerprint regions (a Southern blot; see *Transfer of DNA Fragments: Blotting*, above). The labeled probes reveal the fragments generated from those regions, and the sizes of those fragments constitutes the fingerprint (Figure 5.13). Some regions of the genome vary in that they contain different numbers of back-to-back repeats of the same DNA sequence. For fingerprinting purposes, these regions of the genome can be selectively amplified using PCR, and the amplified products can be examined by hybridization to probes or by simple staining. The lengths of the products correspond to the number of repeats present, and these vary greatly among individuals.

DNA fingerprinting is used for comparisons and to rule out suspects. For example, if blood is found at a crime scene, a DNA fingerprint can be generated from the blood and compared with DNA fingerprints from any suspects. If the crime scene fingerprint does not match a suspect's, that suspect cannot have left the sample and is cleared. In fact, about 30% of the FBI's DNA typing cases have cleared the prime suspect, and DNA typing has cleared people who were in prison serving time for violent crimes.

If the sample fingerprint at the crime scene does match a suspect's DNA fingerprint, the relevant question becomes how likely it is that a person other than the suspect could have left it. Since DNA fingerprinting looks only at a portion of an individual's genome, two people could have very similar fingerprints.

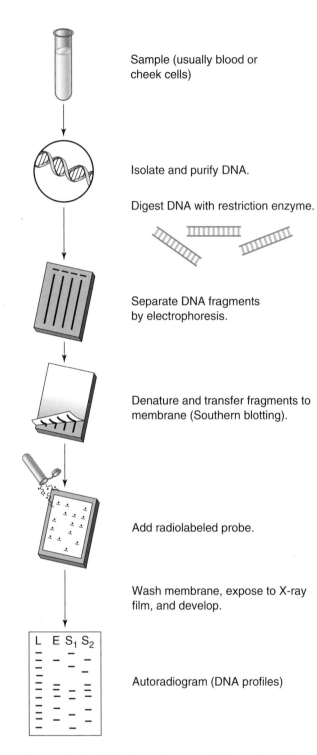

Sample (usually blood or cheek cells)

Isolate and purify DNA.

Digest DNA with restriction enzyme.

Separate DNA fragments by electrophoresis.

Denature and transfer fragments to membrane (Southern blotting).

Add radiolabeled probe.

Wash membrane, expose to X-ray film, and develop.

L E S₁ S₂

Autoradiogram (DNA profiles)

Figure 5.13 DNA fingerprinting by RFLP analysis. Radioactive probes are detected by exposing the membrane to X-ray film after hybridization is complete.

Scientists look at databases of DNA fingerprints and calculate the frequency of the patterns in the fingerprint in question. They then use those numbers to calculate the odds that another individual will have the same DNA fingerprint. Forensic laboratories try to conduct their testing so thoroughly that random matches are extremely unlikely. (For more information, see the activities *Analyzing Genetic Variation: DNA Typing, A Mix-Up at the Hospital, A Paternity Case,* and *The Case of the Bloody Knife.*)

DNA typing is also used to confirm the identity of human remains, as in the recent identification of the Unknown Soldier (see the Reading after chapter 19, *Mitochondrial DNA in DNA Typing*). U.S. soldiers now deposit samples in a DNA data bank as a backup for the metal dog tags they wear in combat.

DNA typing is also used in human rights work. Geneticists are using DNA typing to identify the bodies of people slain during recent political upheavals in Guatemala. The work of these scientists corroborates eyewitness reports and sometimes helps bring the killers to justice. (See chapter 19.)

Although DNA typing gets media coverage for its use in forensic cases, the technique is also widely used in conservation biology and ecological research. DNA typing can reveal the degree of kinship of individual animals. This knowledge can be critical to the success of captive breeding programs for endangered species, for which biologists need to select genetically different individuals as breeding pairs. Knowing degrees of kinship between animals in a living group is also essential to behavioral ecology studies. In many species, it is impossible to determine paternity simply by watching the living group; DNA typing provides a way to solve this problem (see the activity *Analyzing Genetic Variation: DNA Typing*). Analysis of genetic variability can also provide insight into the status of wild populations. Biologists have assumed that vigorous wild populations contain genetically variable individuals rather than genetically identical ones. DNA typing, together with protein comparisons, provides the means for testing this hypothesis.

In another application of DNA typing, scientists were able to solve the mystery of the Mexican loggerhead turtles. Pacific loggerheads nest in Japan and Australia, not Mexico, yet young loggerheads could always be found off the coast of Baja California. Many biologists did not believe the young turtles could have come the 10,000-mile distance from Japan (Australia is even farther), and so the origin of the Baja turtles was a mystery. Now DNA comparisons have established that the Baja population is made up of turtles from both the Japanese and the Australian groups. Apparently the young turtles are carried to Mexico by ocean currents, and they then swim back to Japan or Australia to breed. Some swim!

DNA-Based Detection of Pathogens and Disease Diagnosis

The diagnosis and treatment of a particular disease often require identifying a particular *pathogenic* (disease-causing) microorganism. Traditional methods of identification involve culturing these organisms from clinical specimens and performing metabolic and other tests to identify them. DNA technology is making its presence felt in the clinical laboratory, speeding up and simplifying many identification procedures.

The idea behind DNA-based disease diagnosis of infectious disease is simple: if the pathogen is present in a clinical specimen, its DNA will be present. Its DNA has unique sequences that can be detected. DNA-based diagnosis involves the detection of pathogen DNA, often in the clinical specimen itself.

As you probably expect, detection of pathogen DNA relies on hybridization with probes specific to the pathogen. There are two main approaches to hybridization tests. One is a direct hybridization, in which labeled probe is added to the sample, and hybridization is detected directly. This method may not be useful if the pathogen is present in very low numbers, so that very few molecules of probe can hybridize.

To get around this problem, scientists and clinicians are turning to PCR (see *Amplifying DNA: PCR*, above). PCR also starts with the hybridization of specific DNA sequences. In PCR, however, the oligonucleotide probes are not detected but instead are used as primers for a chain reaction of DNA synthesis. If the primer-probes can hybridize, large amounts of a specific DNA fragment will be synthesized. If the primer-probes do not find complementary DNA sequences, no DNA synthesis occurs. Whether or not a new DNA fragment is synthesized thus provides a means of detecting hybridization. The advantage of the PCR method is that hybridization of probes to even a tiny number of DNA molecules (far too few to be detected by direct methods) yields enough new DNA to be detected easily. PCR is therefore a much more sensitive method of DNA-based diagnosis. (See the activity *The Polymerase Chain Reaction: Paper PCR*.)

DNA technology has proved very useful in the diagnosis of tuberculosis. Until fairly recently (the 1990s), Americans believed that tuberculosis in their country was a disease of the past. However, the spread of AIDS in the 1980s and 1990s created a population of individuals with almost no ability to fight off disease. Many of these individuals lived in the inner city in dire poverty. Inner-city poverty is associated with crowding, which favors the spread of disease, and malnutrition, which also makes people more vulnerable to infection. Tuberculosis reappeared and spread. To make matters worse, drug-resistant strains arose. In immune-compromised individuals, acute drug-resistant tuberculosis can be fatal in a matter of weeks.

The problem with using traditional methods to diagnose tuberculosis is that the organism that causes the disease, *Mycobacterium tuberculosis,* grows extremely slowly in the laboratory. Traditional culturing, followed by testing for antibiotic resistance, takes months. The patients sometimes died before the test results were in. Once scientists identified the genes responsible for drug resistance in *M. tuberculosis,* they made DNA probes that could detect the resistance genes in a sample taken from the patient, eliminating the need for the slow culturing.

Finding genes

Much of biotechnology concerns the manipulation of genes. How are genes identified in the first place? The strategy for finding a particular gene depends on several factors, one of which is whether the protein product for the gene is known. If you know what the protein product of the gene is, you have a handle on finding the gene. It's a different ballgame if you don't even know what the product of an unknown gene does. Let's first consider how to find a gene for a known protein.

Finding Genes for Known Products

Using genetics. If you are looking for a gene in a microorganism, the easiest way to find it may be by using genetics. Let us imagine that we would like to find the gene for a specific enzyme involved in the biosynthesis of the amino acid histidine in *Escherichia coli.* What we need is a mutant *E. coli* that lacks this enzyme. Since literally thousands of *E. coli* mutants have been characterized, it is probably easy to get one, for example, from the *E. coli* Genetic Stock Center at Yale University. Our mutant cannot make histidine and therefore cannot live unless its medium is supplemented with that amino acid. Now we go to our *E. coli* DNA library (see above). The plasmids in this library contain inserted DNA fragments from the entire chromosome of normal *E. coli.* We transform a batch of the mutant, histidine-requiring *E. coli* with this mixture of plasmids. Since transformation occurs at a low frequency, each *E. coli* cell will receive at most one plasmid. Now we look for a transformed *E. coli* that has acquired the ability to make histidine. We can find it easily because it can now grow on histidine-free medium (Figure 5.14). The assumption is that any bacterium now able to make histidine must have received the missing gene on its new plasmid.

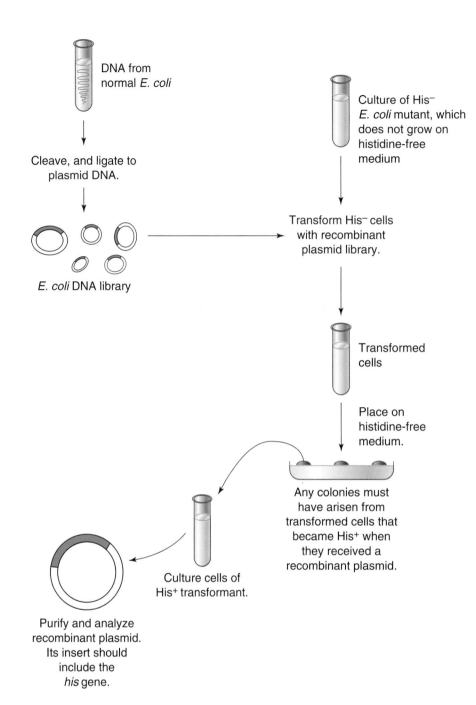

Figure 5.14 Using genetics to find an *E. coli* gene for histidine biosynthesis.

We isolate the plasmid from that bacterium, and, with luck, have found the histidine biosynthesis gene we were looking for within the bacterial DNA insert in the plasmid.

Using the product. The genetic method outlined above is simple and efficient, but is applicable only under limited circumstances. Obviously, if you are looking for an animal gene, you cannot collect a batch of mutant animals, transform them with plasmids, and

look for restoration of a normal phenotype. Instead, we must pull some different techniques from our biotechnology toolbox, as in the example below.

Suppose you wish to locate the gene for polar bear hemoglobin, a known gene product. If you cannot find another scientist who has already made a library of polar bear DNA, you first make one of these. Now you need specific information about the amino acid sequence of the protein, so you obtain a specimen of polar bear blood and isolate the hemoglobin from the red cells. You give the purified hemoglobin to your friend the protein chemist, who analyzes a portion of its amino acid sequence. Then, working backward from the amino acid sequence, you determine the sequence of nucleotides that encodes that portion of polar bear hemoglobin. Next, you go to another friend, a nucleic acid chemist, and have her synthesize an oligonucleotide with that sequence. You now have a probe for the hemoglobin gene.

Returning to your laboratory, you use one of a number of methods to label the probe for detection (such as with a radioactive isotope). Then you hybridize the probe to your polar bear DNA library to identify molecules that have the hemoglobin DNA sequence. Barring complications, you will eventually find the gene (Figure 5.15).

This thumbnail sketch of one approach to finding the gene for polar bear hemoglobin makes it sound like a fairly easy proposition. It is not. Even when you know what you are looking for (polar bear hemoglobin), the procedures are technically demanding, and unexpected difficulties are almost certain to arise. Finding a gene can be one of the most difficult and time-consuming aspects of molecular biology projects.

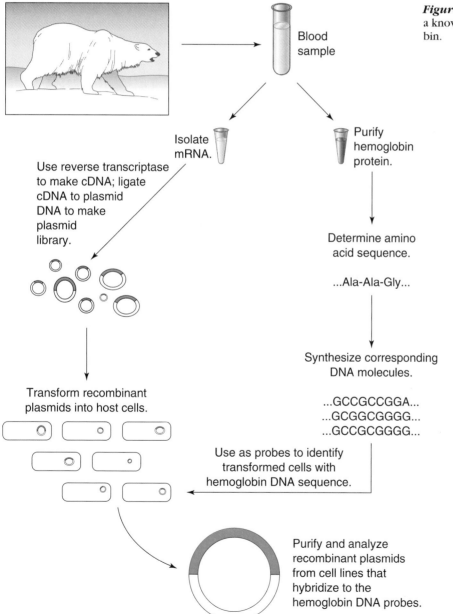

Figure 5.15 Finding the gene(s) for a known product: polar bear hemoglobin.

Finding Genes with Unknown Products

The task of finding a gene when the protein product is not known is much harder than finding a gene for a known protein. Why would anyone want to do it at all? One reason is to find the genes involved in inherited diseases. In many cases, the identity of the defective protein involved in a genetic disease is not known. Identifying the defective gene would be a starting place for understanding the cause of the disorder.

To find a disease gene with an unknown product, scientists must have DNA samples from a large number of related individuals who carry the disease in their family. From the family's medical history, the scientists know which members have the disease and which members are carriers (if the disease is a recessive trait). The DNA samples are digested with many different restriction enzymes, and the digestion products examined for variations between individuals (RFLPs; see *Analyzing genetic variation* above). The goal is to find a restriction fragment or fragments seen only when the disease gene is present. Such a distinctive fragment or pattern is called a **marker.** If a marker is found, it is assumed that the region of the chromosome containing the marker is very close to the disease gene.

After a marker for a disease gene has been identified, a new phase of work begins. Long stretches of DNA (often hundreds of thousands of base pairs) around the marker must be sequenced and searched for sequence patterns that look like genes (with promoters and coding regions). When potential genes are found, their sequences from the healthy and sick family members must be compared. If the sequences of a putative gene from sick individuals are always different from the sequences of the same putative gene from healthy individuals, then the defective gene is probably the culprit in the disease. The gene *BRCA-1,* which is implicated in hereditary breast cancer, was found in this way.

This entire process can take years and is not foolproof. Recently, two well-regarded scientific groups reported that they had found the gene for a particular genetic disease. The only problem was that the two groups had identified completely different genes on different chromosomes. Once a candidate for a disease gene has been identified, further study of more affected people is usually required to confirm that the identification is correct.

Screening for Genetic Diseases

After a genetic marker for a disease has been found and confirmed, that marker can be used as a tool to predict the presence of the disease gene. If an individual wishes to know whether she carries the disease gene, scientists can prepare a sample of her DNA and look for the RFLP that constitutes the genetic marker. If the marker is present, she is likely to carry the disease gene. The laboratory procedure for screening for genetic disease is essentially the same as that for DNA fingerprinting (see above), with the restriction enzymes and the probe selected to reveal the marker region.

Why do we say *likely* to carry the genetic disease? Because genetic markers are usually not the disease-causing mutations themselves but only RFLP patterns that are close to those mutations and tend to be inherited with them, so the presence of a marker is not proof that the mutation is present. Genetic recombination events could separate the marker RFLP from the disease mutation. Likewise, the absence of the genetic marker may not be absolute proof that a disease gene is not present. In addition, there may be many different mutations that can cause a given genetic disease. The vast majority of an affected population may carry one particular mutation and its associated genetic marker. A rare, alternative disease-causing mutation may not be associated with the marker RFLP; in these cases, the marker is not a useful diagnostic tool. Or a marker might be present in some families but not others. However, if the limitations are understood, genetic testing for a disease gene can provide valuable information for physicians and potentially affected individuals.

Once mutant genes are positively identified, more specific probes for diagnosis can be made. Probes for specific disease-causing mutations can be used to screen samples. If many different mutations can cause a specific disease, as in the case of cystic fibrosis, arrays of probes to the different mutations can be bound to a microscope slide. The sample DNA is labeled and hybridized to the slide containing the probe sequences (see *DNA chip technology* in chapter 1). Since more and more disease genes are being identified, direct examination of those genes is becoming increasingly common as a means of diagnosing genetic disease. (See the activity *Human Molecular Genetics.*)

Genetic engineering

The term **genetic engineering** refers broadly to the process of directed manipulation of the genome of an organism. Because genetic engineering usually involves the combining of genes from two or more sources, it is also commonly referred to as **recombinant DNA (rDNA) technology.** Genetic engineering usually involves the manipulation of a specific gene. The goal of genetic engineering, of course, is not to manipulate an organism's DNA per se but to change something about the proteins produced in that organism: to cause it to produce a new protein, to stop producing an old protein, to produce more or less of a protein, and so on. Manipulating the genome is merely the way to influence protein production.

In general, the gene in question must first be identified and cloned into a plasmid or other vector so that it is easy to work with. The cloned gene is manipulated in the laboratory by using the tools and techniques described previously. The altered gene is then inserted into the target organism. The target organism has now been genetically engineered. What are some genetic manipulations that we can make and insert into a host cell?

Regulating the Expression of Existing Genes

Increasing gene expression. Inserting several copies of a useful gene into an organism is one way to increase production of the gene product. This technique has been used by the company Novo Nordisk to increase production of bacterial amylase, an enzyme used in the baking and soft drink industries. For years, the company used the bacterium *Bacillus lichenformis* to produce amylase. Scientists at the company have now given the bacterium extra copies of its own amylase gene, resulting in greatly increased yields of the protein.

Turning genes off. In many cases it is advantageous to decrease the expression of a particular gene. An example can be found in the case of the familiar flavorless grocery store tomato. Grocery store tomatoes lack flavor because they are picked green. Green tomatoes are hard and can be harvested by machine, packaged, and shipped without damage. Ripe tomatoes taste much better but are soft and would very likely be damaged during handling. Scientists discovered that the softening of tomatoes during ripening is caused by the enzyme polygalacturonase. Blocking production of this enzyme would allow a tomato to ripen and develop flavor without becoming soft.

Although there are many approaches to blocking production of an enzyme, scientists at Calgene, Inc., decided to use an exciting new approach to gene control. They introduced into the tomato plant cells a gene encoding an *antisense RNA* for polygalacturonase. Antisense RNA is RNA that is exactly complementary in sequence and opposite in polarity to the normal mRNA of a gene. Its structure allows the antisense RNA to hybridize to the mRNA, creating a

double-stranded RNA molecule. Double-stranded RNA cannot be translated by ribosomes, so expression of the normal gene is reduced (Figure 5.16). After constructing this genetically engineered tomato, Calgene asked the Food and Drug Administration to evaluate its safety. The tomato was approved for sale as food. Calgene test-marketed the tomato but began to run into problems that had nothing to do with genetic engineering and everything to do with the fact that they were inexperienced at growing, harvesting, shipping, and delivering large numbers of tomatoes. The first antisense food foundered on distribution problems.

Many scientists view antisense RNA as an exciting new way to fight disease. The idea is simple: turn off genes specific to the disease-causing organism. A short complementary RNA or DNA molecule should hybridize specifically to the disease organism's RNA and could inactivate expression of proteins essential to the organism without having any undesirable effect on the host. The challenges are to identify the appropriate target RNA sequences and then to deliver the antisense oligonucleotide to the appropriate cells. One antisense drug, Vitravene, is already on the market. Vitravene fights infection caused by cytomegalovirus, a virus that destroys the retinas of many patients with depressed immune systems. At the time of this writing, many other antisense drugs, targeted against cancer, Crohn's disease, psoriasis, colitis, and AIDS, are in various stages of development and clinical trials by several different drug companies.

Adding New Genes

People have long used selective breeding to introduce desirable traits into plants and animals. Selective breeding is, of course, limited by the ability of organisms to mate and by the characteristics of natural variants of the species. Genetic engineering is not limited in either of these ways. By using genetic engineering, we can transfer genes for desirable characteristics from one organism to any other. For example, a bacterial gene can be transferred into a plant.

One of the environmental problems affecting the ability of soil to sustain plant life is aluminum toxicity. More than one-third of the world's soil suffers from aluminum toxicity, and the problem is most severe in the humid tropical climates of many developing countries. Aluminum toxicity is also a consequence of soil acidification as a result of acid deposition, seen in sensitive soils of the United States and other countries. Aluminum ions injure plant root cells, thus interfering with root growth and nutrient uptake. In an effort to make crop plants more resistant to aluminum, Mexican scientists transformed corn, rice, and papaya with a bacterial gene for the enzyme citrate synthase.

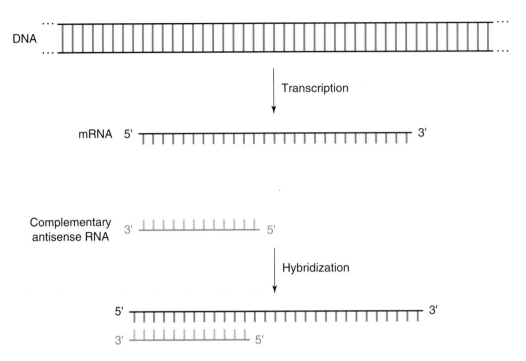

DNA

mRNA 5' ———————————————— 3'

Complementary
antisense RNA   3' ——————————— 5'

↓ Hybridization

5' ———————————————— 3'
3' ——————————— 5'

Figure 5.16 Antisense RNA turns off gene expression by hybridizing to the 5' (front) end of a complementary mRNA. Double-stranded RNA is not recognized by ribosomes and cannot be translated.

The transformed plants released citric acid, which binds to soil aluminum and prevents it from entering plant roots. The genetically modified plants were able to germinate and develop at aluminum concentrations that were toxic to nontransgenic plants.

Another example of adding new genes by genetic engineering is the introduction of a human gene into animals. This kind of genetic engineering is done to obtain large amounts of the gene product. For example, people suffering from type A hemophilia require a protein called factor VIII so that their blood will clot properly. Scientists at Virginia Polytechnic Institute and State University introduced the gene for human factor VIII into pigs, with a promoter that causes the protein to be secreted in the milk. The scientists estimate that 300 to 600 milking sows could produce enough factor VIII to meet the world's demand.

Gene Therapy

A medical application of biotechnology called gene therapy has received a great deal of media attention. And no wonder. Gene therapy holds out the hope that previously incurable, fatal inherited diseases may be treatable. The goal of gene therapy is to treat genetic diseases by giving the patient healthy copies of the defective gene. Why is gene therapy listed under the heading *Genetic engineering* in this chapter? Because gene therapy is actually a form of genetic engi-

neering. Patients' genomes are manipulated so that they can produce a protein they were not previously producing (or were not producing in adequate quantities).

Gene therapy sounds logical and straightforward, but it is not a simple undertaking. First, the defective gene must be identified, and a healthy copy must be cloned. Then the healthy gene must be appropriately delivered to the patient. The delivery is the tricky part: the new gene needs to be delivered to the correct tissue. For example, it does no good to deliver an essential liver receptor gene to blood cells. If you can deliver the new DNA to the appropriate tissue, the gene must still be expressed in those cells. As yet, we have no reliable method for actually replacing a defective gene on a chromosome with a healthy one. Genes can be inserted randomly into the chromosome, with the risk of causing some damage. In addition, genes inserted at random are sometimes not expressed well; the reasons for this are not understood.

The first test of human gene therapy used blood cells. Blood cells were a natural first case because they are easy to obtain and easy to reintroduce into the body. The patient was a little girl with severe combined immunodeficiency disease. This child had an extremely defective immune system and could not defend herself against infection. Her condition was the same as that of the "bubble boy," who lived his short life in a sterile environment. The cause of her condition was a defect in the gene coding for the enzyme adenosine deaminase (ADA). Scientists at the National Institutes of Health removed blood from her, separated the lymphocytes (white blood cells), and used a retrovirus to introduce a healthy copy of the *ada* gene into them. They then reintroduced the lymphocytes into their patient. The altered cells produced the missing enzyme, and at this time, the little girl is healthier than she has ever been.

Scientists are experimenting with ways to deliver genes to other body cells. One such method is to introduce DNA directly, through a method sometimes referred to as DNA vaccination. In this method, the subject receives multiple injections of DNA encoding a protein or proteins the patient lacks. This method has been tested in a colony of hemophilic dogs at the University of North Carolina. The dogs lacked the ability to produce clotting factor IX. Thus, their blood required an extremely long time to clot. When some of these dogs were given multiple intramuscular injections of DNA encoding factor IX, levels of the factor in the dogs' blood approached but did not achieve normal, and their clotting time was dramatically reduced.

Although genetic therapy was originally envisioned as a remedy for inherited diseases, the scope of genetic medicine has already moved far beyond that concept. For example, researchers are looking to antisense RNA as a means of fighting cancer, viral infections, and even baldness. Other scientists are experimenting with genetic engineering of tumors to make them more vulnerable to drug treatments. Still others are using genetic approaches in hopes of stimulating the immune system to fight cancer. In the next few years, gene therapy trials will undoubtedly multiply in number and scope.

Protein engineering

As is probably evident from the foregoing examples, almost any manipulation of DNA sequence is possible. The goal of most of these manipulations is the same: to affect a protein. At present, our efforts are largely centered on protein *production*: blocking it, increasing it, or causing it to happen in a new cell type. A future direction of biotechnology will be to improve the *functioning* of proteins by altering their structure or even to design new proteins to fulfill a specific need. Increasing the use of enzymes and biological processes in manufacturing offers significant advantages in reducing industrial pollution and utilizing renewable rather than nonrenewable resources

(see chapter 1), so protein engineering offers important potential benefits to society.

One desirable way in which industrial enzymes could be modified would be to increase their stability so that they would be less vulnerable to changes in temperature, pH, or other reaction conditions. Hardier enzymes could be used under a greater variety of industrial conditions, and higher temperatures could mean faster reaction rates. Increasing the stability of a protein means increasing the stability of the protein's tertiary and, possibly, quaternary structures. So before a scientist can undertake to engineer a protein for increased stability, he or she must know its three-dimensional structure. Once that structure has been determined, the scientist looks for ways to increase stability without altering the protein's function.

The most obvious way to increase a protein's stability is to introduce disulfide bridges (see *Stability of protein structure*, chapter 4) to hold the tertiary structure together. These bridges have to be geometrically compatible with the protein's structure and must not alter crucial amino acids. The bacteriophage enzyme lysozyme has been engineered in this way. This enzyme has two domains. Using recombinant DNA technology, scientists introduced codons for cysteines in positions within the gene that were selected after careful study of the protein's structure. The altered gene was reintroduced into cells, which duly produced the protein with the new cysteines. The cysteines combined to form three disulfide bridges that tied the domains together in their appropriate three-dimensional configuration (Figure 5.17). These alterations significantly increased the thermal stability of the enzyme. The engineered protein withstood heating to a temperature $23^{\circ}C$ higher than the native protein could endure.

Scientists are already looking beyond simply modifying existing proteins to creating new ones from scratch. Significant advances in protein design will not be achieved until we gain a greater understanding of how amino acids determine protein structure and exactly how that structure determines activity. Gaining this understanding is a difficult undertaking, considering the enormous variety in both protein structure and functions. However, progress is being made. With the improving technologies of NMR spectroscopy and X-ray crystallography, determining the actual three-dimensional structure of a protein molecule is becoming more common (although not easy). Knowing the structure of a protein provides valuable clues about how the protein actually works. Some day we hope to be able to predict the three-dimensional structure of a protein from its amino acid sequence

Figure 5.17 Protein stability engineering. Three engineered disulfide bridges (gold) tie the two domains of bacteriophage T4 lysozyme in the proper configuration. The two domains are blue and gray and are connected by the stippled helix. The numbers are the positions of the cysteines in the 164-amino-acid protein. The cysteine at position 54 was changed to a threonine to keep it from interfering with proper formation of the engineered bridges. (Adapted from M. Matsumara, G. Signor, and B. W. Matthews, *Nature* 342:291, 1989, with permission.)

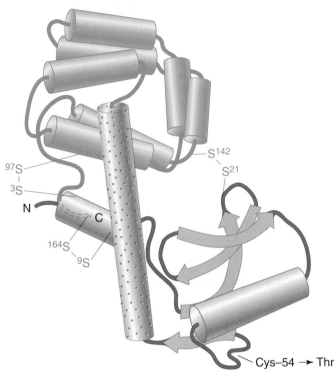

and understand exactly how that structure interacts with other compounds to produce the protein's activity. This knowledge could then be used to design proteins for specific tasks, such as breaking down toxic chemicals or manufacturing products from renewable resources that are currently manufactured from petroleum. The possibilities are limitless.

Has anyone already produced an "artificial" enzyme? Yes. A group of researchers in Denver, Colorado, designed a hypothetical protein-digesting enzyme. The design was assisted by computer programs that make predictions about protein structure. When the protein was chemically synthesized, it had the predicted enzyme activity. This achievement, published in 1990, is the first clear example of synthesis of an artificial protein with catalytic activity.

The Denver group's protein was made by chemical synthesis. Protein engineers of the future will probably use recombinant DNA technology to produce their designer enzymes. Once the desired amino acid sequence of the new enzyme has been determined, they could synthesize DNA encoding that amino acid sequence, clone the DNA into a plasmid, and transform an appropriate host. The host cell would then synthesize the protein. This process of starting with a protein sequence and making a gene for it turns the linear "central dogma" of biology (DNA to RNA to protein) into a circle. It is amazing that only 5 decades

after the discovery of the structure of DNA, we have learned so much.

Summary

As our understanding of the genetic machinery and protein biochemistry of cells grows, so does the number of technologies based on that understanding. We learned the chemistry of the DNA molecule; now we can synthesize it ourselves. We learned how genetic variation is determined by DNA sequence variation; now we look directly at variation in DNA for identification. We learned how nature copies, recombines, and transfers genes; now we copy, recombine, and transfer genes ourselves. When we understand the forces that determine protein structure and function, we will create our own custom enzymes.

Where is this new technology taking us? In many cases, to clear-cut benefits. More precise identification techniques will minimize mistakes in paternity and criminal cases. Faster diagnostics will result in more effective disease treatments. Genetic engineering will give us crops that are more nutritious and disease resistant. We will use biological systems to diagnose and then clean up environmental pollution. Gene therapy will alleviate much suffering.

There are also perplexing questions about the appropriate use of these technologies. Decisions about appropriate use will inevitably (and should) involve values and moral judgments (see Part III of this book). An understanding of the technologies themselves will provide some factual basis for making these important judgments.

Apart from the social implications of DNA-based technologies, their development offers us an exciting, unfolding story of science. The more we learn, the more we find that nature is vastly more complicated than we originally imagined. The intricacies of heredity and gene expression dwarf our first conceptions of the processes. New surprises are constantly announced. Regardless of the practical applications that may come from our increasing understanding of life's basic processes, the wonder engendered by the incredible complexity of the processes must surely inspire and enrich all of us.

Selected Readings

Anderson, W. F. 1995. Gene therapy. *Scientific American* 273:124.

Bayley, H. 1997. Building doors into cells. *Scientific American* 277:62.

Cohen, J. S., and M. E. Hogan. 1994. The new genetic medicines. *Scientific American* 271:76–82.

de la Fuente, J. M., and L. Herrera-Estrella. 1997. Aluminum tolerance in transgenic plants by alteration of citrate synthesis. *Science* 276:1566–1568.

Hall, S. 1995. Protein images update natural history. *Science* 267:620–624.

Kahn, P., and A. Gibbons. 1997. DNA from extinct hominids. *Science* 277:176–178.

Kidd, G. 1995. Low-tech woes derail Calgene's high-tech tomato. *Bio/Technology* 13:540–541.

Making gene therapy work. 1997. *Scientific American* 276(5). This special edition contains several articles about topics in gene therapy, including overcoming obstacles to gene therapy, nonviral strategies, gene therapy for cancer and AIDS, and what cloning means for gene therapy.

Mullis, K. 1990. The unusual origin of the polymerase chain reaction. *Scientific American* 262:56–65. How the Nobel Prize-winning surfer-scientist came up with the PCR technique.

Paabo, S. 1993. Ancient DNA. *Scientific American* 269:86–92. Isolating and studying DNA from ancient materials. Anyone with questions about the feasibility of Jurassic Park should read this article.

Ronald, P. 1997. Making rice disease-resistant. *Scientific American* 277:100.

Roush, R. 1997. Antisense aims for a renaissance. *Science* 276:1192–1193.

Saiki, R. K., S. Scharf, F. Faloona, K. Mullis, G. Horn, H. Erlich, and N. Arnheim. 1985. Enzymatic amplification of beta-globin genomic sequences and restriction site analysis for diagnosis of sickle cell anemia. *Science* 230:1350–1354.

Sarinaga, M. 1997. Making plants aluminum tolerant. *Science* 276:1497.

Shreeve, J. 1999. Secrets of the gene. *National Geographic* 196:42–75.

Watson, J., M. Gilman, J. Witkowski, and M. Zoller. 1992. *Recombinant DNA*. Scientific American Books, W. H. Freeman & Co., New York, N.Y.

Weiner, D., and R. Kennedy. 1999. Genetic vaccines. *Scientific American* 281:50.

White, R., and J. Lalouel. 1988. Chromosome mapping with DNA markers. *Scientific American* 258:40–48.

PART II
Classroom Activities

This part contains activities that will help you understand some basics of recombinant DNA technology and its applications. The activities are grouped into four sections with specific themes. Section A teaches concepts basic to molecular biology: the structure and function of DNA. Section B illustrates methods of manipulating and analyzing DNA, concluding with activities that illustrate how these methods are applied in DNA typing. Section C focuses on the transfer of genetic material between organisms. Gene transfer is necessary for cloning DNA in the laboratory but also occurs frequently in nature, where it has important effects on our health and disease. In this section, you will transfer genes between bacterial strains and also from a bacterium to a plant, and you will learn about some of the implications of natural gene transfer. In Section D, you will focus on a single protein common across the five biological kingdoms: the enzyme amylase. You will look for its activity in a variety of samples, visualize it in electrophoresis gels, use a portion of its amino acid sequence to construct an evolutionary tree, and, finally, use it as a gateway to explore the online bioinformatics resources. Section E makes the connection between the fields of genetics and molecular biology. It illustrates how an understanding of biological processes at the molecular level explains genetic observations we make at the organism level, using coat color in Labrador retrievers as a specific example. This section concludes with information on human molecular genetics and on cancer.

A. DNA Structure and Function

A basic principle of biology is that structure must serve function. We can see examples of this principle when we consider the structure of a bird's wing in relation to flight or a human eye in relation to sight. The same relationship between structure and function holds true at the molecular level. Nowhere is this relationship more apparent than in the DNA molecule. Its structure is wonderfully suited to its two functions: replication and information translation. It also permits easy regulation through a number of mechanisms.

These activities focus on the DNA molecule and how its information is replicated and translated into proteins. You will make models, do simulations, and have discussions to explore the structure, packaging and storage, and replication of DNA; the processes of transcription and translation; and aspects of gene regulation, including the exciting field of antisense technology. At the end of the section are simple wet laboratory procedures that you can use to extract DNA from *Escherichia coli,* onion, animal tissue, and store-bought baker's yeast.

Classroom Activities

Constructing a Paper Helix

6

Introduction

DNA is called the blueprint of life. It got this name because it contains the instructions for making every protein in your body. Why are proteins important? Because they are what your muscles and tissues are made of; they synthesize the pigments that color your skin, hair, and eyes; they digest your food; they make (and sometimes are) the hormones that regulate your growth; they defend you from infection. In short, proteins determine your body's form and carry out its functions. DNA determines what all of these proteins will be.

The DNA molecule is a double helix. Think of it as a ladder that has been twisted into a spiral. The outside of the ladder is made up of alternating sugar and phosphate groups. The sugar is called deoxyribose. The rungs of the ladder are made up of nitrogen-containing bases. There are four different nitrogen-containing bases in DNA: adenine, guanine, cytosine, and thymine. These four bases are of two types: purines and pyrimidines. Purines are large double-ring structures. Adenine and guanine are purines. Pyrimidines are smaller single-ring structures. Cytosine and thymine are pyrimidines. For more information on DNA and its component parts, see chapter 4.

Inside the DNA ladder, two bases pair up to make a "rung." One base sticks out from each sugar-phosphate chain toward the inside of the ladder. It forms a pair with a base sticking out from the opposite sugar-phosphate chain. Only three rings can fit between the two sugar-phosphate chains, so a pyrimidine (one ring) and a purine (two rings) form a pair. Because of the chemical structures of the bases, adenine always pairs with thymine, and cytosine always pairs with guanine.

Activity

The goal of this lesson is to construct a paper model of a DNA helix. You will do so by making the fundamental unit of DNA. This unit, called a nucleotide, consists of one sugar molecule, one phosphate group, and one nitrogenous base. Each member of the class will make nucleotides and then join them to form the ladderlike helix.

Although making models is fun, building models is also a technique scientists use to help them figure out how things are put together or how they might work. In fact, Watson and Crick, who discovered the structure of the DNA molecule, used cut-and-paste paper models to help them.

Procedure

1. Cut out the pattern for the nucleotide(s) assigned to you.
2. Place the pattern on construction paper of the appropriate color.
3. Label your pieces as the pattern is labeled.
4. Glue the nitrogen base to your sugar molecule by matching up the dots.
5. Glue the phosphate group onto your model by matching up the stars.
6. Your teacher will join the nucleotides together to form a helix.

Questions

1. What base does adenine pair with?

2. What base does guanine pair with?

3. What is the smallest unit of DNA called?

4. What is the shape of the DNA molecule?

5. Which bases are purines?

6. Which bases are pyrimidines?

7. Why must a purine pair with a pyrimidine?

8. What is the name of the sugar in the DNA backbone?

Advanced Questions

9. Suppose you know that the sequence of bases on one DNA strand is AGCTCAG. What is the sequence of bases on the opposite strand?

10. Referring to question 9, suppose that the 5′-most base on the given strand is the first A from left to right. What would be the 5′-most base on the opposite strand?

11. Assume that a 100-base-pair DNA double helix contains 45 cytosines. How many adenines are there?

Classroom Activities

DNA Polymerase: The Replicator

The accurate copying of DNA is one of the most important jobs an organism must do during its life. Why do you think this statement is true? What would happen to you if your ancestors' cells had not taken the trouble to make accurate copies of their DNA?

For such an important task, cells employ not one but a whole team of enzymes. However, the star of the team is the enzyme DNA polymerase. This protein builds the new daughter strand from nucleotides in the cell and checks the new base pairs for accuracy. The other protein team members help DNA polymerase do its job.

All bacteria and higher organisms and many viruses have their own DNA polymerase proteins, which are encoded by their DNA. All of these DNA polymerases work in a similar way and even resemble one another. Scientists therefore often study the complicated process of DNA replication in simple systems, such as bacterial and animal viruses, and apply what they have learned as they look at higher organisms. From these simple systems, many general facts about DNA polymerases have emerged. Let's look at some of these facts and then consider how they were determined.

1. DNA polymerases use deoxynucleoside triphosphates as precursors for the synthesis of DNA. These molecules are held at a special binding site on the polymerase before they are incorporated into a new DNA molecule. Nucleotides with only one or two phosphate groups will not bind there and are not incorporated into new DNA.

2. DNA polymerases cannot begin synthesizing DNA without a starting point called a **primer.** A primer is a piece of DNA or RNA base paired to the template strand so that the template strand sticks out past the 3′ end of the primer (see the figure). DNA synthesis begins at the 3′ end of the primer, where DNA polymerase attaches the first new nucleotide.

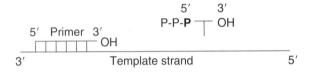

3. DNA polymerases synthesize DNA in one direction only. They start by attaching the 5′ phosphate group of the new nucleotide to the 3′ OH group of the last nucleotide in the primer, and they continue in this manner. They cannot connect the 3′ OH group of a new nucleotide to the 5′ phosphate group of a primer. DNA synthesis is therefore said to occur in the 5′-to-3′ direction (relative to the new strand).

How were these facts deduced? By looking at reactions with purified DNA polymerase proteins in vitro. ("In vitro" means "in glass"; it refers to the use of glass test tubes in the laboratory and basically means "in the test tube," even though almost everyone uses disposable plastic tubes for this kind of work now). A scientist could take purified DNA polymerase protein, add a DNA template, and then experiment by adding deoxynucleoside monophosphates, diphosphates, or triphosphates and looking for DNA synthesis. By using radioactive nucleotides, a scientist can detect DNA synthesis by the appearance of radioactive DNA strands. In tests like this, it is clear that only nucleoside triphosphates are incorporated into DNA. It is also possible to mix DNA polymerase with radioactive nucleotides and ask whether the radioactivity becomes associated with the protein. In this manner, it is possible to detect binding of nucleotides to DNA polymerase.

Besides looking at the effects of different forms of nucleotides, scientists looked at different forms of DNA templates. They constructed different types of DNA molecules and asked whether DNA synthesis occurred when polymerase and deoxynucleoside triphosphates were added. Through these experiments they learned that DNA polymerase must have a primer on the template strand before it can synthesize DNA. Finally, by examining the DNA that was synthesized, they learned that DNA polymerase can add nucleotides only in the 5′-to-3′ direction. The enzyme could never synthesize a complementary strand to a portion of the template that was "upstream" (on the 5′ side) of the primer.

DNA polymerases are now important tools in molecular biology research and in biotechnology applications. They are used in cloning, copying, and sequencing DNA. In later lessons, you may have the opportunity to learn how DNA polymerases can help reveal the base sequence of a piece of DNA or how they are used to make a "DNA Xerox machine" that is extremely useful in detecting organisms that cause disease, among other applications. When you do these activities, notice that in each case, the DNA polymerase is given a primer and that DNA synthesis occurs in the 5′-to-3′ direction.

Questions

1. Given what you have learned about DNA polymerase, on which of the DNA template molecules shown below could the enzyme synthesize a new strand if given nucleoside triphosphates? Show where DNA synthesis would begin on each molecule and in what direction it would proceed.

2. Given what you have learned about DNA polymerase, what is wrong with the simple model of DNA replication that you used earlier?

A

B

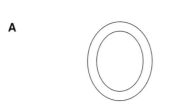

C

D

From Genes to Proteins

You Are Your Proteins

You have probably heard many times that "your DNA determines your characteristics." Did you ever wonder how DNA does that? DNA determines all your characteristics (and the characteristics of every plant, animal, fungus, bacterium, etc., in the world) by determining what proteins your cells will synthesize.

Why are proteins so important? Because of the many things they do. There are many types of proteins, and each of these types performs an important kind of job in your body. For example, *structural proteins* form the "bricks and mortar" of your tissues. Two of them, actin and myosin, enable your muscles to contract. Another structural protein, keratin, is the basic component of hair. *Carrier proteins* transport important nutrients, hormones, and other critical substances around your body. One of these proteins is hemoglobin, which carries oxygen through your blood to your tissues.

Another large class of proteins is the enzymes. Enzymes are the body's workhorses. They carry out chemical reactions in your body. Enzymes digest your food, synthesize fats so your body can store energy, and carry out the work of making new cells. They make molecules and perform activities necessary for life.

In fact, if you want to sum up the importance of proteins in your body, you could say: "*Nearly every biological molecule in my body either is a protein or is made by proteins.*" So by telling your cells what proteins to synthesize, DNA controls your characteristics.

What Proteins Are

What exactly is a protein? A protein is a biological molecule made of many small units linked together in a chain (rather like beads on a string). The units are amino acids, small molecules composed of carbon, hydrogen, oxygen, and nitrogen. Twenty different amino acids can be used in making your proteins.

Imagine sitting down with 20 containers, each filled with a different type of bead, to make a string of beads. You may use as many or as few of each kind of bead as you wish, you may string them in any order that you wish, and you may use any number of beads you wish. This situation is a reasonable representation of the possibilities for protein synthesis within a cell. The beads represent amino acids, and any string of them represents a protein. You can get an idea of the infinite variety of proteins that could be made!

However, in the cell, one thing is different from our example. The cell is handed an instruction sheet that tells it which beads (amino acids) to string to make a particular chain. The instructions come from the cell's DNA.

How DNA Directs Protein Synthesis

DNA contains a *genetic code* for amino acids, with each amino acid being represented by a sequence of three DNA bases. These triplets of bases are called codons. The order of the codons in a DNA sequence is reflected in the order of the amino acids assembled in a protein chain. The complete stretch of DNA needed to determine the amino acid sequence of a single protein is called a gene.

In your cells, the DNA is located within the chromosomes in the nucleus. DNA contains all the instructions for making every protein in your body. However, your cells make proteins at the ribosomes, located in the cytoplasm. An individual ribosome makes only one protein at a time. So when your cell needs to make a protein, a "working copy" of the instructions for that one protein is copied from the DNA and sent to the ribosome for use. This working copy is messenger RNA (mRNA).

After the base sequence of a gene (DNA) is copied into mRNA, the mRNA travels to the ribosome, where its code is translated into protein. The translation step is carried out by a second type of RNA called transfer RNA (tRNA). The tRNA matches the correct amino acids to the codons in the mRNA. The amino acids are linked together to make the new protein.

Wait a Minute— Isn't Protein a Kind of Food?

Most of us have heard that protein is part of a healthy diet. In fact, most people think of food when they hear the word "protein." So what's going on? How does protein in the diet fit into a discussion of genes?

The protein you eat is composed of individual protein molecules made by plant and animal cells. Animal muscle tissue (meat) is particularly rich in protein (remember actin and myosin, mentioned above?). When you eat protein, regardless of its source, your digestive enzymes break the individual protein molecules down into amino acids (like taking the beads off the string). The individual amino acids are used by your cells to make your proteins—a form of biological recycling—or can be further broken down for energy.

Classroom Activity

Today, you and your classmates will act out the steps involved in translating the DNA code into proteins. You will see the roles played by mRNA and tRNA and may also be introduced to some of the "traffic signals" that direct their action.

Questions

1. DNA is double stranded. One strand is the coding strand, and the other is the noncoding strand. The noncoding strand is used as the template to make the mRNA. What is the relationship between the base sequences of the coding and noncoding strands?

2. What is the relationship between the base sequence of the coding strand and the base sequence of mRNA?

3. What is the relationship between the base sequence of the noncoding strand and the base sequence of mRNA?

4. Would there be a problem if the RNA polymerase transcribed the wrong strand of DNA and the cell tried to make a protein?

5. A frameshift mutation is caused by the insertion or deletion of one or two DNA bases. What would be the effect on the amino acid sequence of a protein if one extra base were inserted into the gene near the beginning? Use an example to show what you mean.

6. Suppose an individual has a nutrient deficiency due to poor diet and is missing a particular amino acid. How would transcription and translation be affected?

7. Given below are some tRNA anticodon-amino acid relationships and a stretch of imaginary DNA. Fill in the empty boxes in the chart by writing the DNA sequence of the coding strand and the correct mRNA codons, tRNA anticodons, and amino acids. What are the similarities between the noncoding DNA sequence and the tRNA sequence? Use the following tRNA-amino acid relationships:

GGC	UUA	CAG	CUC	GAU	AGG	CCG
Pro	Asn	Val	Glu	Leu	Ser	Gly

DNA Coding Noncoding											
	TAC	AGG	GGC	CTC	TTA	CAG	CTC	GAT	AGG	CCG	ATC
mRNA											
tRNA											
Amino acid	start										stop

8. A new and exciting branch of biotechnology is called protein engineering. To engineer proteins, molecular biologists work backward through the protein synthesis process. They first determine the exact sequence of the polypeptide they want and then create a DNA sequence to produce it. Use the rules of transcription and translation to "engineer" the peptide sequence below. Fill in the rows for tRNA anticodons, mRNA codons, and the two DNA strands. Use the tRNA-amino acid relationships given in question 7.

DNA Coding Noncoding											
mRNA											
tRNA											
Amino acid	Met start	Leu	Val	Pro	Gly	Asn	Ser	Glu	Glu	Pro	Val

Does Antisense Make Sense?

Imagine what you might be able to do if you could prevent or decrease the expression of any gene. Prevent cancer from developing? Prevent viral diseases? Control certain genetic diseases?

Biotechnologists are pondering these and other questions largely because of a new technology for gene regulation: antisense. What is antisense technology, and how does it work? Antisense technology, like the rest of biotechnology, is based on a natural phenomenon. In antisense gene regulation, a cell synthesizes a very short piece of RNA that is exactly complementary to the ribosome recognition region (usually the 5′ or "front" end) of the messenger RNA (mRNA) of the gene to be regulated. Because of its complementary base sequence, the antisense RNA can base pair with the mRNA. When it does this, the 5′ end of the mRNA becomes double stranded. The ribosome can no longer recognize and translate it. By preventing translation of mRNA, antisense RNA can decrease gene expression (see figure).

How antisense regulation works. The gene to be regulated is transcribed normally, producing an mRNA molecule. The antisense oligonucleotide is exactly complementary in sequence to the area of the message where the ribosome normally binds. When the antisense molecule base pairs to the mRNA, the ribosome cannot bind to its recognition site, and translation does not occur. (The bacterial gene sequence shown is taken from the *Escherichia coli lacZ* gene.) RRS, ribosome recognition site; ic, initiation codon; aRNA, antisense RNA.

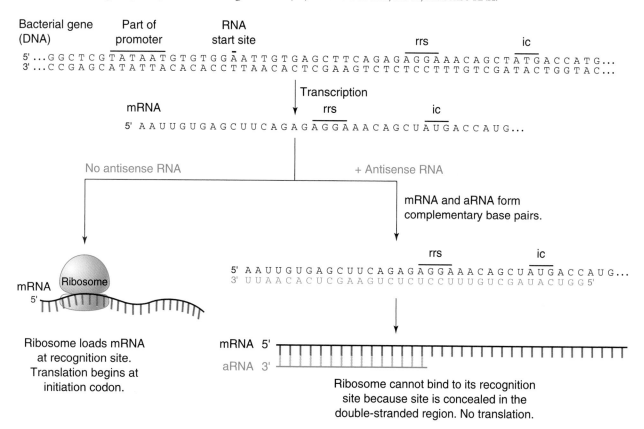

Antisense technology copies this natural approach except that the antisense molecule can be RNA, DNA, or even a chemically modified version of either one, just as long as it will base pair with the target molecule. A general term used to refer to these short molecules is "antisense oligonucleotides" (the prefix "oligo" means "several"; "nucleotide" can refer to the building blocks of DNA or RNA). It is theoretically possible to reduce the expression of any gene by introducing an appropriate antisense oligonucleotide into the cell.

Regulation by antisense is very precise because of the specificity of base pairing. For any oligonucleotide, the chance that its complementary sequence will occur randomly is 1 in 4^n, where n is the number of bases in the oligonucleotide. So the chance of a 15-nucleotide antisense molecule accidentally pairing with an unintended target is 1 in 4^{15}, which is less than 1 in 1 billion (multiply it out for yourself). For longer antisense molecules, the odds are even more remote.

Is antisense technology being used for any practical applications? Yes. In 1999, the first antisense drug went on the market in the United States. It is called Vitravene, and it is manufactured by Isis Pharmaceuticals. Vitravene targets the replication protein of cytomegalovirus (CMV). CMV is very common; the odds are excellent that you have been infected by it. CMV infection doesn't usually present any problems to people with intact immune systems, but in people with impaired immunity, such as AIDS patients, CMV infection can be devastating. One of the most debilitating forms of CMV infection is retinitis, in which the virus gradually destroys the retina. CMV retinitis is the most common cause of blindness in people with AIDS. Vitravene fights CMV by preventing expression of its replication protein and thereby blocking viral replication.

Questions

1. What kinds of things did the Isis scientists need to know to develop Vitravene? What kinds of things would they probably have tested during the development of the drug?

2. Can you think of a disease or a type of disease that would probably not be treatable by antisense drugs?

3. Imagine that you are part of a research team at a pharmaceutical company. You have been asked to prepare a proposal to develop an antisense drug to control AIDS. What kinds of things would you need to know to begin to design your antisense molecule? Find the information you need, and prepare a proposal for your company's management. Explain to them why blocking expression of the target gene should be an effective strategy. Of course, you don't know for certain if it will work—it's a research proposal—but you should be able to justify your idea.

Sizes of the *Escherichia coli* and Human Genomes

Introduction

DNA is stored in cells in the form of chromosomes and plasmids. The amount of DNA required to store the information necessary for making even a simple organism such as a bacterial cell is very large. One of the wonders of biology is that cells are able to store and access the great lengths of DNA needed to encode their hereditary information.

The bacterium *Escherichia coli* is estimated to have about 2,000 to 3,000 genes in its genome; the average bacterial gene is considered to be 1,200 base pairs. The DNA of *E. coli* is about 4.6 million base pairs long and includes some noncoding DNA. A typical plasmid is about 3,000 base pairs long and encodes just a few genes.

What are the physical sizes of these DNA molecules? The *E. coli* chromosome consists of one large circular DNA molecule that if stretched out would be approximately 10^{-3} m (1 mm) long. By comparison, the *E. coli* cell is only 1×10^{-6} to 2×10^{-6} m long. The *E. coli* DNA molecule is thus 1,000 times longer than the cell! Even a lowly plasmid of 3,000 base pairs would be 10^{-6} m long, or approximately the length of the cell, if it were linear. Yet the chromosome of *E. coli* constitutes only 2 to 3% of the cell's weight and occupies only 10% of its volume.

DNA occupies such a small fraction of the cell's volume (considering the enormous length of the DNA molecule) because it is an extremely slender molecule. It is capable of a high degree of folding and coiling, an essential feature for packing it into the cell. Although the degree of folding required to fit the DNA of *E. coli* into the bacterium is impressive, the folding necessary for packaging DNA into a human cell is even more remarkable.

The human cell is approximately 2×10^{-5} m in diameter. The human genome is estimated to consist of about 3 billion base pairs. If the DNA of a single human cell DNA were stretched out, it would be about 2 m long, or 100,000 times longer than the cell! Yet all this DNA not only fits into the cell but is also accessible to the cell's enzymes for information transfer and replication.

This activity uses models that are 10,000 times life size to demonstrate the relationship between the sizes of an *E. coli* cell, its chromosome, its plasmid, and a single gene. The ×10,000 *E. coli* is represented by a 2-cm gelatin capsule, the ×10,000 *E. coli* chromosome is represented by a 10-m length of thread, the plasmid is represented by 10 mm of thread, and a gene is represented by 4 mm of thread.

Materials

Obtain from your teacher a letter-size envelope containing the following items.

- A 2-cm (20-mm) gelatin capsule
- 10 m of thread
- An index card with 4 mm of thread labeled "Average length of bacterial gene"
- An index card with 10 mm of thread labeled "Length of a typical bacterial plasmid"

At the direction of your teacher, form groups of four for the next steps.

1. Remove the gelatin capsule from the envelope. It represents a single *E. coli* bacterium that has been enlarged 10,000 times.
2. Remove the two index cards from the envelope. The lengths of thread or string on the cards represent the length (and *not* the diameter) of an *E. coli* gene and plasmid magnified 10,000 times.
3. Remove the thread from the envelope and stretch it out. The thread represents the bacterial chromosome magnified 10,000 times.
4. Let two people make a circle with the threads. The *E. coli* chromosome is circular and is attached to the cell membrane.
5. A third person can now hold up the index card with the ×10,000 bacterial gene next to the DNA

loop. The average bacterial gene contains about 1,200 base pairs. Remember that a gene is composed of all the segments of DNA that instruct the cell to make a single protein, whether those segments are continuous or not.

6. With the bacterial chromosome and bacterial gene models still in view, the fourth person can hold up the index card with the ×10,000 bacterial plasmid. Plasmids carry one or a few genes necessary for their own replication and stability and often carry genes that give the bacterium important characteristics such as antibiotic resistance. You can see that the plasmid is tiny in comparison to the chromosome.

7. Now that you have compared the sizes of the chromosomes, plasmids, and genes, try to reconstruct the bacterium by inserting the "chromosome" into the capsule. It isn't easy! The real *E. coli* chromosome occupies about 10% of the cell volume.

Questions

1. How could two 10-m lengths of thread represent the *E. coli* chromosome more accurately?

2. How many bacterial genes would fit on your DNA circle (formed in step 4)?

3. Is the thread that you tried to stuff in the capsule too thick to represent the DNA's actual thickness? What is the reason for your answer? (Hint: What percentage of "bacterial cell" volume does the thread occupy in your model, and what is the actual volume that DNA occupies in *E. coli?*)

4. In this activity, how many meters of thread did it take to represent the *E. coli* genome? If the human genome is 1,000 times longer than the *E. coli* genome, how many meters would it take to represent the human genome? How many miles of thread would that be?

Mathematical Calculations

- The distance between DNA base pairs is 3.4×10^{-10} m.
- The *E. coli* chromosome contains about 3×10^6 base pairs.
- The average *E. coli* gene contains 1,200 base pairs.
- A typical plasmid contains about 3,000 base pairs.

Using the information given above, calculate the following.

5. How long (in meters) is the *E. coli* chromosome?

6. How long (in meters) is the average *E. coli* gene?

7. What is the circumference (in meters) of a typical *E. coli* plasmid?

8. If *E. coli* were magnified 10,000 times, how long (in meters) would its chromosome be?

9. If *E. coli* were magnified 10,000 times, how long would its average gene be?

10. If *E. coli* were magnified 10,000 times, how long would a typical plasmid be?

The human genome can be related to a length of railroad track. The railroad ties represent the base pairs, and the rails represent the sugar-phosphate backbone of the DNA molecule. The railroad ties are 2 ft apart.

11. The human genome contains 3×10^9 base pairs. How many miles of track will it take to represent the human genome?

12. The circumference of the earth is 24,000 miles. How many times would the railroad track representing the human genome wrap around the earth at the equator?

13. Another way to represent the size of the human genome is to relate the base pairs to characters

on a page in a book. Calculate how many of these books it would take to represent the human genome in the following manner.

- Choose a page in your text that is mostly print.
- Count the number of characters on five randomly selected lines. Find the average number of characters per line (C). Record C.
- Count the number of lines on the page (L). Record L. Calculate the average number of characters (N) on a page by multiplying C times L. Record N.

- Calculate the number of characters in your text (T) by multiplying N by the number of pages in your text. Record T.
- To determine how many books like your text it would take to represent the human genome if every character represented a base pair, divide the number of base pairs in the human genome (3×10^9) by the number of characters in your text (T). How many books would be required?

Extraction of Bacterial DNA

Introduction

In this activity, you will extract a visible mass of DNA from bacterial cells.

The preparation of DNA from any cell type involves the same general steps: (1) breaking open the cell (and nuclear membrane, if applicable), (2) removing proteins and other cell debris from the nucleic acid, and (3) doing a final purification. These steps can be accomplished in several different ways, and the method chosen generally depends on the purity needed in the final DNA sample and the relative convenience of the available options.

If a cell is enclosed by a membrane only, the cell contents can be released by dissolving the membrane with detergent. Cell membranes are made of proteins and fats. Just as detergent dissolves fats in a frying pan, a little detergent dissolves cell membranes. (The process of breaking open a cell is called cell lysis.) As the cell membranes dissolve, the cell contents flow out, forming a soup of nucleic acid, dissolved membranes, cell proteins, and other cell contents that is referred to as a cell lysate. Additional treatment is required for cells with walls, such as plant cells and many bacterial cells. These treatments can include enzymatic digestion of the cell wall material or physical disruption by means such as blending or grinding.

After cell lysis, the next step in a DNA preparation usually involves purification by removing proteins from the nucleic acid. Treatment with protein-digesting enzymes (proteinases) and/or extractions with the organic solvent phenol are two common methods of protein removal. Proteins dissolve in phenol, but DNA does not. Furthermore, phenol and water, like oil and water, do not mix but instead form separate layers. If you add phenol to an aqueous (water-based) DNA-protein mixture like a cell lysate and mix well, the protein dissolves in the phenol. After you stop mixing, the phenol separates from the aqueous portion, carrying the protein with it. The DNA remains in the aqueous layer. To remove the protein, simply remove the phenol layer. Following removal of protein, DNA is usually subjected to additional purification.

In this activity you will not attempt any DNA purification: your goal is simply to see the DNA. You will lyse *Escherichia coli* with detergent and layer a small amount of alcohol on top of the cell lysate. Because DNA is insoluble in alcohol, and it will form a white, weblike mass (precipitate) at the interface of the alcohol and water layers. By moving a glass rod up and down through the layers, you can collect the precipitated DNA. This DNA is very impure; the mass contains cellular proteins and other debris along with the stringy fibers of DNA.

Before you begin the DNA isolation, make sure you know whether to follow procedure 1 or procedure 2. They are essentially the same but differ in the volume of cells and the volumes and nature of the reagents you will use.

Procedure 1

1. Obtain from your teacher 2.5 ml of *E. coli* cells in a salt solution. Add 250 μl (1/4 ml) of 10% sodium dodecyl sulfate (SDS), and mix well. SDS is a detergent and an ingredient of many detergents we buy at the store, such as Woolite.
2. Your teacher will provide a 60 to 70°C water bath. Place each tube into the water bath for 15 min. *Note:* Maintain the water bath temperature above 60°C but below 70°C. A temperature higher than 60°C is needed to destroy the enzymes that degrade DNA.
3. Cool the tube (on ice if you have it) until it reaches room temperature.
4. For the DNA to be visible, it must be taken out of solution, or precipitated. Watch your teacher demonstrate the following technique. Use a pipette to carefully layer 6 ml of 95% ethanol (or isopropanol) on top of the suspension in each tube. The alcohol should float on top and not mix. (It *will* mix if you stir it or squirt it in too fast, so be careful.) Water-soluble DNA is insoluble in al-

cohol and precipitates when it comes in contact with it.

5. A weblike mass (precipitate) of DNA will float at the junction of the two layers (the interface). Push a rod through the alcohol into the soup and turn the rod. The rod carries a little alcohol into the soup and makes DNA come out of solution onto the rod. Keep moving the rod through the alcohol into the cell soup, and more DNA will appear. *Do not totally mix the two layers.*

Observe and draw the tube. Label the different substances in the tube. Answer the questions.

Procedure 2

1. Obtain from your teacher 4 ml of *E. coli* cells and 4 ml of broth in test tubes. Label the tubes. Shake your *E. coli* culture gently to resuspend the cells. Add to each labeled tube 3 ml of a 50% solution of dishwashing detergent in water. (Your teacher may substitute some other detergent.) Shake each tube to ensure complete mixing.

2. Your teacher will provide a 60 to 70°C water bath. Place each tube into the water bath for 15 min. *Note:* Maintain the water bath temperature above

60°C but below 70°C. A temperature higher than 60°C is needed to destroy enzymes that degrade DNA.

3. Cool the tubes to room temperature (on ice if you have it).

4. For the DNA to be visible, it must be taken out of solution, or precipitated. Watch your teacher demonstrate the following technique. Use a dropper to carefully layer 3 ml of 95% ethanol on top of the suspension in each tube. The alcohol should float on top and not mix with the cell lysate. (It *will* mix if you stir or squirt it in forcefully, so be careful.) Water-soluble DNA is insoluble in alcohol and precipitates when it comes in contact with it.

5. A weblike mass (precipitate) of DNA will float at the junction of the two layers (the interface). Push a rod through the alcohol into the soup and turn the rod. The rod carries a little alcohol into the soup and makes DNA come out of solution onto the rod. Keep moving the rod through the alcohol into the cell soup, and more DNA will appear. *Do not totally mix the two layers.*

Observe and draw both tubes. Indicate the substances in each tube. Answer the questions.

Questions

1. What information storage advantage(s) lies in DNA's double helix structure?

2. Why does the alcohol stay on top of the cell suspension and the broth in step 3?

B. Manipulation and Analysis of DNA

In the last 4 decades, our knowledge of DNA structure and function and of the biochemical processes cells use to modify the structure and carry out the functions has grown explosively. This extensive new knowledge has led to the ability to manipulate and analyze DNA outside the cellular environment. For example, we can cut DNA into specific pieces, separate and isolate those pieces, join pieces, copy DNA, and determine its base sequence. We are using our skills at manipulating genes to gather even more knowledge about how basic life processes are carried out and to make useful products for our daily lives.

In the activities in this section, you will use models and simulations to learn techniques scientists use to analyze DNA and make recombinant molecules: restriction digestion, ligation, gel electrophoresis, hybridization analysis, DNA sequencing, and the polymerase chain reaction. Puzzles, problems, and information are included throughout to illustrate how these techniques are applied to answer various kinds of questions. Also in this section are wet laboratory procedures that you can use to analyze DNA molecules yourself. The section concludes with an activity that shows how these techniques are currently used in DNA typing, followed by three short puzzles in which you will use DNA data to solve real-life mysteries.

Classroom Activities

DNA Scissors

Background Reading

Genetic engineering is possible because of special enzymes that cut DNA. These enzymes are called *restriction enzymes* or *restriction endonucleases*. Restriction enzymes are proteins produced by bacteria to prevent or restrict invasion by foreign DNA. They act as DNA scissors, cutting the foreign DNA into pieces so that it cannot function.

Restriction enzymes recognize and cut at specific places along the DNA molecule called restriction sites. Each different restriction enzyme (and there are hundreds, made by many different bacteria) has its own type of site. In general, a restriction site is a 4- or 6-base-pair sequence that is a palindrome. A DNA palindrome is a sequence in which the "top" strand read from 5′ to 3′ is the same as the "bottom" strand read from 5′ to 3′. For example,

5′ GAATTC 3′
3′ CTTAAG 5′

is a DNA palindrome. To verify this, read the sequences of the top strand and the bottom strand from the 5′ ends to the 3′ ends. This sequence is also a restriction site for the restriction enzyme called *Eco*RI. The name *Eco*RI comes from the bacterium in which it was discovered, *Escherichia coli* strain RY 13 (*Eco*R), and I, because it was the first restriction enzyme found in this organism.

*Eco*RI makes one cut between the G and A in each of the DNA strands (see below). After the cuts are made, the DNA is held together only by the hydrogen bonds between the four bases in the middle. Hydrogen bonds are weak, and the DNA comes apart.

```
                      ↓
Cut sites:   5′  GAATTC 3′
             3′  CTTAAG 5′
                      ↑

Cut DNA:     5′  G          AATTC 3′
             3′  CTTAA          G 5′
```

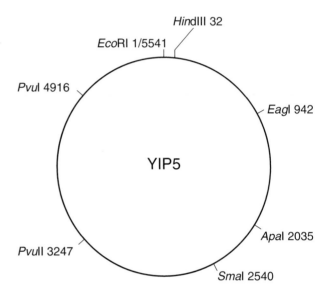

Figure 11.1   Restriction map of YIP5, a 5,541-base-pair plasmid. The number after each restriction enzyme name indicates at which base pair the DNA is cut by that enzyme.

The *Eco*RI cut sites are not directly across from each other on the DNA molecule. When *Eco*RI cuts a DNA molecule, it therefore leaves single-stranded "tails" on the new ends (see the example just given). This type of end has been called a "sticky end" because it is easy to rejoin it to complementary sticky ends. Not all restriction enzymes make sticky ends; some cut the two strands of DNA directly across from one another, producing a blunt end.

When scientists study a DNA molecule, one of the first things they do is to figure out where many restriction sites are. They then create a restriction map, showing the location of cleavage sites for many different enzymes. These maps are used like road maps to the DNA molecule. A restriction map of a plasmid is shown in Figure 11.1.

The restriction sites of several different restriction enzymes, with their cut sites, are shown on the next page.

```
              ↓                              ↓
EcoRI:  5′ GAATTC 3′    HindIII:  5′ AAGCTT 3′
        3′ CTTAAG 5′              3′ TTCGAA 5′
                  ↑                          ↑


              ↓                              ↓
BamHI:  5′ GGATCC 3′    AluI:     5′ AGCT 3′
        3′ CCTAGG 5′              3′ TCGA 5′
                  ↑                        ↑


              ↓                              ↓
SmaI:   5′ CCCGGG 3′    HhaI:     5′ GCGC 3′
        3′ GGGCCC 5′              3′ CGCG 5′
                  ↑                        ↑
```

Which ones of these enzymes would leave blunt ends? Which ones would leave sticky ends? Refer to this list of enzyme cut sites as you do the activity.

Exercises and Questions

Exercise 1

Cut the DNA sequence strips (Appendix A) along their borders. These strips represent double-stranded DNA molecules. Each chain of letters represents the phosphodiester backbone, and the vertical lines between base pairs represent hydrogen bonds between the bases.

1. You will now simulate the activity of *Eco*RI. Scan along the DNA sequence of strip 1 until you find the *Eco*RI site (refer to the list above for the sequence). Make cuts through the phosphodiester backbone by cutting just between the G and the first A of the restriction site on both strands. Do not cut all the way through the strip. Remember that *Eco*RI cuts the backbone of each DNA strand separately.

2. Now separate the hydrogen bonds between the cut sites by cutting through the vertical lines. Separate the two pieces of DNA. Look at the new DNA ends produced by *Eco*RI. Are they sticky or blunt? Write *Eco*RI on the cut ends. Keep the cut fragments on your desk.

3. Repeat the procedure with strip 2, this time simulating the activity of *Sma*I. Find the *Sma*I site, and cut through the phosphodiester backbones at the cut sites indicated above. Are there any hydrogen bonds between the cut sites? Are the new ends sticky or blunt? Label the new ends *Sma*I, and keep the DNA fragments on your desk.

4. Simulate the activity of *Hin*dIII with strip 3. Are these ends sticky or blunt? Label the new ends *Hin*dIII, and keep the fragments.

5. Repeat the procedure once more with strip 4, again simulating *Eco*RI.

6. Pick up the "front-end" DNA fragment from strip 4 (an *Eco*RI fragment) and the "back-end" *Hin*dIII fragment from strip 3. Both fragments have single-stranded tails of 4 bases. Write down the base sequences of the two tails, and label them *Eco*RI and *Hin*dIII. Label the 5′ and 3′ ends. Are the base sequences of the *Hin*dIII and *Eco*RI tails complementary?

7. Put down the *Hin*dIII fragment, and pick up the back-end DNA fragment from strip 1 (cut with *Eco*RI). Compare the single-stranded tails of the *Eco*RI fragment from strip 1 and the *Eco*RI fragment from strip 4. Write down the base sequences of the single-stranded tails, and label the 3′ and 5′ ends. Are they complementary?

8. Imagine that you have cut a completely unknown DNA fragment with *Eco*RI. Do you think that the single-stranded tails of these fragments would be complementary to the single-stranded tails of the fragments from strip 1 and strip 4?

9. An enzyme called **DNA ligase** re-forms phosphodiester bonds between nucleotides. For DNA ligase to work, two nucleotides must come close together in the proper orientation for a bond (the 5′ side of one must be next to the 3′ side of the other). Do you think it would be easier for DNA ligase to reconnect two fragments cut by *Eco*RI or one fragment cut by *Eco*RI with one cut by *Hin*dIII? What is your reason?

Exercise 2

Figure 11.1 is a restriction map of the circular plasmid YIP5. This plasmid contains 5,541 base pairs. There is an *Eco*RI site at base pair 1. The locations of other restriction sites are shown on the map. The numbers after the enzyme names tell at which base pair that enzyme cleaves the DNA. If you digest YIP5 with *Eco*RI, you will get a linear piece of DNA that is 5,541 base pairs long.

10. What would be the products of a digestion with the two enzymes *Eco*RI and *Eag*I?

11. What would be the products of a digestion with the two enzymes *Hin*dIII and *Apa*I?

12. What would be the products of a digestion with the three enzymes *Hin*dIII, *Apa*I, and *Pvu*I?

13. If you took the digestion products from question 10 and digested them with *Pvu*II, what would the products be?

DNA Goes to the Races

You have already learned about restriction enzymes and how they cut DNA into fragments. You may have even looked at some DNA restriction maps and figured out how many pieces a particular enzyme would produce from that DNA. But when you actually perform a restriction digest, you put the DNA and the enzyme into a small tube and let the enzyme do its work. Before the reaction starts, the mixture in the tube looks like a clear fluid. Guess what? After the reaction is finished, it still looks like a clear fluid! Just by looking at it, you can't tell that anything has happened.

For restriction digestion to mean much, you have to be able to see the different DNA fragments that are produced. There are chemical dyes that stain DNA, but obviously it doesn't do much good to add these dyes to the mixture in the test tube. In the laboratory, scientists use a process called **gel electrophoresis** to separate DNA fragments so that they can look at the results of restriction digests (and other procedures).

Gel electrophoresis takes advantage of the chemistry of DNA to separate fragments. Under normal circumstances, the phosphate groups in the backbone of DNA are negatively charged. In electrical society, opposites do attract, so DNA molecules are very much attracted to anything that is positively charged. In gel electrophoresis, DNA molecules are placed in an electric field (which has a positive and a negative pole) so that they will migrate toward the positive pole.

The electric field makes the DNA molecules move, but to cause the molecules to separate and be easy to look at later on, the whole process is carried out in a gel (obviously the source of the name *gel* electrophoresis). If you have ever eaten Jell-O, you have had experience with a gel. The gel material in Jell-O is gelatin; different gel materials are used to separate DNA. One gel material often used for electrophoresis of DNA is called *agarose,* and it behaves much like Jell-O but without the sugar and color. To make a gel for DNA (called *pouring* or *casting* a gel), you dis-

solve agarose powder in boiling water, pour it into the desired dish, and let it cool. As it cools, it hardens (sound familiar?).

Since the plan for agarose gels is usually to add DNA to them, scientists place a device called a *comb* in the liquid agarose after it has been poured into the desired dish and let the agarose harden around the comb. Imagine what would happen if you stuck the teeth of a comb into liquid Jell-O and let it harden. Afterwards, when you pulled the comb out, you would have a row of tiny holes in the solid Jell-O where the teeth had been. This is exactly what happens with laboratory combs. When the comb is removed from the hardened agarose gel, a row of holes in the gel remains (look at Figure 12.1). The holes are called *sample wells.* DNA samples are placed into the wells before electrophoresis is begun.

For electrophoresis, the entire gel is placed in a tank of salt water (not table salt) called buffer. An electric current is applied across the tank so that it flows through the salt water and the gel. When the current is applied, the DNA molecules begin to migrate through the gel toward the positive pole of the electric field (Figure 12.2). Figure 12.3 shows a scientist loading a DNA sample into an agarose gel sitting in an electrophoresis tank.

At this point, the gel does its most important work. All of the DNA in the gel migrates through the gel toward the positive pole, but the gel material makes it more difficult for larger DNA molecules to move than smaller ones. So in the same amount of time, a small DNA fragment can migrate much farther than a large one. You can therefore think of gel electrophoresis as a DNA footrace, where the "runners" (the molecules being separated) separate just like runners in a real race (Figure 12.4). The smaller the molecule, the faster it runs. Two molecules the same size run exactly together.

After a time, the electric current is turned off, and the entire gel is placed into a DNA staining solution. After

A   B

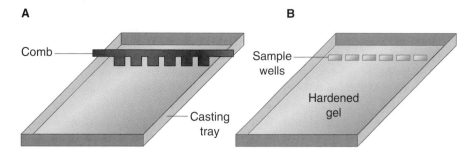

Comb ——   Sample ——
                                           wells

                                                    Hardened
                                                    gel
                          —— Casting
                             tray

Figure 12.1 Casting an agarose gel. (A) To make a gel, hot liquid agarose solution is poured into a casting tray (any shallow container), and the comb is put in place. (B) After the agarose cools and hardens, the comb is removed, leaving behind pits in the gel called sample wells. Samples are loaded into the wells prior to electrophoresis.

staining, the DNA can be seen. The pattern looks like a series of stripes (bands) in the gel; each separate band is composed of one size of DNA molecule. There are millions of actual molecules in the band, but they are all the same size (or very close to it). At any rate, after a restriction digest, there should be one band in the gel for each different-size fragment produced in the digest. The smallest fragment will be the one that has migrated farthest from the sample well, and the largest will be closest to the well, as shown in Figure 12.5.

Activity

In Appendix A, you have three representations of a DNA molecule and the outline of an electrophoresis gel. The representations show the cut sites of three different restriction enzymes on the same DNA molecule. You will simulate the digestion of this DNA with

each of the three enzymes and then simulate agarose gel electrophoresis of the restriction fragments.

1. Cut out the three pictures of the DNA molecule.
2. Simulate the activity of the restriction enzyme *Eco*RI on the DNA molecule that shows the *Eco*RI sites by cutting across the strip at the vertical lines representing *Eco*RI sites. You have now digested the molecule with *Eco*RI. Put your "restriction fragments" in a pile apart from the other two DNA strips.

Figure 12.3 A scientist is using a micropipette to load a DNA sample into an agarose gel. The gel is in an electrophoresis chamber full of buffer. The power supply for the chamber is on the laboratory bench behind the chamber.

Figure 12.2 In electrophoresis, the gel is placed in a tank of salt solution, and an electric current is applied. The DNA migrates toward the positive pole.

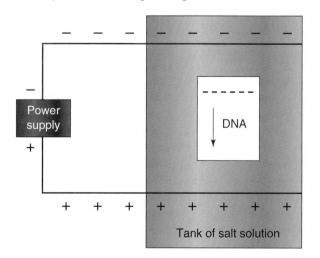

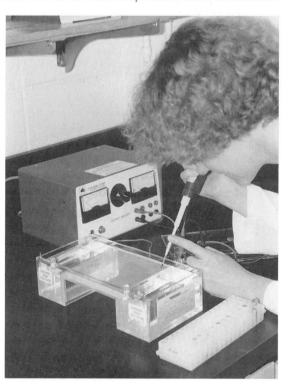

Figure 12.4 In electrophoresis races, the small DNA always wins!

3. "Digest" the second DNA strip with *Bam*HI. Put the *Bam*HI fragments in a separate pile.

4. Now "digest" the remaining DNA molecule with *Hin*dIII. Put these fragments in a third pile.

5. In our imaginary gel electrophoresis, you will separate the *Eco*RI, *Bam*HI, and *Hin*dIII fragments as if you had loaded the three sets of fragments into separate but adjacent sample wells. Arrange your fragments as they would be separated by agarose gel electrophoresis. Designate an area on your desk as the end of the gel with the sample wells. Starting with the *Eco*RI fragments, arrange them from longest to shortest, with the longest one closest to the well.

6. Next, separate the *Bam*HI fragments, and place them adjacent to the *Eco*RI fragments. Be sure to order the fragments correctly by size with respect to other *Bam*HI fragments and to the *Eco*RI fragments you have already laid out.

7. Repeat the same procedure for the *Hin*dIII fragments. You should now have all three of your sets of fragments arranged in order in front of you.

8. Look at the outline of the electrophoresis gel provided in Appendix A. Notice that it has a size scale in base pairs on the left-hand side and that sample wells are drawn in. Using the outline and the size scale as a guide for where to draw your fragments, draw the pattern your restriction di-

Figure 12.5 Gel electrophoresis is used to separate products of restriction digestion. (A) Restriction map, with fragment sizes in base pairs; (B) view of gel after electrophoresis.

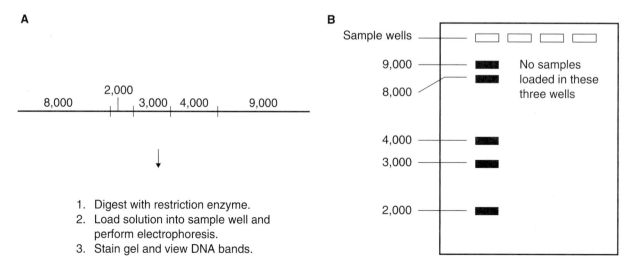

gest would make in the gel. Use the *Eco*RI sample well for the *Eco*RI fragments, and so on.

9. After you draw the bands representing the restriction fragments, use the size information on the paper DNA strips to label the bands on the gel with the sizes of the fragments in base pairs.

10. Use the fragment sizes as a check for your work.

Are all the smaller fragments across all the gel "lanes" in front of all the larger fragments? Did you notice that the size scale doesn't seem to have regular intervals? The size scale looks the way it does because agarose gels separate fragments that way.

Classroom Activities

Restriction Analysis of Lambda DNA

13

Introduction

In this activity, you will be carrying out a series of steps that molecular biologists and biotechnologists use very frequently in their work. You will separate restriction fragments of DNA by electrophoresis through an agarose gel, stain the fragments so that you can see them, and analyze the products of the digest. You will either use predigested DNA samples or set up and carry out the digestions yourself. The process from restriction digestion to analysis of gel patterns is called restriction analysis.

The procedure has been broken into several parts. The first part is setting up the restriction digests. If you are using predigested samples, you will skip this step. The remaining steps are preparing and loading the agarose gels, staining the gels, and analyzing the data. If you were in a laboratory all day (as scientists are), you could carry out the analysis in a single day. As it is, you will carry out a different part of the procedure each day for several days.

The DNA you will analyze is from bacteriophage lambda. Lambda is a virus that infects *Escherichia coli* and sometimes destroys it. Lambda DNA contains about 48,500 base pairs. You will cut lambda DNA with the enzymes *Eco*RI, *Hin*dIII, and, possibly, *Bam*HI. These enzymes will cut the DNA into a number of different-size pieces called restriction fragments.

You will separate the fragments by agarose gel electrophoresis. The mixture of fragments is loaded into a

well in the gel, then an electric current is applied. Because DNA is negatively charged, the fragments will migrate toward the positive electrode in the gel box. The shorter the fragment, the faster it can progress through the agarose. When you stop the electrophoresis (by turning off the current), the smallest fragment will be closest to the positive electrode, and the largest will be furthest away. The intervening fragments will be sorted by their sizes. You will stain the DNA so that you can see the fragments and analyze your results by comparing them with a restriction map of lambda (Figure 13.1).

Day 1: Restriction Digests

Each laboratory team will need

- A device for measuring microliter volumes
- Four sterile microcentrifuge tubes
- Waterproof marking pen or other way to label the tubes
- One holder or rack for microcentrifuge tubes

Obtain from your teacher microcentrifuge tubes containing

- 10× restriction buffer
- Uncut lambda DNA
- Sterile water

1. Label your four sterile microcentrifuge tubes *Eco*RI, *Hin*dIII, *Bam*HI (if you are using *Bam*HI), and Control. Set up the restriction digests according to the following instructions. Use Table

Table 13.1   Guide for setting up restriction digests

| Component | Amount (µl) in tube with: | | | |
	*Eco*RI	*Hin*dIII	*Bam*HI	Control
Water	4	4	4	5
10× buffer	1	1	1	1
Uncut DNA	4	4	4	4
Enzyme	1	1	1	None

13.1 to check off each component of the reaction as you add it to each tube.

2. Carefully add 4 μl of sterile water (from the microcentrifuge tube you got from your teacher) to the *Eco*RI, the *Hin*dIII, and the *Bam*HI tubes. When you add a droplet to the tube, touch the pipette tip to the side of the tube to deposit a small bead on the inside of the tube.

3. Add 5 μl of sterile water to the Control tube.

4. Using a fresh tip or capillary tube, carefully add 1 μl of 10× restriction buffer to each of the four tubes, checking them off in Table 13.1 as you go.

5. Using another fresh tip or capillary tube, carefully add 4 μl of uncut lambda DNA to each of the four tubes, checking them off as you go.

6. Your digests are ready for the enzymes. Your teacher has the enzymes on ice. When you are ready to add the enzymes, prepare your measuring device with a fresh tip or capillary tube and immediately and carefully add 1 μl of *Eco*RI to the *Eco*RI tube. Close the tube.

7. Change to a fresh tip. Immediately and carefully add 1 μl of *Hin*dIII to the *Hin*dIII tube. Close the tube.

8. Change to a fresh tip. Immediately and carefully add 1 μl of *Bam*HI to the *Bam*HI tube. Close the tube.

9. Mix the reagents in the closed tubes either by microcentrifuging them for a few seconds or by tapping the bottoms of the tubes gently on the top of your desk. Tap the bottoms of the tubes gently with your finger to ensure good mixing. Make sure the enzyme is mixed into the DNA solution.

10. Place the tightly closed tubes in a 37°C water bath or dry incubator. Incubate for 1 to several hours. Your teacher will put the tubes in the freezer for you.

Day 2: Gel Electrophoresis

Each laboratory team will need

- 30 to 50 ml of agarose solution
- Gel-casting tray
- Gel electrophoresis chamber
- Electrophoresis power supply (one per two groups)
- Gel-loading device
- Your own DNA digests and loading dye *or* precut DNA samples
- Microliter measuring device if you prepared your own restriction digestions
- Precut lambda DNA standard
- 1× TBE buffer

1. If appropriate, remove your frozen restriction digestions from the freezer, and let them thaw while you prepare the gel.

2. Cast the gel.

- Seal the ends of the tray with masking tape. Place the comb at one end of the tray, making sure it does *not* touch the bottom of the tray but is close to it.

- Pour enough agarose into the gel tray to cover the lower third of the comb. Allow the agarose to cool (it will become whitish and opaque).

- Remove the tape from the gel tray without damaging the ends of the gel. Do not remove the comb at this point. Place the gel tray into the electrophoresis chamber with the comb nearest the *negative* electrode end (black leads) of the chamber.

3. Fill the electrophoresis chamber with 1× TBE buffer. The buffer must completely cover the gel.

4. Carefully remove the comb from the gel. The holes left in the gel are the wells that you will fill with your restriction digests.

5. If you prepared your own restriction digests, add 2 μl of loading dye to each of your four tubes. Mix by microcentrifuging briefly and tapping the bottom of the tube (or by tapping the tube on your desk).

6. Load each sample into a separate well. *Be sure to draw a diagram of the gel and label which well contains which sample.* Also load one lane with 10 μl of the precut standard (obtain this from your teacher). Record its position.

7. Plug in the electrophoresis chamber to the power supply by connecting the red (positive) lead to the red electrode and the black (negative) lead to the black electrode. Be sure that the end of the gel with the DNA is connected to the negative pole of the power supply (even if you have to do the colors backwards).

8. Turn on the current to the level indicated by your teacher. Allow the second marker dye to migrate 2.5 to 3.5 cm into the gel. Depending on the voltage applied to the gel, this could take as little as 40 min (170 V) and as much as 24 h (10 to 12 V).

Day 3: Staining

Each laboratory team will need

- One gel-staining tray
- Methylene blue solution

1. Carefully remove your gel from the electrophoresis chamber, and place it in the staining tray.

2. Add enough methylene blue solution to the tray to just cover your gel. The stain solution will stain

your clothes as well as DNA, so be careful. If you get any on your hands, it will wear off in a few hours. There is no hazard from skin contact with blue DNA stains. Soak the gel in the stain for the amount of time indicated by your teacher.

3. Using a funnel, return as much as possible of the stain solution to the stain solution container. Rinse the gel carefully with dechlorinated tap water or deionized or distilled water.

4. Cover the gel with fresh dechlorinated tap water or deionized or distilled water and soak it for 3 to 5 min. Pour off the water.

5. Repeat step 4 until you can see the stained DNA bands clearly or until you run out of time. You will probably need to soak your gel overnight in a volume of dechlorinated, deionized, or distilled water sufficient to cover it.

Day 4: Data Analysis

Each laboratory team will need either photography equipment or centimeter-ruled graph paper, plastic wrap, and a needle or pin.

Make a permanent record of your data in one of three ways.

1. Photograph the gel with Polaroid film, and fasten the picture in your laboratory notebook.

2. Use the graph paper method to make an accurate reproduction of your gel.

3. Lay a transparent film over your gel, and carefully trace the bands.

If you photograph the gel, follow your teacher's instructions exactly. If you use the graph paper method,

1. Tape a piece of centimeter-ruled graph paper to your desk.

2. Cover the graph paper smoothly with plastic wrap. Tape down the plastic wrap.

3. Lay your gel on the plastic wrap, and line up the leading edge of the wells with a line on the graph paper.

4. Use the needle or pin to pierce through the gel at the center of the leading edge of each band. Also pierce through the leading edge of each well you used. The goal is to generate a pattern of pinpricks on the graph paper that will mark exactly the locations of the bands in the gel.

5. Remove the gel and plastic wrap. Using the pinpricks as a guide, draw the band patterns on the graph paper. Keep the graph paper in your laboratory notebook as a permanent record of your data.

Look at the restriction map of lambda (Figure 13.1). List in descending order (from largest to smallest) the sizes of the DNA fragments that should be generated in the following digests: *Eco*RI, *Hin*dIII, and *Bam*HI (if you used *Bam*HI). The ideal gel pattern is shown in Figure 13.2. Now sketch the pattern of bands from your gel. Label each gel lane with the name of the enzyme used in that digest. Compare the pattern of bands in your *Eco*RI lane with the expected set of fragments you just listed. Starting with the largest band in your gel, label each band with the size of the fragment in base pairs. Remember that two fragments very similar in size may not resolve and that very small fragments may have run off the gel or may not be detectable because they don't bind sufficient stain for you to see. Do the same for the *Hin*dIII fragments and the *Bam*HI fragments, if you used *Bam*HI. Discuss your band assignments with your laboratory teammates and your teacher.

When you are satisfied that you have made the best possible assignment of sizes to the bands, resketch your diagram with the size labels and put it in your laboratory notebook with the permanent record of your gel.

Questions

1. After labeling the fragment sizes in your gel pattern, compare your fragment patterns with the ideal fragment patterns shown in Figure 13.2. Account for differences in separation and band intensity between the experimental gel pattern and the ideal pattern.

2. Can you predict the fragment pattern for a digest using *Eco*RI and *Hin*dIII at the same time with only the information (the location of the bands) from the gels you ran? Why or why not?

(continued on page 170)

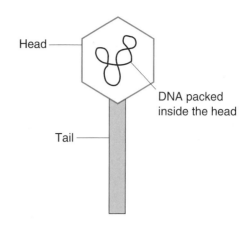

Head

DNA packed
inside the head

Tail

*Hin*dIII restriction map, with fragment sizes in base pairs

23,130			9,416			6,557	4,361

2,027   2,322   564 125

*Eco*RI map

21,226	4,878	5,643	7,421	5,804	3,530

*Bam*HI map

5,505	16,841	5,626	6,527	7,233	6,770

Figure 13.1 Bacteriophage lambda is the best known of the phages. It has a linear chromosome of approximately 48,500 base pairs and about 45 known genes. Its restriction map is shown.

Figure 13.2 Ideal gel pattern. The 125-base-pair *Hin*dIII fragment will never be seen. The 564-base-pair *Hin*dIII fragment is also usually not visible. It is not shown in the example.

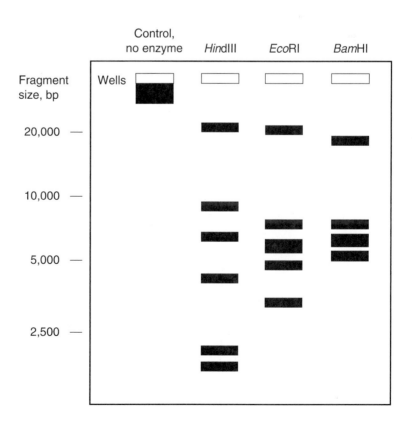

Classroom Activities

3. Use the lambda restriction map in Figure 13.1 to predict the size of fragments that would be generated in a double digest using *Eco*RI and *Hin*dIII.

4. If you used precut DNA, propose an experiment to test whether the restriction enzymes actually cut DNA. Be sure to describe exactly what you would use for control and experimental treatments. Assume that you have access to the restriction enzymes you wish to test.

5. If you did the restriction digest, what was the purpose of the control tube?

6. If you did the restriction digest, why did the control tube receive 5 μl of water instead of 4 μl?

7. The lambda restriction map in Figure 13.1 indicates that six *Eco*RI fragments should be generated, yet you probably saw only five in your gel. Why? If you used *Bam*HI, you expected six fragments, yet you probably saw only four. Why? Can you suggest a way to see all of the expected fragments?

Classroom Activities

Recombinant Paper Plasmids

Background Reading

Some of the most important techniques used in biotechnology today involve making recombinant DNA molecules. A recombinant object has been re-assembled from parts taken from more than one source. You could make a recombinant bicycle by dis-assembling two bicycles and reassembling them in a new way: putting the wheels of one on the frame of the other, for example. Your genome is recombinant in that part of it came from your mother and part came from your father. Recombinant DNA molecules are pieces of DNA that have been reassembled from pieces taken from more than one source of DNA. Often, one of these DNA sources is a plasmid.

Plasmids are small, circular DNA molecules that can reside in cells. Plasmids are copied by the cell's DNA replication enzymes because they contain a special sequence of DNA bases called an **origin of replication.** DNA replication enzymes assemble at this special sequence to begin synthesizing a new DNA molecule. As you might expect, bacterial and other chromosomes have replication origins, too. Replication origins are essential to heredity; if a DNA molecule does not have a replication origin, it cannot be copied by the cell and will not be transmitted to future generations.

Plasmids often contain genes for resistance to antibiotics. Antibiotics are natural substances produced mostly by soil microorganisms. Antibiotic production allows these microorganisms to kill off competing microbes. Antibiotic resistance is also a natural phenomenon; at the very least, the antibiotic producers must be resistant to the antibiotics they make! We will be working with genes for resistance to the antibiotics ampicillin and kanamycin.

In this activity, we will assemble plasmids carrying genes for ampicillin and kanamycin resistance and then recombine the two plasmids. We call the plasmid

with ampicillin resistance pAMP, the plasmid with kanamycin resistance pKAN, and the recombinant plasmid pAMP/KAN. We will use paper plasmid DNA models to go through the process that scientists use when making recombinant DNA. Scissors will substitute for restriction enzymes. The enzyme DNA ligase, which forms phosphodiester bonds between pieces of DNA, is represented by Scotch tape. Our result will be a model of a recombinant DNA molecule. Scientists place real recombinant plasmids back into bacteria where they multiply. The bacteria also multiply, making millions of copies of the recombinant DNA molecule and the proteins it encodes.

Construction of the pAMP and pKAN Plasmids

Locate the three strips of DNA code on the worksheet (Appendix A) marked "Paper pAMP plasmid model." On each strip, the two rows of letters indicate the nucleotide bases, and the solid horizontal lines indicate the sugar-phosphate backbone of the DNA molecule. The hydrogen bonds between the base pairs are located in the white space between the rows of letters.

1. Use your scissors to cut carefully along the *solid* lines. Cut out each strip, leaving the solid lines intact. Make a vertical cut to connect the open end of the box formed by the solid lines. This cut will remove the 5′ and 3′ from the strip.
2. After all three strips are cut out, glue or tape the "1" end to the "paste 1" area, covering the vertical lines. Connect "2" to "paste 2" and "3" to "paste 3" until you complete the circular model of the pAMP plasmid.
3. Using the page (Appendix A) marked "Paper pKAN plasmid model," cut out and paste together a plasmid containing a kanamycin resistance gene. The procedure is exactly the same as for the pAMP plasmid.

Constructing a Recombinant pAMP/KAN Plasmid

You have now prepared a pAMP plasmid and a pKAN plasmid. In this part of the exercise, you will use them as starting materials to make a recombinant plasmid. You will cut pAMP and pKAN with two specific enzymes, *Bam*HI and *Hin*dIII. You will ligate together fragments that come from each plasmid, creating a pAMP/KAN plasmid.

1. First, simulate the activity of the restriction enzyme *Bam*HI. Reading from 5′ to 3′ (left to right) along the top row of your pAMP plasmid, find the base sequence GGATCC. This is the *Bam*HI restriction site. Notice that it is a palindrome. Cut through the sugar-phosphate backbone between the two G's, stopping at the center of the white area containing the hydrogen bonds (don't cut all the way through the paper). Do the same on the opposite strand. Cut through the hydrogen bonds between the two cut sites, and open the plasmid into a strip. Each end of the strip should have a single-stranded protrusion with the sequence 3′ CTAG 5′. Mark the ends of the strip *Bam*HI.

2. Next, simulate the activity of the restriction enzyme *Hin*dIII. Reading from 5′ to 3′ (left to right) along the top row of the pAMP strip, find the sequence AAGCTT. This is the *Hin*dIII restriction site. It is also a palindrome. Cut the sugar-phosphate backbone between the two A's, stopping at the center of the white space containing the hydrogen bonds. Repeat the cut between the two A's on the opposite strand of the restriction site. Cut the intervening hydrogen bonds. This time you should have two pieces with single-stranded protrusions. One protrusion on each piece is the *Bam*HI end; the other should have the sequence 3′ TCGA 5′. Mark each of these two new ends *Hin*dIII. You now have two pieces of pAMP DNA. Set them aside.

3. Using your pKAN plasmid model, repeat steps 1 and 2. The pKAN plasmid is now in two pieces with labeled ends. Including the two pieces of pAMP, you now have four pieces of plasmid.

4. Take the piece of plasmid with the ampicillin resistance gene in it, and connect it to the piece containing the kanamycin resistance gene by using DNA ligase (Scotch tape). Be sure that complementary bases are paired where you make the ligations. Notice that the *Bam*HI end will not pair with the *Hin*dIII end but will pair with another *Bam*HI end. Likewise, the *Hin*dIII end must pair with another *Hin*dIII end.

Remember that DNA is three-dimensional. In our model, the letters representing the bases can look upside down, but in real DNA, it doesn't matter. So in this paper simulation, the letters representing the bases do not need to be right side up. All that matters is that the 5′-to-3′ directions match within a strand and that the base pairs are correct.

You have now created a recombinant pAMP/KAN plasmid.

Transformation

It is possible to introduce plasmids into bacterial cells through the process of transformation. Bacteria that can be transformed (can take up DNA) are called competent. Some bacteria are naturally competent. Others can be made competent by chemical and physical treatment. After the bacteria absorb the plasmid DNA, they express the new antibiotic resistance genes as instructed by the new DNA. Bacteria that express new proteins in this way are said to be transformed. New copies of the plasmid are synthesized by the cell's DNA replication enzymes and passed to daughter cells as the bacteria multiply. Because many identical copies of the new genes are generated in this process, you are said to have cloned the genes.

After transformation by plasmids containing antibiotic resistance genes, transformed bacteria can be detected by plating on media containing the antibiotic(s). Only bacteria expressing the new antibiotic resistance genes (the transformed bacteria) can form colonies on the antibiotic plates. This method of selecting transformants (because they are the only ones that can grow on the media) is a big advantage, because transformation is usually very inefficient. In a typical experiment, less than 1 cell in 1,000 will become transformed.

Questions

1. A *Bam*HI cut site will ligate only to the matching end of another *Bam*HI site. The same is true for *Hin*dIII cut sites and most other restriction enzyme cut sites. If you use two fragments at a time, how many possible different combinations could be formed from the four fragments you made in

exercise 3? How many of these could you detect in colonies of transformants on antibiotic-containing media?

2. Assume you did the activity with real DNA and attempted to transform bacteria with your new pAMP/KAN plasmid. Describe an experimental procedure for growing the transformed bacteria on plates to find out if the bacteria have actually been transformed by pAMP/KAN. Include controls that would tell you whether any of the experimental bacteria are alive and whether the antibiotics are effective.

3. Referring to question 2, describe an experimental procedure for finding experimental bacteria that were transformed by the other DNA molecules that formed during the ligation step (you described these molecules in question 1).

4. Scientists often combine an antibiotic resistance gene with whatever gene they are trying to clone. The desired gene is then associated with the antibiotic resistance gene. Any bacterium that contains the desired gene is then resistant to an antibiotic. In most cases, the scientists have no use for the protein that destroys the antibiotic. Why would scientists combine genes in this way?

Restriction Analysis Challenge

15

The combination of restriction digestion and gel electrophoresis is often called restriction analysis, since the information obtained from these procedures can be used to analyze the structure of a DNA molecule. Restriction analysis is especially important in checking the structure of recombinant DNA molecules and in analyzing an unknown DNA. The following examples show you some typical restriction analysis problems encountered by scientists in the laboratory.

Restriction Analysis Challenge I

You wish to insert a 1,500-base-pair (bp) *Eco*RI restriction fragment into the plasmid vector shown in Figure 15.1. You digest the plasmid DNA with *Eco*RI, stop the digest, add the 1,500-bp fragment, and ligate.

You know that this procedure will give you a mixture of regenerated starting plasmid and the recombinant molecule you desire. Outline a restriction analysis procedure you could use to distinguish between regenerated vector and the desired product. State the predicted fragment sizes. Use the gel outline provided to show the predicted products from your analysis (one product type per lane). Label each lane with the DNA molecule type (vector or recombinant) and the expected fragment sizes.

Restriction Analysis Challenge II

You wish to remove the 1,500-bp *Eco*RI fragment from the starting plasmid shown in Figure 15.2 and replace it with a *different* 1,500-bp *Eco*RI fragment

Figure 15.1 Information and gel outline for restriction analysis challenge I.

Gel outline

Starting plasmid vector

Distance between restriction sites is shown in base pairs.

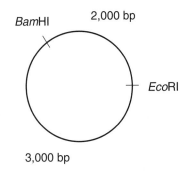

Fragment to be inserted into vector

*Eco*RI   1,500 bp   *Eco*RI

Starting plasmid

EcoRI fragment to be inserted

Distances between restriction sites are shown in base pairs.

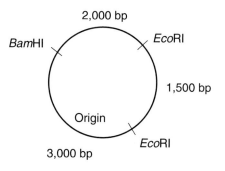

Figure 15.2   Information for restriction analysis challenge II.

(also shown). You digest the starting plasmid with *Eco*RI, stop the digest, add the new fragment to the mixture, and religate.

A. Draw the products you could generate that would have *one replication origin* and be *less than 7,000 bp in size.* (Hint: what are all the fragments present in the ligation reaction?) Label the products A, B, etc., and indicate distances between restriction sites.

B. Design a restriction analysis procedure that will let you identify as many of your proposed products as possible after a single digest. You may use more than one restriction enzyme in the digest. State the predicted fragment sizes for each one of your products from IIA.

C. Use the gel outline in Figure 15.3 to draw the predicted gel pattern from your proposed analysis of each of the products shown in IIA. Label the gel lanes with the product molecule (A, B, etc.). Label

Figure 15.3   Gel outline for restriction analysis challenge IIC.

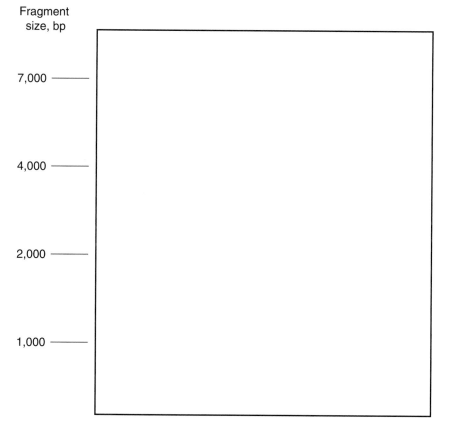

the bands with the fragment sizes. (A size scale has been provided.)

Restriction Analysis Challenge III

When scientists study a new piece of DNA, often one of the first things they do is put together a restriction map of that piece. The restriction map is then used as a road map when they study the genes encoded by the DNA. How is a restriction map put together? Digests with single enzymes and combinations of enzymes are performed. Observing the sizes of fragments produced, scientists determine restriction site locations that would give them the patterns they see. Here is one such puzzle to try for yourself.

A *linear* (not circular) piece of DNA is digested with the restriction endonuclease *Eco*RI and gives fragments of 3,000, 3,600, and 3,400 bp. When digested with *Bam*HI, the DNA molecule gives fragments of 4,500, 3,000, and 2,500 bp. A double digest with *Eco*RI and *Bam*HI gives fragments of 2,500, 500, 3,600, 3,000, and 400 bp.

Draw a restriction map of the starting piece of DNA, showing the locations of the *Eco*RI and *Bam*HI restriction sites. Indicate the distances between the restriction sites in base pairs, and label the sites *Eco*RI or *Bam*HI as appropriate. Also indicate the distance from each end of the starting piece to the nearest restriction site.

Classroom Activities

Detection of Specific DNA Sequences Part I. Fishing for DNA

Imagine this: you are a scientist staring at a test tube full of DNA that you just prepared. You want very much to know whether that DNA contains the gene for cystic fibrosis. You know a little about the cystic fibrosis gene, so how can you tell if it is in your sample?

Scientists often need to fish for particular pieces of DNA such as the cystic fibrosis gene. How do they do it? Once they know a little about the DNA they are looking for, it is not very hard.

When a scientist fishes for a particular piece of DNA, she uses a special DNA "hook" called a **probe**. A probe is usually a form of DNA called single-stranded DNA. Think of pictures of DNA or of models of DNA you have made. The base pairs are in the middle, and the two backbones are on the outside. Single-stranded DNA is only half of this structure: one backbone with bases sticking out from it (rather like a messenger RNA molecule; look at the picture below). If a piece of single-stranded DNA finds another piece of single-stranded DNA with the right base sequence, it will pair with it to make a regular DNA helix. Scientists use a special term for the pairing of two single DNA strands to make a double helix: **hybridization** (see Diagram 16.I).

To see whether a gene is present in a DNA sample, our scientist uses a probe that has the same sequence as part of the gene, and a label, such as a radioactive isotope, that allows the probe to be detected. She adds the probe to the DNA sample to see whether it will hybridize with anything. How does she know when it hybridizes? If the probe hybridizes to the sample, it will stick to the DNA in the sample. If not, it can be easily rinsed away. After the scientist rinses her sample, she exposes it to photographic film to determine if the probe has stuck to it. If any radioactive probe has stuck to the sample, it will expose the photographic film at that position. By examining the film, our scientist knows that the sequence of DNA in the probe is also present in her sample. Since the probe has the sequence of part of a gene, that gene is most likely in the sample DNA.

Activity

On Worksheet 16.I (see Appendix A), you have a 379-base-pair (bp) DNA sequence representing your sample. Every 10th base is in boldface type, and the base at the beginning of each row is numbered. You also have a short sequence that is a probe for the cystic fibrosis gene. You will use the probe to determine whether the cystic fibrosis gene is in your sample DNA.

Do you notice anything funny about your long DNA sequence? It is only one strand. To test whether a probe will hybridize to a DNA sample, it is necessary to separate the strands of the sample DNA so that the probe can find complementary base pairs. Your sample DNA has already been prepared.

1. Cut out the probe sequence.
2. Scan the sample DNA sequence to see whether the probe will hybridize to it.
3. Is the cystic fibrosis gene present in your DNA sample? If you found a place where the probe could hybridize, mark it on the DNA sequence.

Diagram 16.I   Hybridization of a probe to a sequence within single-stranded sample DNA.

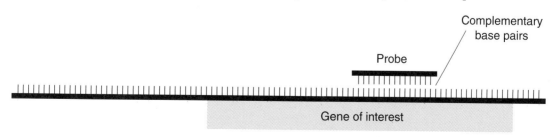

Questions

1. What is DNA hybridization?

2. How can you use hybridization to tell whether a certain DNA sequence is present in a DNA sample?

3. Below is a DNA sequence. Write out the sequence of a 10-bp probe that would hybridize to it.

5′ AATGCAGGCCCTATATGCCTTAACGGCATATGCAATGTACAATGCAAGTCCAACCGG 3′

Detection of Specific DNA Sequences Part II. Combining Restriction and Hybridization Analysis

Introduction

The hybridization analysis you learned about in *Fishing for DNA* can indicate whether a given DNA sequence is present in a sample. Sometimes this simple piece of information is all that is needed. However, a positive hybridization test does not give any information about where the sequence of interest is located within the sample DNA molecule.

To get information about the presence and location of a particular DNA sequence, scientists combine restriction analysis with hybridization analysis. Basically, the sample DNA is digested with a restriction enzyme (or enzymes), and the fragments are separated by electrophoresis. After electrophoresis, hybridization analysis is performed on the fragments. Only the fragment or fragments to which the probe can base pair will be detected (Diagram 16.II).

There are a few important technical details about this whole process. First, it doesn't work just to soak an agarose gel in a probe solution. In 1975, a scientist named Southern figured out a way to transfer DNA fragments from a gel directly to a membrane so that the exact pattern of the fragments in the gel was preserved. After the fragments were stuck on the membrane, they could be tested for hybridization to a probe. This method of transfer is called Southern blotting, and the combination of restriction analysis, transfer to membrane, and hybridization to probe is called Southern hybridization analysis.

Activity: Southern Hybridization Analysis

1. Take the sample DNA sheet from the previous exercise (Worksheet 16.I, Appendix A), and cut out the DNA strips by cutting along the dotted lines. Tape strip 1 to strip 2 to strip 3 through strip 10 to form one long linear molecule. This molecule could be a chromosome like one of yours.
2. Now simulate the activity of the restriction enzyme *Sma*I. This enzyme cuts the DNA sequence

5′ CCCGGG 3′ between the C and G in the middle of the sequence. Digest your chromosome by cleaving at every *Sma*I site. (Your fragments are single stranded in anticipation of the hybridization. In reality, the restriction digestion and electrophoresis are performed on double-stranded DNA, and then the two strands of the molecules are separated.)
3. Next, simulate electrophoresis of your fragments. Sort them by size, and lay them on your desk top as if they were in a gel.
4. Using the outlines provided on Worksheet 16.II (see Appendix A), draw the arrangement of fragments in your gel.
5. On your drawing of your gel, mark with an asterisk any band(s) that would hybridize to the probe.
6. Keep in mind that the pattern of DNA fragments in your gel will be exactly transferred to a membrane. On the membrane, the fragments will be tested for hybridization with the probe. Afterwards, only the fragment(s) that hybridizes to the probe will be detected (as in the accompanying diagram of Southern hybridization analysis). Draw what will be detected after hybridization in the box marked "Results of hybridization analysis."

Diagram 16.II Southern hybridization analysis shows which restriction fragments hybridize to a probe.

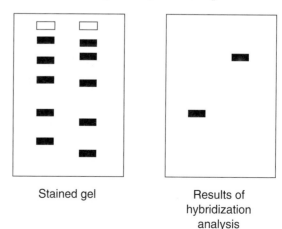

Stained gel

Results of hybridization analysis

Classroom Activities

Question

1. It is possible to cut a slice from an electrophoresis gel, purify the DNA in that slice, and clone it. Suppose that you have a probe for a viral gene that you would like to clone, but you do not know where it is within the 40,000-base-pair virus chromosome. How could you use Southern hybridization analysis to help you clone your gene?

Detection of Specific DNA Sequences Part III. Southern Hybridization

Problem A

You are analyzing the chromosome of a newly discovered virus, X. You have already constructed a restriction map of the 26,500-base-pair (bp) linear chromosome, shown in Worksheet 16.IIIA (see Appendix A). Since you are interested in DNA polymerase enzymes, you conducted a hybridization analysis on virus X DNA using as a probe a DNA polymerase gene from another virus you believe is closely related to X. To your delight, the probe hybridized to the DNA from virus X.

Now you would like to determine where the virus X DNA polymerase gene is located, so you perform a Southern hybridization analysis. You digest viral DNA with *Eco*RI and *Bam*HI separately, separate the fragments in a gel, transfer them to a membrane, and hybridize with the same DNA polymerase gene probe. Shown on Worksheet 16.IIIA is a picture of the fragments in the electrophoresis gel and the pattern seen after the hybridized probe is detected on the membrane.

Problem B

On Worksheet 16.IIIB (see Appendix A) is a restriction map of bacteriophage lambda. You digest some lambda DNA with the enzymes *Bam*HI and *Hin*dIII separately, load the fragments into an agarose gel, and perform electrophoresis.

Next, you transfer the fragments to a membrane and carry out hybridization analysis using the 4,878-bp *Eco*RI lambda fragment (refer to the map) as a probe.

1. Draw a picture of the electrophoresis gel, using the stained electrophoresis gel outline in Worksheet 16.IIIB (the two smallest *Hin*dIII fragments will run off the gel).
2. Indicate which fragments will hybridize to the 4,878-bp probe. (Hint: Even if there is only a partial overlap of the probe with a restriction fragment, the probe will still hybridize to that fragment.)
3. In the second outline, draw what would be seen after detection of the probe on the membrane.

Questions (Problem A)

1. Label the fragments on the gels with their sizes in base pairs.

2. Indicate on the restriction maps the region of the virus X chromosome in which the DNA polymerase gene is located.

3. How far to the left in terms of restriction sites could the gene lie? How far to the right?

4. Why would a DNA polymerase gene from another virus hybridize to DNA fragments from virus X?

DNA Sequencing: The Terminators

Determining the base sequence of a piece of DNA is a critical step in many applications of biotechnology. How is it done? The method most commonly used today employs compounds called chain terminators, a term that describes the lethal effect these compounds have on DNA synthesis, and nature's sequence readers, the DNA polymerase enzymes. These enzymes read the sequence of a single template DNA strand and synthesize a complementary strand. The chain terminators allow us to "look over the shoulder" of the DNA polymerase enzyme: we can see the order in which it adds bases to a new DNA strand and therefore deduce the sequence of the template strand.

What are chain terminators, and how do they block DNA synthesis? Chain terminators are molecules that closely resemble normal nucleotides but lack the essential 3′ hydroxyl (OH) group (Figure 17.1). In DNA replication, a nucleotide complementary to the tem-

plate base is brought into position. The DNA polymerase then adds this nucleotide to the growing DNA strand by forming a bond between the 5′ phosphate group of the new nucleotide and the 3′ OH group of the previous nucleotide (see Figure 17.2). DNA polymerases cannot synthesize DNA without a preexisting 3′ OH group to use as a starting point (this is referred to as a requirement for a primer; see chapter 7). If a DNA polymerase mistakenly adds a chain terminator instead of a normal deoxynucleotide, no further nucleotides can be added, and DNA synthesis is terminated.

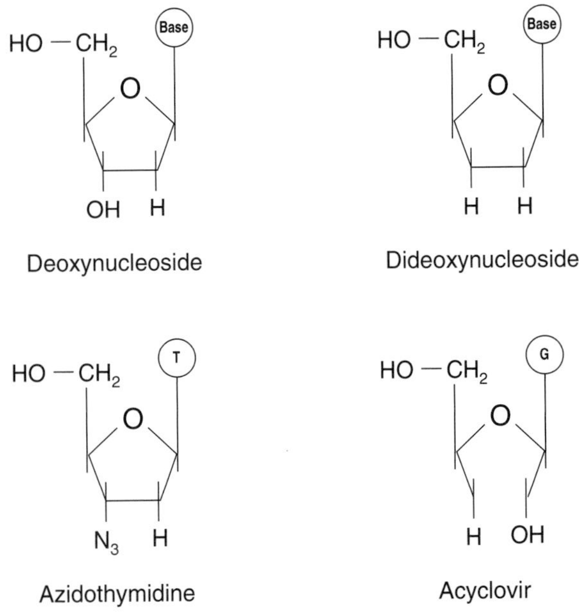

The chain terminators used in DNA sequencing are the dideoxynucleotides (Figure 17.1). How do dideoxynucleotides let us see the base sequence of a DNA molecule? To set up a sequencing reaction, the template DNA molecule to be sequenced is mixed with primers and radioactive normal deoxynucleotides (deoxyadenosine [dA], dG, dC, dT). This master mix is divided into four batches, and each batch is then "spiked" with a different dideoxynucleotide: dideoxyadenosine (ddA), ddG, ddC, or ddT. DNA polymerase enzyme is added, and it synthesizes new DNA strands on the template molecules. Occasionally, however, a dideoxynucleotide chain terminator is inserted in place of the analogous normal deoxynucleotide, and the synthesis of that DNA molecule is terminated. In the reaction mixture containing ddA (the A reaction), some percentage of the new molecules will get a ddA at each place there is a T in the template. The result is a set of new DNA molecules in which some of the molecules terminate at each T in the template. Similarly, in the G reaction, some of the new molecules will terminate at each C in the template. In the C reaction, some of the new molecules terminate at each G in the template, and in the T reaction, molecules terminate at each template A.

After the synthesis reactions are complete, the A-, G-, C-, and T-reaction mixtures are denatured by heating and loaded separately into adjacent lanes of a gel, and a current is applied (Figure 17.3). The newly synthesized molecules are separated by size and then visual-

Figure 17.1 Normal deoxynucleoside shown with chain terminators. All are incorporated into DNA from their triphosphate forms.

Deoxynucleoside

Dideoxynucleoside

Azidothymidine

Acyclovir

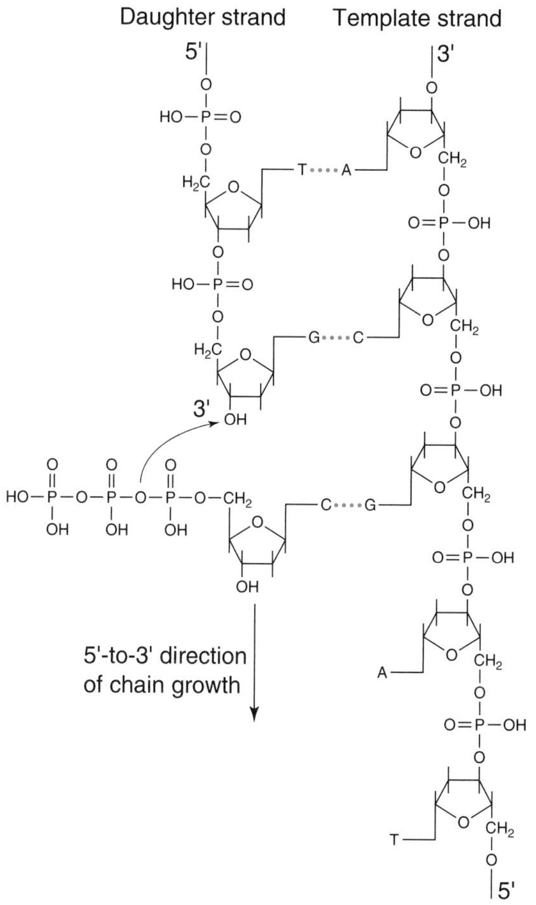

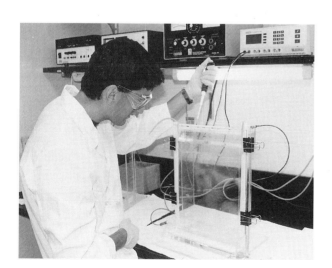

Daughter strand   Template strand

5'-to-3' direction
of chain growth

Figure 17.2   DNA replication. Base pairing between an incoming deoxynucleoside triphosphate and the template strand of DNA guides the formation of a new complementary strand.

Figure 17.3   A student loading a sequencing gel. A sequencing gel is a vertical gel: the DNA runs from top to bottom. It is tall and quite thin (less than 1 mm thick). Sequencing gels are made with acrylamide instead of agarose. Acrylamide forms a much tighter mesh than agarose, making it possible to separate DNA molecules.

ized by exposing the gel to photographic film (remember that the new nucleotides are radioactive). The A-reaction lane will show bands that correspond in length to the sites of the T's in the template. The G-reaction lane shows bands whose lengths correspond to the sites of template C's, and so on. The sequence of the new DNA strand is "read" from the sequencing gel by starting at the bottom (the shortest new molecule) and reading upward (see below). Figure 17.4 shows an autoradiogram of a sequencing gel.

If this written explanation seems confusing, don't worry. The sequencing simulation you will do shows how the whole approach works. After you complete the simulation, go back and reread the introductory material, and then answer the questions.

Procedure

1. You will be assigned the base A, G, C, or T. You will conduct the sequencing reactions with a terminator that substitutes for your base. Each base will

Figure 17.4   Autoradiogram of a sequencing gel. The scientist who did this sequencing procedure was screening several plasmids from transformants to see if she had gotten the clone she was trying to construct. She did. In fact, every one of the plasmids had the sequence she wanted.

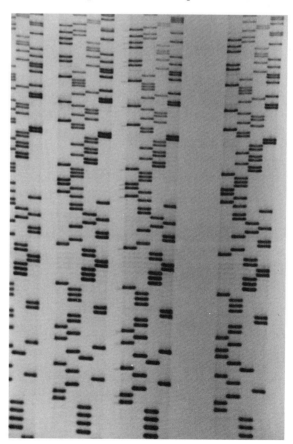

be assigned a specific color, and the terminator will be assigned a color.

2. Obtain four paper cups and label them A, G, C, and T. Put 20 paper clips of the correct color in each cup *except* for the one containing your assigned base. In that cup, put 16 paper clips of the correct color and 4 of the terminator color and mix them thoroughly. The terminator color represents dideoxy molecules.

3. Line up the four template molecules on your desk, with the paper cups in easy reach. Be sure the "spiked" paper clips (the ones with dideoxynucleotides) are well mixed. Begin to synthesize new DNA from the 3′ end of the paper primer. Since the first template nucleotide is an A, draw out a complementary nucleotide (a T) from the correct paper cup and bond it to the paper primer (by piercing the paper). Do this four times, once for each template molecule. The drawings must be done at random (eyes closed). If you have the T reaction and pull out a terminator paper clip, add the terminator paper clip to the primer and then set that template aside. No more nucleotides can be added to it.

4. Repeat step 3 for the next base in the template (the second A), linking the second T paper clip to the first. Again, if you randomly select a terminator paper clip, that paper clip should be added to

the growing DNA strand, but then that template must be set aside and no more nucleotides can be added to it. The terminator paper clips represent dideoxynucleotides that lack the 3′ hydroxyl group for forming a new bond.

5. Continue down the template to the end.

6. Now it is time to load the gel and perform electrophoresis. DNA sequencing reactions are denatured before loading, so "denature" your molecules by cutting the paper primer away from the paper template but leaving the paper primer attached to the paper clips.

The A-reaction products will be loaded and run in the A lane, the G-reaction products in the G lane, and so forth. Imagine that you have loaded your products into the appropriate lane and that electrophoresis has occurred. You will place your reaction products in the correct gel lane according to their sizes.

7. When all the reaction products are arranged in the gel, read the sequencing gel. Starting at P+1, identify the lane the band is in (T). Continue with P+2, P+3, etc., writing the sequence as you go. Compare the base sequence you write with the sequence of the newly synthesized DNA strand. Compare it with the sequence of the paper template.

Questions

1. Would the long template molecule (represented by the paper DNA strand) show up in the sequencing gel?

2. If a cell culture were "fed" some dideoxynucleotides in media, what cellular process, if any, would be most affected?

3. (Optional: refer to the additional reading.) What chain terminator is currently used in DNA sequencing and as an anti-AIDS drug?

Chain Terminators as Antiviral Drugs

Chemicals that were first used as investigative tools in research laboratories often find their way into the pharmacy. Compounds that affect fundamental biological processes are important to basic research and sometimes turn out to be therapeutically useful as well. The chain terminators are a good example of such compounds. As shown in the previous activity, chain terminators are essential to the now-favorite method of DNA sequencing. In addition, three of the currently available anti-AIDS drugs and the best available antiherpes drug belong to this class of compounds.

How are chain terminators used in fighting AIDS and herpesvirus infections? They fight viral infections in the same way they assist in DNA sequencing: by terminating DNA synthesis. For a virus to establish an infection, it must replicate its nucleic acid. Blocking this replication is potentially a very effective way to fight the spread of a virus, but blocking DNA synthesis could be as harmful to the patient as to the virus invader. Herpesviruses and the AIDS virus are good candidates for antireplication drugs because they encode their own DNA-synthesizing enzymes that have unique properties that can be exploited.

The AIDS virus (human immunodeficiency virus [HIV]) is an RNA virus. It encodes a special enzyme, reverse transcriptase, that synthesizes DNA by using the viral RNA as a template. This step is essential to HIV infection. Chain terminators fight the spread of HIV in the body by interfering at this stage. Reverse transcriptase is a good target for chain terminator drugs because it is a sloppy enzyme: it is much more likely to incorporate an incorrect nucleotide or a chain terminator than is the human polymerase. Furthermore, reverse transcriptase lacks the ability to proofread its work, so it cannot remove an incorrect nucleotide once it is incorporated.

There are more issues involved in developing a successful chain terminator drug than simply the sensitivity of the viral replication enzymes. The drug must survive in the body and be absorbed by the proper cells. The active triphosphate form of the chain terminators (shown in Figure 17.2 for the nucleotide being added to the growing chain) is not absorbed by cells, so the drugs are given in unphosphorylated form (as shown in Figure 17.1). Once these compounds enter the cell, the human enzymes must add the phosphate groups (phosphorylate the molecules) to activate them. Different compounds are absorbed and phosphorylated with different efficiencies. So the development of a new drug depends not only on its toxicity to the virus and harm to the host but also on how it is metabolized in the human system.

Three chain terminators that are currently employed against the AIDS virus are azidothymidine (AZT), dideoxycytosine (ddC; the same compound used in DNA sequencing), and dideoxyinosine (ddI) (Figure 17.1). AZT, an analog of thymidine, is incorporated into a growing DNA molecule in place of normal thymidine. Likewise, ddC is incorporated in place of cytosine. The compound inosine is a nucleoside analog that is identical to adenosine except for the absence of one amino group. Cells synthesize inosine and convert it directly to adenosine. When ddI is given as a drug, it enters cells and is rapidly converted to ddA. The ddA is incorporated by reverse transcriptase in place of normal adenosine.

These drugs have toxic side effects in patients. Although human DNA polymerase is less sensitive to the drugs than reverse transcriptase is, the drugs do affect DNA replication in normal human cells. AZT is particularly toxic to the bone marrow, while ddC and ddI affect the peripheral nerves.

Another class of anti-AIDS drugs that has been developed in recent years targets a second HIV enzyme: protease. HIV infects cells as an RNA genome packaged with the enzymes reverse transcriptase and integrase inside a protein envelope. After infection, reverse transcriptase makes a double-stranded DNA copy of the RNA genome. The integrase enzyme integrates the DNA into the host cell's chromosome. There, the host cell's RNA polymerase transcribes the

HIV genes, and the host ribosomes translate the mRNA into protein.

The transcribed protein, however, is not the active form. The HIV protease must cleave the protein into its active pieces for them to function. Several drugs have been developed that block the function of this protease enzyme. These drugs are referred to as protease inhibitors. Current AIDS therapy usually involves two nucleoside analogs and at least one protease inhibitor. This therapeutic regimen has extended the life expectancy of AIDS patients and improved its quality.

The herpesviruses are DNA viruses with relatively large genomes. They encode many of their own DNA replication enzymes, including a DNA polymerase. After the initial infection, herpesviruses remain in the body in an inactive (latent) state from which they can be activated and can then cause subsequent outbreaks of disease. Herpes simplex viruses 1 and 2 cause fever blisters and genital herpes, respectively. The herpesvirus varicella-zoster causes chicken pox when it first infects a person and shingles in subsequent outbreaks.

Herpesvirus diseases can now be treated with the chain terminator acyclovir (Figure 17.1). Acyclovir (marketed under the name Zovirax) is relatively nontoxic to the human host because human cells take up acyclovir but do not phosphorylate it, so the drug remains inactive. Herpesviruses, however, encode an enzyme that does phosphorylate acyclovir. Therefore, acyclovir becomes an active chain terminator only in herpesvirus-infected cells. Acyclovir is an analog of guanidine and is incorporated opposite cytosine residues in the template. Once incorporated, it terminates further DNA synthesis and inhibits the activity of the herpes polymerase.

The Polymerase Chain Reaction: Paper PCR

Introduction

One of the difficulties scientists often face in the course of DNA-based analysis is a shortage of DNA. A forensic scientist may have only a tiny drop of blood or saliva to test. An evolutionary biologist may want to analyze DNA from a museum specimen without destroying the specimen. Even if ample amounts of tissue or numbers of sample cells are available, it takes some work to purify specific DNA fragments in large quantities.

In 1985, a new technique that changed the whole picture was introduced. This technique essentially allows a scientist to generate an unlimited number of copies of a specific DNA fragment. It was invented by biotechnology industry scientist Kary Mullis, who had the initial inspiration one night in 1983 as he was driving and thinking about a technical problem he faced at work.

The essence of Mullis's idea was this. If you could set up a test tube reaction in which DNA polymerase duplicated a single template DNA molecule into 2 molecules, then duplicated those into 4, then duplicated those into 8, then 16, then 32, etc., you would soon have a virtually infinite number of copies of the original molecule. Each round of DNA synthesis would yield twice as many molecules as the previous round: a chain reaction producing specific pieces of DNA. Mullis's new technique was called the polymerase chain reaction, or PCR.

Of course, Mullis did more than just realize that DNA polymerase can copy one DNA helix into two. You already knew that, too. What he did was figure out how to generate a chain reaction in a test tube and how to get the reaction to copy the DNA segment of the scientist's choosing.

Mullis's approach relies on the characteristics of DNA polymerase enzymes and on the process of hybridization. Recall that DNA polymerases must have a primer base paired to a template DNA strand so that they can synthesize the complement to the template strand. Also remember that hybridization is the spontaneous formation of base pairs between two complementary single strands: you can separate the two strands of a helix by heating, but if you then allow the mixture to cool, the base pairs between the strands will re-form.

Now, how did Mullis use DNA polymerase and hybridization to get a chain reaction of DNA synthesis? Refer to the diagram in Figure 18.1 during this explanation.

First, you decide what DNA segment you wish to amplify (scientists say *amplify* instead of duplicate, because they are making so many copies). Then you synthesize two short single-stranded DNA molecules that are complementary to the very ends of the segment. These two short molecules must have specific characteristics. Look at the first panel in Figure 18.1 under round 1. It shows a double-stranded parental DNA molecule with two copies each of the two short, single-stranded DNAs. Each of the single-stranded molecules is complementary to only one strand of the parental DNA, and each one is complementary to only one end of the segment. Furthermore, if you imagine these short molecules base paired to the complementary regions in the duplex, their 3′ ends would point toward each other. These short, single-stranded molecules are the *primers*.

To begin the chain reaction, a large number of primers are mixed with the template molecule in a test tube containing buffer and many deoxynucleotide triphosphates. (What are they for?) This mixture is heated almost to boiling, so that the two strands of the parental molecule denature (Denaturation in Figure 18.1).

Next, the mixture is allowed to cool. Ordinarily, the two strands of the parental DNA molecule would eventually line up and re-form their base pairs. However, there are so many molecules of primers in the mixture that the short primers will find their complementary sites on the parental strands before the two

Round 1

Double-stranded parental DNA

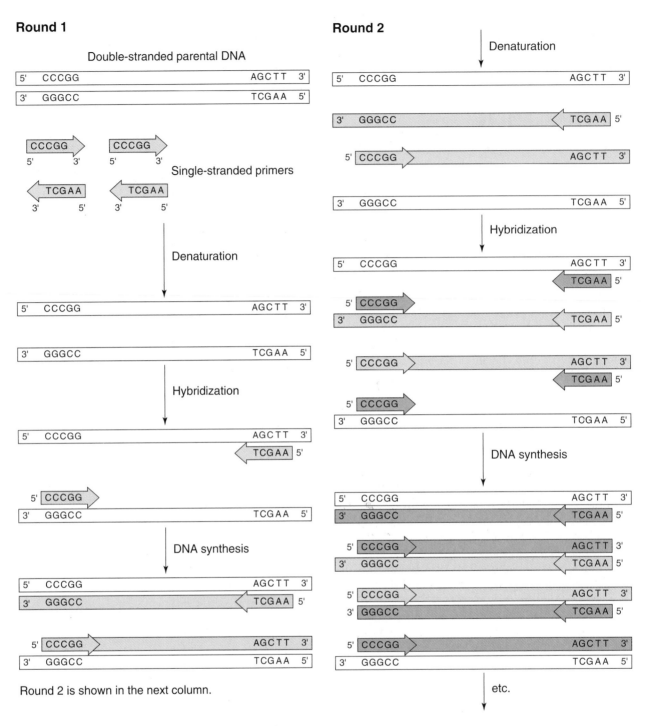

Round 2 is shown in the next column.

Round 2

Figure 18.1   PCR.

parental strands can line up correctly for base pairing. So a primer molecule hybridizes to each of the parental strands (Hybridization in Figure 18.1).

Now DNA polymerase enzyme is added. The primers hybridized to the single-stranded parental molecules meet the requirements of the enzyme for DNA syn-

thesis. DNA polymerase begins adding the correct deoxyribonucleotides to the 3′ ends of the primers, forming new complementary strands (DNA synthesis in Figure 18.1).

After a short time, the mixture is heated up again. Now the two new double-stranded molecules denature, leaving four single strands (Denaturation, round 2). The mixture is cooled, and the abundant primer molecules hybridize to the single-stranded molecules (Hybridization, round 2). DNA polymerase is added again, and new deoxynucleotides are added to the 3′ ends of the hybridized primers, yielding four double-stranded molecules (DNA synthesis, round 2). Notice that two of the newly synthesized strands begin and end at the primer hybridization sites.

This process of denaturation, hybridization, and DNA synthesis is repeated over and over, often 25 to 30 times, yielding huge numbers of molecules. The overwhelming majority of the newly synthesized molecules reach exactly from one primer hybridization site to the other. So by choosing the primers, a scientist controls which segment of the parental molecule is amplified. PCR is now used routinely for many different purposes: to amplify a specific fragment of DNA for cloning, to generate a DNA fingerprint from a minute sample, even to diagnose diseases.

One technical improvement to the process outlined previously has made performing PCR even easier. Did you notice that we added DNA polymerase before each DNA synthesis step? That is because PCR was first carried out with *Escherichia coli* DNA polymerase, which is rendered inactive at the high denaturation temperature. So more enzyme had to be added for each round of synthesis. However, some organisms inhabit the very hot waters of hot springs and thermal ocean vents. The DNA polymerase enzymes from these organisms are not inactivated by the temperatures required for DNA denaturation. Now PCR is carried out with heat-resistant DNA polymerase, so the enzyme can be added only once at the beginning of the reaction cycles.

When PCR was first developed, scientists preset three water baths to the temperatures required for denaturation, hybridization, and DNA synthesis. They performed PCR by simply moving the reaction tube from one water bath to another. It wasn't long before enterprising biotechnology companies manufactured incubators that rapidly cycled between the desired temperatures, eliminating the need for manually moving tubes. These *thermal cyclers* have further simplified the performance of PCR. PCR can now be carried out by mixing parental DNA, primers, buffer, deoxynucleotides, and heat-resistant polymerase in a reaction tube; placing the tube in the thermal cycler; programming the thermal cycler to the desired time and temperature specifications; and waiting for the cycles to finish. It usually takes a few hours for many cycles of amplification. Figure 18.2 shows a scientist placing a reaction tube into a thermal cycler in his laboratory.

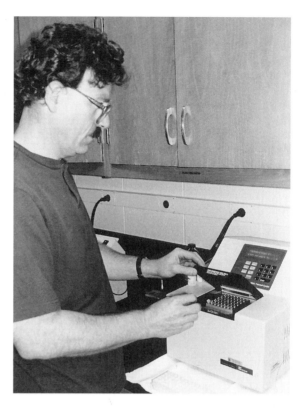

Figure 18.2 A scientist is loading a reaction tube into an automated thermal cycler (PCR machine). The digital keypad on the machine is for programming the desired temperatures, times, and number of reaction cycles.

When we (the authors writing this material) and other scientists first learned about PCR, the idea made perfect sense. However, we did not completely get it until we had worked through several reaction cycles ourselves, drawing the parental DNA molecules, the primers, the products, etc. Because we really learned PCR by working through it, we have provided a paper simulation for you to do the same thing. The template and primers are included in this chapter. Your instructor has directions, though you may be able to figure it out yourself. We have also included materials to simulate a PCR-based diagnostic test.

If you do not do the paper simulation, we highly recommend that you get some clean paper and draw through four rounds of PCR yourself. Figure 18.1 should get you started. We like using different colors of pens to keep track of the parental DNA, the primers, and the newly synthesized molecules (remember, primers will form part of the new molecules).

Here are some questions to think about.

Questions

1. What would a scientist have to know before she could design a PCR-based diagnostic test for virus X?

2. How could a scientist help ensure that her primers will hybridize only to the DNA she wishes to detect?

3. Write an expression that predicts the number of product molecules generated from a single double-stranded DNA molecule after n rounds of synthesis.

4. Predict the number of product molecules that are *not* the short primer-to-primer strands, that would be generated from one double-stranded DNA molecule after n rounds of synthesis.

5. Could you amplify DNA given only one primer? What would the products be after one round? Two rounds? Four rounds?

Analyzing Genetic Variation: DNA Typing

Introduction: Studying Variation

We all recognize that everyone is different. Every day, we rely on differences in appearance, sound, or gait to recognize our friends and family. At work, school, and play, we observe people and note their great variety. In an informal way, we all study human diversity.

Some people literally make a career out of studying human diversity and similarity. These people may be scientists seeking to understand human evolution, physicians looking at the prevalence of genetic diseases, social workers trying to identify the father of a child, or law enforcement agents trying to determine the identity of a violent criminal. All of them need precise ways to measure differences between individuals. How do they do it? Over the years, the techniques they use have evolved from a reliance on general appearance to measurement of specific phenotypes to, finally, examining the genotype itself. Let's look at what that means.

In the past, investigators of human diversity had to rely completely on outward traits such as eye color, ear shape, and the like. As you can imagine, information like this would not usually be precise enough to prove that a particular man fathered a child or to predict the occurrence of genetic diseases and would not even be available in the case of an unwitnessed murder.

As we learned more about biology, it became possible to get a closer view of individuals by looking at specific proteins rather than overall appearance. For example, we learned that although everyone has blood, not everyone's blood is the same. Certain proteins on blood cells are different in different people, giving rise to blood types, and these blood types can be determined precisely. In recent years, an examination of proteins such as those that cause blood types has been very helpful in studying the amount of similarity in different groups of people, in ruling out suspects in crime cases, and in ruling out potential fathers in paternity cases.

Now we know that individuals have even more differences in their DNA than show up in their proteins. For example, an individual with blood type A could have the genes AA or AO. Even more important, we have learned that most of the differences in people's DNA may not even be in their protein-encoding genes but seem to be in the noncoding regions between their genes. Two individuals with blood type A might seem the same on that criterion, but an examination of their blood group genes and the surrounding DNA could reveal many differences.

It is very clear that the bottom line on inherited differences is in DNA. As a result of developments in the field of biotechnology, we can now look directly at DNA to study differences in various populations, to screen for disease genes, to determine paternity, and to identify individuals.

Analyzing differences at the DNA level

Everyone's DNA is different (with the exception of identical siblings). Therefore, it is easy to talk about identifying people by looking at their DNA, but when you begin to think about actually doing that, many questions arise. For example, humans have 3 billion base pairs of DNA. Obviously we cannot look at all of them. Which ones should we look at? Are differences between people scattered uniformly over the chromosomes? Are some sections of the human genome likely to be more dissimilar from person to person than others? How much DNA do we need to look at to be sure that two samples are really the same? Just how different is the DNA from different people anyway?

Based on comparisons of many DNA sequences from many individuals, the average number of differences is estimated to be one for every 300 to 1,000 base pairs. Using the conservative end of this estimate, we can determine that a genome of 3 billion base pairs would show 3 million differences between two unrelated individuals.

Where can the differences be found? Again, studies show that the differences in people's DNA are not evenly distributed. Some regions of the genome are very similar between individuals, and other regions are highly variable. In general, these highly variable regions seem to be in noncoding DNA.

What is the nature of the differences? One type of difference is simply base changes, such as C to T or A to G. These can be used in DNA-based identification, as described below. A second kind of difference is also widely used in DNA-based identification. Human DNA contains end-to-end repeats of different short DNA sequences (from as few as 2 bases up to 30+ bases long). In some regions of the genome, the number of repeats varies highly from individual to individual. The abbreviations VNTR (variable number tandem repeat) and STR (short tandem repeat) are used to refer to these areas. DNA-based identification, or DNA typing, focuses on highly variable regions of the genome, whether the variety comes from many base changes or from a tandemly repeated sequence. In DNA typing, several of these highly variable regions from an individual are characterized, and a DNA "profile" is generated. The DNA profile can then be compared with other profiles (such as from potential fathers or from crime scene evidence). Let's see how the profiles are generated.

DNA typing

How do you examine DNA for typing? The immediate answer might seem to be, "Determine the DNA sequence." At this time, however, DNA sequencing is not a practical approach because of the time it takes and the cost involved. Instead, scientists use other methods to look at differences in people's DNA. You have already learned about the laboratory techniques involved. One of the approaches uses restriction enzymes and Southern hybridization; the other uses the polymerase chain reaction (PCR), usually followed by hybridization.

The basic approach to DNA typing by Southern hybridization analysis is to identify regions of the human genome that when cut with a particular restriction enzyme generate varying sizes of fragments from individual to individual. When different sizes of restriction fragments are generated from the same region of the genome in different people, we say we have found a restriction fragment length polymorphism, or RFLP ("poly" = many; "morph" = form; therefore, "polymorphism" = many forms).

An RFLP can arise in one of two ways. First, in a chromosome region with many changes in the base sequence, restriction sites can be created or destroyed

by some of those changes. For example, the enzyme *Hae*III cuts the sequence GGCC between the GG and CC. Any base change in the sequence would destroy the site. Conversely, if the A in the sequence GACC were mutated to a G, a site would be created. Gain or loss of restriction sites would obviously change both the number and the sizes of restriction fragments generated in a digest.

The second way that different-size fragments can arise from the same region of a chromosome (from different people) is through a VNTR or STR. Imagine a VNTR region with a *Hae*III site on either side of it. Obviously, the length of the *Hae*III fragment containing the VNTR depends on how many repeats are there. The more repeats, the longer the fragment.

Because there are many VNTR and STR regions throughout the human genome, RFLP analysis based on these repeated regions is widely used for DNA profiling. Exercise 1 shows an example of an RFLP created by a VNTR.

To type DNA by using RFLP analysis, a laboratory worker digests the DNA sample with restriction enzymes, runs a gel to separate the fragments, and then performs Southern hybridization with a DNA probe that hybridizes to the highly variable region. The probe reveals the pattern of restriction fragments from that region. Often, the analysis is performed with multiple probes that reveal RFLPs from several different highly variable regions. The result is a DNA profile. If many highly variable regions of the genome are analyzed per sample, it is highly unlikely that two profiles from unrelated individuals will ever match by chance (see below).

Why not just look at the band pattern revealed by staining and skip the hybridization? Remember the size of the genome: 3 billion base pairs. If an average restriction fragment were 1,000 base pairs long, there would be 3 million of them. There is no way to look at individual bands by staining: you would only see, all the way down the gel lane, a smear caused by hundreds, thousands, or millions of overlapping fragments. DNA typing depends on the use of methods like hybridization and PCR, which allow you to look only at specific regions of the genome.

DNA typing by PCR is based on VNTRs and STRs. PCR primers that hybridize to the DNA sequence on each side of a tandem repeat are used to amplify the area containing the repeats. The PCR cycle produces millions of copies of the VNTR region. The PCR products are separated in an electrophoresis gel and usually visualized by hybridization. (If the amount of background sample DNA is low, PCR products can sometimes be seen by simple staining, since they are present at such high concentration.) The lengths of the PCR products are determined by the number of repeats present, so the sizes of products vary from person to person. Several sets of primers can be used to look at several VNTR or STR regions, generating a unique profile for the individual.

Applications of DNA typing

Forensics

When most people think of DNA typing, they think of solving crimes or paternity cases. If a blood or tissue sample not belonging to the victim is found at the scene of a violent crime, it can be used to link a suspect to the crime scene through blood analysis and/or DNA typing. Most important, DNA typing can clear suspects who are innocent. In about 30% of DNA typing cases thus far, the results have proved that the prime suspect could not have left the incriminating sample. In some cases, men who were convicted of terrible crimes have been released after DNA testing of old crime scene samples proved these men could not have left these samples.

DNA typing is used to exclude suspects. If a suspect's DNA profile does not match the crime scene sample, he or she is cleared. If a suspect's DNA profile does match the crime scene sample, the question becomes how many other people in the population might have this same DNA profile just by chance. Studies of profiles of large numbers of people have made it possible to calculate the frequency of patterns seen with different probes.

For example, if a suspect's DNA profile with probe A matches the crime scene sample but 10% of the population also has that profile, then 1 in 10 people could also have left the sample. As a forensic scientist, what do you do? You use another probe. Suppose the suspect's profile with probe B also matches the crime scene sample profile. Let's say that particular profile is found in 5% of the general population. If the profiles from probe A and probe B are independent, we now have a (0.1)(0.05) = 0.005 or 0.5% probability of finding the two matches in a random person. Another way of looking at it is that 1 person in 200 would be expected to have both of these profiles.

In real crime scene analyses, many probes are used. For example, the FBI's Combined DNA Identification System (CODIS) is based on 13 STR locations. The odds of two individuals matching by chance with this many probes are very small, though people argue about exactly how small. Originally, many scientists

were concerned that individuals belonging to distinct ethnic groups might be shortchanged by this type of calculation. Specifically, they worried that the DNA profiles of all the members of a particular ethnic group might be more similar to each other than the DNA profiles of the general population would be (just as Scandinavians have blond hair and blue eyes more frequently than the general population). DNA identification laboratories, whether governmental or private, have responded to this criticism by building databases of DNA profiles of individuals from various ethnic groups to serve as comparison data. Studies of DNA profiles from more than 70 ethnic groups have not shown great differences between the profiles in terms of frequencies. If an individual's DNA profile matches a crime scene profile exactly after analysis with several different probes, the odds that a different person left the sample are extremely small.

Biological Applications

Forensics is only one area in which DNA typing has become an important tool. Determining genetic similarity and analyzing kinship through DNA typing is useful in many areas of biological research such as conservation biology, evolutionary biology, taxonomy, and behavioral biology. For example, primatologists would like to understand how kinship affects the social behavior of chimpanzees. However, it is impossible to establish which chimpanzee is the father of an offspring simply by watching the troop, because both male and female chimpanzees mate promiscuously. If primatologists wanted to determine the father of a chimpanzee, they had to tranquilize the animals, take blood samples, and analyze them. This procedure could traumatize the animals and disrupt the very behavior the primatologists wanted to study.

DNA typing using PCR has provided a solution to this problem. The PCR method is so sensitive that it requires only a few hairs (there is DNA in the follicle cells) for analysis. Chimpanzees sleep alone at night and change sleeping sites each night. In the morning, without disturbing the animals at all, researchers can collect shed hairs from the sleeping sites after the chimps have left them. Now that researchers can determine kinships among the chimps, the scientists can analyze whether the degree of relatedness affects the social behavior of the chimps.

In an example of tying biological specimens to crime, DNA typing of a tree led to a murder conviction. A woman's body was found buried under a paloverde tree in Arizona. The police had a prime suspect of whose guilt they were certain, but there was no physical evidence linking him to the site. However, the police found paloverde tree pods in the bed of his

Classroom Activities

pickup truck. The prosecuting attorney wondered whether it might be possible to determine whether those pods had come from the tree under which the body was found. He called a nearby university and asked.

The scientists agreed to try the experiment. First, they had to develop a profiling method. Using PCR, they tried out many random sets of primers until they were able to find some that yielded unique patterns of products when tested on several different paloverde trees. They were then ready to test the crime scene evidence. They generated profiles from the pods from the suspect's truck, from the tree under which the body was buried, and from about 15 other paloverde trees. The profile from the pods matched the tree at the crime scene. None of the other trees had the same pattern. The prosecutor had the evidence link-ing the suspect to the crime scene, and he obtained a conviction.

Exercise 1:
STRs Can Cause an RFLP

Use with worksheet 1.

Suppose that we are taking a close-up look at a region of chromosome 8 from two individuals named Bob and Mary. Laboratory workers are focusing on a re-gion of chromosome 8 in which they have found an STR. Shown on worksheet 1 are the sequences from that region of Bob's and Mary's chromosome 8s. The laboratory will perform Southern hybridization analy-sis using the restriction enzyme *Hae*III and the probe shown on worksheet 1.

Questions

1. What is the tandemly repeated sequence in this chromosome region?

2. What size fragments would be generated from the region containing the repeats by digesting Bob's maternal chromosome with *Hae*III? His paternal chromosome? How large would the fragments from Mary's chromosomes be?

3. Suppose you digested samples of Mary's and Bob's DNA with *Hae*III and performed a Southern analy-sis using the probe shown on worksheet 1. Draw the results on the outline provided.

4. If you were analyzing DNA prepared from Bob and Mary's white blood cells, why couldn't you simply look at the stained gel pattern and skip the hybridization step?

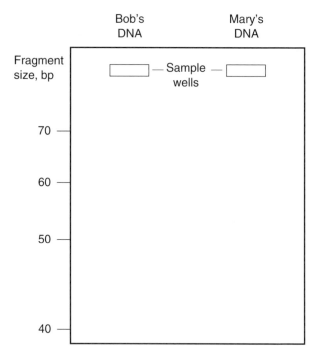

Exercise 2: PCR Can Reveal Differences between Chromosomes

Use with worksheet 2.

Workers at another laboratory wish to look at differences between Bob's and Mary's DNA. They will also look at the highly variable region of chromosome 8 shown on worksheet 2 (the same region is also shown on worksheet 1), but they will use PCR instead of Southern hybridization analysis.

Laboratory workers prepare DNA from samples of Bob's and Mary's blood and then mix a tiny sample of DNA from each person with DNA polymerase enzyme, deoxynucleotides, and the primers shown on worksheet 2. The reaction is allowed to proceed through 30 replication cycles.

Questions

1. Write the sequence of the major PCR product from Bob's maternal chromosome.

2. Write the sequence of the major PCR product from Bob's paternal chromosome.

3. Write the sequence of the major PCR products from Mary's maternal and paternal chromosomes (label them "maternal" and "paternal").

4. Draw what the products would look like in a stained gel on the outline provided.

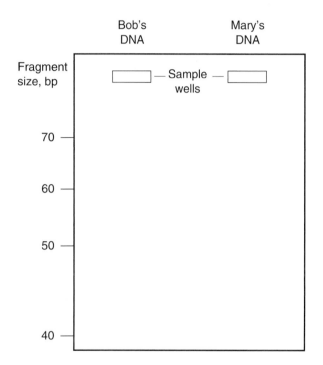

Because PCR yields so many copies of the major product molecule, it is possible to see the major products by simple staining, even when there is too little original sample DNA to see in a stained gel. PCR products can also be visualized by hybridization to a probe, as in Southern analysis.

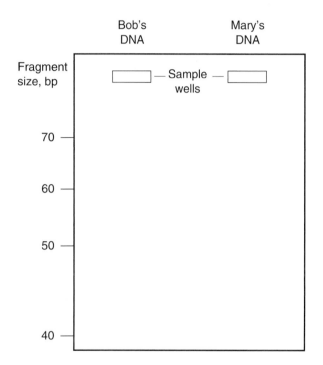

Worksheet 1

Use with Exercise 1: STRs Can Cause an RFLP

Shown below are the probe sequence and the DNA sequence from the highly variable region of Bob's and Mary's chromosome 8 homologs. Use the sequence information to answer the questions in exercise 1.

The restriction enzyme *Hae*III recognizes the sequence 5' GGCC 3' and cuts between the G and the C.

Probe sequence: `3'-GGAGATCCTGTACGATTT-5'`

Highly variable region of Bob's chromosome 8s:

Maternal chromosome:

`5'-AGGCCTCTAGGACATGCTAAAGCTAGCTAGCTAGCTAGCTAAGGCCTAGGTGCGAT-3'`

`3'-TCCGGAGATCCTCTACGATTTCGATCGATCGATCGATCGATTCCGGATCCACGCTA-5'`

Paternal chromosome:

`5'-AGGCCTCTAGGACATGCTAAAGCTAGCTAGCTAGCTAGCTAGCTAGCTAGCTAAGGCCTAGGTGCGAT-3'`

`3'-TCCGGAGATCCTGTACGATTTCGATCGATCGATCGATCGATCGATCGATCGATTCCGGATCCACGCTA-5'`

Highly variable region of Mary's chromosome 8s:

Maternal chromosome:

`5'-AGGCCTCTAGGACATGCTAAAGCTAGCTAGCTAGCTAGCTAGCTAGCTAGCTAGCTAAGGCCTAGGTGCGAT-3'`

`3'-TCCGGAGATCCTGTACGATTTCGATCGATCGATCGATCGATCGATCGATCGATCGATTCCGGATCCACGCTA-5'`

Paternal chromosome:

`5'-AGGCCTCTAGGACATGCTAAAGCTAGCTAGCTAGCTAGCTAAGGCCTAGGTGCGAT-3'`

`3'-TCCGGAGATCCTGTACGATTTCGATCGATCGATCGATCGATTCCGGATCCACGCTA-5'`

Worksheet 2

Use with Exercise 2: PCR Can Reveal Differences between Chromosomes

Shown below are the sequences of the PCR primers and the highly variable regions from Bob's and Mary's chromosome 8 homologs. Use the information to answer the questions in exercise 2.

PCR primers:

5'-GGCCTCTAGGACATGTAAAGC-3' and 3'-TCGATTCCGGATCCACGC-5'

Remember to hybridize the 5'-to-3' primer to the 3'-to-5' DNA strand and vice versa.

Highly variable region of Bob's chromosome 8s:

Maternal chromosome:

5'-AGGCCTCTAGGACATGCTAAAGCTAGCTAGCTAGCTAGCTAAGGCCTAGGTGCGAT-3'

3'-TCCGGAGATCCTCTACGATTTCGATCGATCGATCGATCGATTCCGGATCCACGCTA-5'

Paternal chromosome:

5'-AGGCCTCTAGGACATGCTAAAGCTAGCTAGCTAGCTAGCTAGCTAGCTAGCTAAGGCCTAGGTGCGAT-3'

3'-TCCGGAGATCCTGTACGATTTCGATCGATCGATCGATCGATCGATCGATCGATTCCGGATCCACGCTA-5'

Highly variable region of Mary's chromosome 8s:

Maternal chromosome:

5'-AGGCCTCTAGGACATGCTAAAGCTAGCTAGCTAGCTAGCTAGCTAGCTAGCTAGCTAGCTAAGGCCTAGGTGCGAT-3'

3'-TCCGGAGATCCTGTACGATTTCGATCGATCGATCGATCGATCGATCGATCGATCGATCGATTCCGGATCCACGCTA-5'

Paternal chromosome:

5'-AGGCCTCTAGGACATGCTAAAGCTAGCTAGCTAGCTAGCTAAGGCCTAGGTGCGAT-3'

3'-TCCGGAGATCCTGTACGATTTCGATCGATCGATCGATCGATTCCGGATCCACGCTA-5'

Mitochondrial DNA in DNA Typing

Human mitochondria contain a circular genome of about 17,000 base pairs (Figure 19.1). Mitochondrial genes encode some of the proteins that are associated with the electron transport chain and some of the proteins required for mitochondrial protein synthesis. There are hundreds of mitochondria in a typical cell.

Mitochondrial DNA plays a special role in DNA typing because of its inheritance pattern and its high mutation rate. Except in rare instances, mitochondrial DNA is inherited from the mother. The mitochondria present in the ovum become the mitochondria of the zygote after fertilization. In addition, the region of the mitochondrial genome around the replication origin is highly variable and has a high mutation rate. Therefore, two people with the same DNA sequence in that region are highly likely to share a recent female ancestor (Figure 19.2).

Taken together, this information means that an analysis of mitochondrial DNA can show very clearly whether two people are related through the female line. Analysis of mitochondrial DNA can be helpful in reuniting families and in identifying the dead. A particularly poignant application of mitochondrial DNA typing occurred in Argentina. During the Argentine military's brutal rule (1976 through 1983), many families were torn apart. Often, parents were murdered, and their children given away or sold. In other cases, parents were dragged away to prison, unwillingly leaving their babies to uncertain fates. When the dictatorship was overthrown, the relatives of these kidnapped or disappeared children tried desperately to find them. Many of the relatives were women whose children were murdered and who sought their missing grandchildren.

An American scientist, Dr. Mary-Claire King, was instrumental in helping Argentine families find their lost relatives. Dr. King used mitochondrial DNA in her analyses. Because the parents of the lost children had often been murdered, DNA from more distant suspected relatives was usually the only evidence available for com-

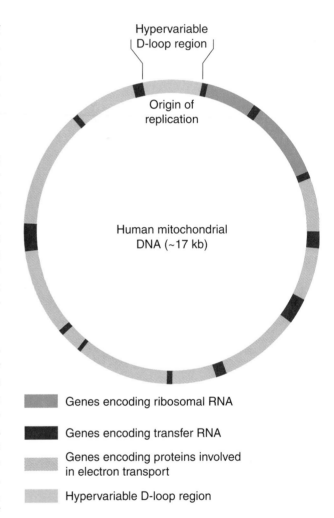

Figure 19.1 Human mitochondrial DNA. The D-loop region around the origin of replication is the hypervariable region.

parison. Since mitochondrial DNA is passed on through the females in a family lineage, a child's mitochondrial DNA profile exactly matches the profile from her mother's mother, all of her mother's siblings, and the children of her mother's sisters (Figure 19.2). This approach was successful in identifying many of the missing children. The story of Dr. Mary-Claire

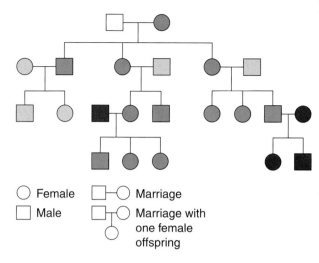

○ Female □─○ Marriage

□ Male □─○ Marriage with
                    one female
        ○   offspring

Figure 19.2 Inheritance of mitochondrial DNA is matrilineal. That is, the mitochondrial genotype is inherited from the mother. Mitochondrial genotype is symbolized here by the shading in the male and female symbols.

King's work in Argentina is told in the article "Genes of war" (*Discover,* October 1990, p. 46-52).

Mitochondrial DNA typing was also used to identify the skeletal remains of the royal family of Russia, murdered by Bolshevik soldiers in 1918 and buried in an anonymous grave. Because many of the members of various European royal families are related through Queen Victoria, they share mitochondrial DNA sequences. Czar Nicholas's remains were identified through comparison with mitochondrial DNA sequences of living relatives.

The U.S. Armed Forces have an ongoing program to identify previously unidentified remains from military conflicts. Although this mission may seem strange at first, to the families of missing soldiers it is very important. In 1999, there were approximately 2,200 soldiers unaccounted for from the Vietnam War, 8,000

from the Korean War, and 78,000 from World War II. Many of these soldiers' bodies were found but could not be identified and were buried as unknowns.

The U.S. Armed Forces DNA Identification Laboratory uses polymerase chain reaction (PCR) amplification of mitochondrial DNA to identify human skeletal remains from previous conflicts. Because there are so many copies of mitochondrial DNA per cell, the chances are greater that it can be amplified even from fragmented remains.

When the laboratory workers attempt to identify a set of remains, they first collect all available information about the remains: where they were found, what unit was fighting there, which soldiers were unaccounted for from that engagement, and so on. This results in a set of possible identities for the remains. Laboratory personnel then contact family members and, with permission, obtain DNA samples for comparison. Under extremely strict conditions designed to minimize contamination, PCR amplifications of mitochondrial DNA are performed, and the products are sequenced. The sequences are compared with sequences from the candidate families. If there is a match, the remains are identified.

In 1998, the remains buried in the Tomb of the Unknown Soldier were brought to the Armed Forces DNA Identification Laboratory under full military honor guard. With great care, the mitochondrial DNA sequence of the remains was determined. The sequence was compared with sequences from seven candidate families, and it matched only one family. The Unknown Soldier was identified as Michael J. Blassie, an Air Force pilot who was shot down in Vietnam in 1972.

The Armed Forces DNA Identification Laboratory now generates a nuclear DNA profile for all soldiers, in the hopes of eliminating future unknown remains.

A Mix-Up at the Hospital

On June 6 at approximately 1:00 p.m., Mrs. Smith, Mrs. Stevenson, and Mrs. Jones each delivered a healthy baby boy at Metropolitan General Hospital. At 1:20 p.m., the hospital's fire alarm sounded. Nurses and orderlies scrambled to evacuate patients, and the three new babies were rushed to safety. After the danger had passed, the hospital staff was distressed to find that in the confusion, they had forgotten which baby was which! Since the babies were rescued before receiving their identification bracelets, there was

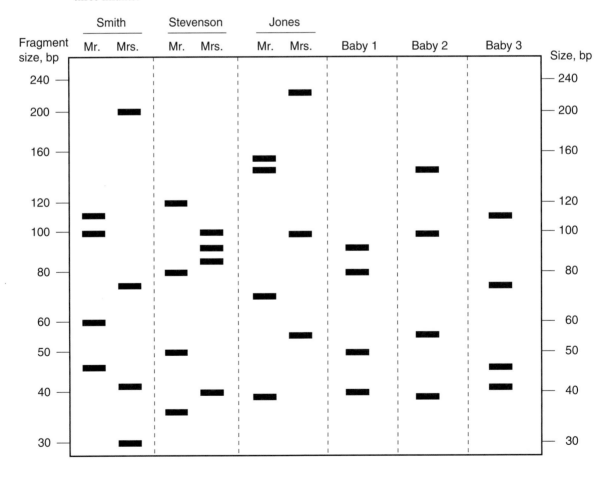

Figure 20.1 DNA profile data from the Smith, Stevenson, and Jones parents and the three infants.

no easy way to identify them. Dr. Anne Robinson, head of pediatrics, ordered that DNA typing be performed on the babies and their parents.

The DNA typing laboratory looked at two different highly variable chromosome regions. The DNA profiles are shown in Figure 20.1. Your job is to decide which baby belongs to which set of parents. To assign a baby to a set of parents, every band in the baby's profile should match a band from either the mother or the father. Not all of the bands in the mother's or father's profiles will have a counterpart in the baby's DNA profile. Hint: Use a ruler or straightedge to help you line up the bands.

Which baby belongs to which couple? Show which bands each baby inherited from its mother and from its father by marking the bands M and F.

A Paternity Case

Mr. I. M. Megabucks, the wealthiest man in the world, recently died. Since his death, three women have come forward. Each woman claims to have a child by Megabucks and demands a substantial share of his estate for her child. Lawyers for the estate have insisted on DNA typing of each of the alleged heirs. Fortunately, Megabucks anticipated trouble like this before he died, and he arranged to have a sample of his blood frozen for DNA typing.

Laboratory technicians used restriction digestion and Southern hybridization analysis to look at four highly variable VNTR (variable number tandem repeat) chromosome regions. The results of the blots are shown in Figure 21.1. Your job is to analyze the data and

determine whether any of the children could be Megabucks' heir.

Remember that every person has two of each chromosome, one inherited from the mother and one from the father. Half of every person's DNA comes from the mother and half from the father, so some of the DNA bands showing in the Southern blots of the children will come from their mothers, and the rest will come from their fathers. The question is, could that father be Megabucks?

1. For the first child, identify the bands in the DNA profile that came from the mother. (Remember that not all of the mother's DNA is transmitted to

Figure 21.1 Results of hybridization analysis.

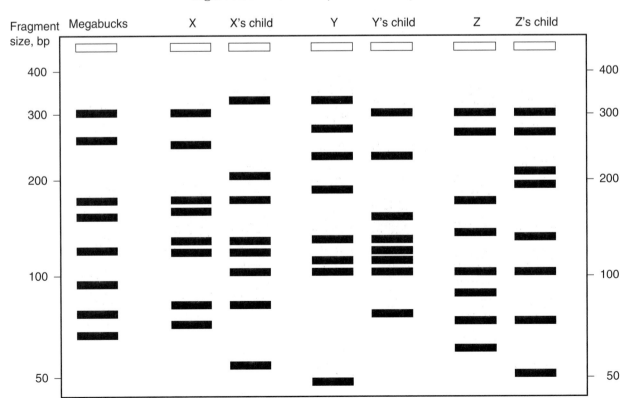

the child; just one of each pair of chromosomes is transmitted.) Mark the bands that came from the mother with an M. Circle the remaining bands.

2. Compare the remaining bands with the DNA profile from Megabucks. If he is the father, then all of the circled bands in the child's profile should have a corresponding band in his profile. Use a straightedge to help you line the bands up accurately. (Remember that only half of the father's chromosomes are transmitted to a child, so not every band from the father would match the child's profile.)

3. Repeat the analysis for the other alleged heirs. Could any of them be Megabucks' children?

Question

Why are there eight bands in every lane, and why do four of each child's bands match corresponding bands in the mother's profile?

The Case of the Bloody Knife

Late one April night, government agents received an anonymous tip that the National Art Museum was about to be robbed of a priceless jewel collection. When they arrived at the museum, they saw they were too late: the jewels were gone. Lying facedown on the floor next to the empty jewel case was the body of a man the chief inspector recognized as the international jewel thief Heinrich Milhouse. Milhouse had been shot in the chest at close range; his clothes were saturated with blood. Underneath the body, the inspectors found a bloody knife.

At the airport the next day, police apprehended Englewood Smink, the murdered thief's occasional partner in crime. Smink denied all knowledge of the murder and the theft. When asked about the fresh cut on his hand, Smink said that he had had an accident in the kitchen that morning.

Suspicious, the chief inspector ordered DNA tests on the victim, the blood on the victim's clothes, the blood on the end of the knife found under the victim, and Smink. Police laboratory technicians used the polymerase chain reaction to look at two different chromosome regions that contain a variable number tandem repeat. They used one set of primers for each region. The chromosome regions, primers, and results of the tests are shown in Figure 22.1.

What is your interpretation of the data? State your reasons. Should Smink be released? Should other tests be performed?

Figure 22.1 Results of polymerase chain reaction analysis.

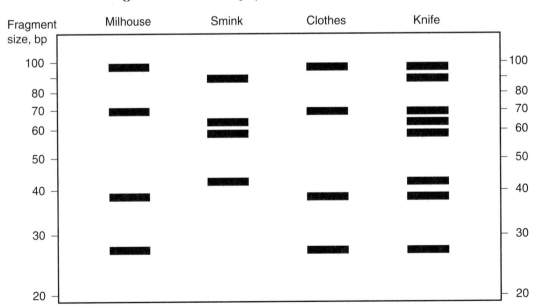

Chromosome region 1:

```
5'-TCCGAGCTGGACGTGCAG...variable number of TAGA repeats...GTTACACGCCTGAGTTACGGT-3'
3'-AGGCTCGACCTGCACGTC...variable number of ATCT repeats...CAATGTGCGGACTCAATGCCA-5'
```

Primer set 1: `5'-CCGAGCTGGACGTGCAG-3'` + `3'-AATGTGCGGACTCAATG-5'`

Chromosome region 2:

```
5'-CGACGCTTAGCATGTCCAG...variable number of CCAGT repeats...CGCTAGTCGACGCCATC-3'
3'-GCTGCGAATGCTACAGGTC...variable number of GGTCA repeats...GCGATCAGCTGCGGTAG-5'
```

Primer set 2: `5'-GACGCTTACGATGTCC-3'` + `3'-GCGATCAGCTGCGGTA-5'`

C. Transfer of Genetic Information

The previous set of classroom activities focused on the manipulation and analysis of DNA. Most of the methods we discussed depend on the availability of large numbers of identical DNA molecules. We get these identical molecules primarily through DNA cloning. The activities in the previous section illustrate part of the cloning process: the construction of a recombinant DNA molecule and its analysis by restriction and DNA sequencing. The other essential part of the cloning process is gene transfer: the introduction of new genetic information into an organism.

We use gene transfer in cloning as a means of propagating recombinant DNA molecules. The recombinant molecule is replicated inside the host cell as the cell multiplies, producing an essentially unlimited number of copies. We also use gene transfer to deliberately change the genotype of an organism so that the organism will have a new trait or make a useful product. For example, we have made corn resistant to certain caterpillar pests by introducing a gene for an insecticidal bacterial protein. We have introduced the gene for clotting factor VIII into pigs so that they secrete the protein in their milk. We have induced *Escherichia*

coli to make human insulin by introducing a complementary DNA copy of the human messenger RNA.

The main reason we have learned how to transfer genes between organisms is because gene transfer is a natural process—it happens in many different ways in nature, and biotechnologists have taken advantage of those natural mechanisms. Natural gene transfer also has a significant impact on human life. For example, the spread of antibiotic resistance in disease-causing organisms, an issue of increasing concern to public health officials, takes place through natural gene transfer. This topic is of such importance that we have included a Reading on it in this section. Other examples of the medical importance of natural gene transfer are given in the individual activities.

The first three activities in this section are wet laboratories that illustrate natural methods of gene transfer in *E. coli*: conjugation, transformation, and transduction. Following the *E. coli* laboratories is an activity in which you can witness genetic alteration of a living plant by a natural genetic engineer, the bacterium *Agrobacterium tumefaciens*.

Transformation of *Escherichia coli* 23

Introduction

Transformation is the uptake and expression of foreign DNA by bacterial cells. *Escherichia coli* is one of the many bacterial strains that do not undergo transformation naturally. However, in 1970, a process for increasing the ability of *E. coli* cells to be transformed was developed. Rapidly growing cells were suspended in cold calcium chloride and exposed to high concentrations of plasmid DNA. The cells were then briefly incubated at a relatively high temperature. After this treatment, some of the cells expressed genes encoded on the plasmid: the cells had been transformed. This laboratory exercise is based on that procedure.

Transformation occurs in nature: some bacteria take up DNA from their environment. Transformation was discovered in an organism, *Streptococcus pneumoniae,* that does this. *S. pneumoniae* causes pneumonia, and other members of the *Streptococcus* genus cause other diseases. Transformation is thought to be an important means by which streptococci undergo genetic change. And streptococci are undergoing genetic changes that are important to humans: they are becoming increasingly resistant to antibiotics. More information about the spread of antibiotic resistance among streptococci and other organisms is provided in the reading *Gene Transfer in Nature*, which follows chapter 24.

In this activity, you will cause *E. coli* to take up plasmid DNA. A plasmid is a small, circular DNA molecule that acts like a minichromosome in a cell. The cell's DNA replication enzymes duplicate the plasmid DNA just as they duplicate the regular chromosome, so plasmid DNA molecules are inherited by both daughter cells when the bacterium divides. This means that if a single bacterial cell takes up a plasmid molecule, all its descendents will contain the molecule, too. Their genetic makeup will include the genes encoded by the plasmid DNA.

Scientists use plasmids as convenient vehicles for introducing new genes into cells. It is easy to isolate

plasmid DNA. New genes can be added to the purified plasmid DNA by using restriction enzymes and DNA ligase. Finally, the new recombinant plasmid can be introduced into host cells by transformation. When plasmids are used this way, they are called vectors. A vector is any DNA molecule used to deliver new genes to cells.

Here is a problem for you to think about. Even the most carefully conducted transformations of *E. coli* are very inefficient. Only one cell in thousands or millions takes up plasmid DNA. How could you find the few transformed cells among the many untransformed ones?

When scientists do recombinant DNA work, they usually use plasmids that carry marker genes. Marker genes are genes that produce an easily detected phenotype such as resistance to an antibiotic or a color change under certain conditions. The plasmid you will be using in this activity carries an antibiotic resistance marker gene. How could you take advantage of this marker to detect cells that take up your plasmid DNA?

Procedure

1. Label one sterile 1.5-ml microcentrifuge tube as + (with plasmid) and another as – (without plasmid). Plasmid DNA (pAMP) will be added to the + tube; none will be added to the – tube, which is our control.
2. Using a 100- to 1,000-μl pipette and a sterile tip (or other appropriate device), add 250 μl of 0.1 M calcium chloride solution to each labeled tube (+ and –).
3. Place both tubes in an ice-filled cup.
4. Transfer one or two large (3-mm-diameter) colonies from an agar plate to the + tube as follows.

   - Sterilize an inoculating loop in a Bunsen burner flame until the loop glows red-hot. Continue to pass the lower third of the shaft through the flame.
   - Stab the loop into the agar several times to cool the loop. *Do not touch the bacterial colonies*

until you have cooled the loop. If you touch the cells with a hot loop, you will kill them.

- Scrape into the loop one or two 3-mm-diameter bacterial colonies, but be careful not to transfer any agar. Impurities in the agar can inhibit transformation and ruin your experiment.
- Immerse the filled loop in the calcium chloride solution in the + tube, and *vigorously* tap the loop against the tube's wall to dislodge the cells. Hold the tube up to the light to observe whether the cell mass fell off the loop.

5. Spin the loop in the tube to suspend the cells. Hold the tube up to the light, and inspect it carefully to ensure that the suspension is homogeneous. It should be milky white. You cannot hurt the bacterial cells by being too vigorous. *It is important that no visible clumps of cells remain.*

- Reflame the loop before setting it down.

6. Return the + tube to ice.
7. Transfer to the – tube and suspend a second mass of cells, as described in steps 4 and 5.
8. Return the – tube to the ice. Incubate both tubes on ice for 10 to 15 min.
9. Use the 0.5- to 10-μl pipette and a sterile tip (or other sterile device) to add 5 μl of plasmid DNA to the + tube. Tap the tube with your finger to mix the contents. Avoid making bubbles in the suspension or splashing the suspension up the sides of the tube.
10. Return the + tube to the ice. Incubate both tubes on ice for an additional 15 min.
11. When the incubation period is nearly over, prepare the heat shock bath. Go to the sink and fill your second Styrofoam cup with water that is at 43°C. Use the thermometer to determine the temperature of the water. You will need the water to be at 42 to 43°C at the end of the incubation period.
12. When the incubation on ice is over, heat shock the bacteria. *It is essential that the cells receive a sharp and distinct shock.*

- Make sure that your heat shock bath is 42 to 43°C.
- Remove both tubes from the ice bath, and immerse them in the hot-water bath for exactly 90 s.
- Immediately return the tubes to the ice bath, and let them stand for at least one additional minute.

13. Set the tubes in a rack at room temperature.
14. Use a 100- to 1,000-μl pipette to add 250 μl of broth to each tube. Gently tap the tubes to mix the contents.

The cells can be plated immediately if you are looking for ampicillin resistance, or they can be left until the next day. If you are looking for kanamycin resistance, incubate the cells for at least 30 min before plating them or inoculating them into liquid media. If cells are to be left overnight, incubate them for about 1 h at 37°C or longer at room temperature, and then refrigerate them.

15. Clean up responsibly. Put all waste that has come in contact with bacterial cells in the designated biological waste containers.

Detection of transformants

Obtain from your teacher the following materials.

- Two plates of tryptic soy agar (TSA) or other nutrient medium, such as Luria broth agar
- Two plates of TSA with ampicillin (TSA-AMP)
- One spreader (bent glass rod) *or* four sterile cotton swabs and sterile broth *or* some sterile glass beads

1. Label one TSA plate and one TSA-AMP plate as +; label the other two plates as -.
2. Use the matrix below as a checklist as you spread the + tube and - tube cells on each type of plate.

	TSA plate	TSA-AMP plate
Control cells (- tube)	100 μl	100 μl
Transformed cells (+ tube)	100 μl	100 μl

3. Use a 100- to 1,000-μl pipette and sterile tip (or a sterile transfer pipette) to add 100 μl of cell suspension from the - tube to the TSA- plate and another 100 μl to the TSA-AMP plate. Do not let the suspension sit on the plate too long before spreading it; if too much liquid is absorbed by the agar, the cells cannot be evenly distributed.
To avoid inhaling or splashing in your eyes any aerosol that might be created, keep your face away from the tip end of the pipette while you are pipetting the suspension culture.
4. Spread the cells with a sterile spreader. The object is to spread the cells out evenly and isolate them from each other on the agar surface so that each cell can give rise to a distinct colony.

If you are spreading with cotton swabs, use the following procedure.

Open the package at the handle end. Withdraw a swab from the package. Have a laboratory partner open the container of sterile broth and hold the lid by the outside. Without touching any other surface, dip the cotton swab into the broth. Remove the wet swab, and use it to spread cells on one plate. Repeat the procedure with a fresh swab for each additional plate.

If you are using glass beads, use the following procedure.

Pour a few sterile beads onto each plate and gently shake the plate back and forth to spread your cells. Let the plates sit a minute or two, and then dump the beads into an appropriate container.

If you are using a spreader, use the following procedure.

- Dip the spreader in ethanol, and then pass it through a Bunsen flame only long enough to ignite the alcohol. Remove the spreader from the flame. (The spreading rod will become too hot if left in the flame.)
- Allow the alcohol to burn off. Lift the lid from the TSA– plate, but do not set the lid down on the laboratory bench.
- Cool the spreader by touching it to the agar surface away from the 100-μl cell suspension. It is essential to cool the spreader before touching the cells.

- Touch the spreader to the cell suspension, and gently drag it back and forth across the surface of the agar.
- Rotate the plate a quarter turn, and repeat the spreading motion. Remember, the object is to spread the cells out as evenly as possible.

5. Repeat step 4 with the TSA-AMP– plate, and spread the cell suspension.
6. Use a new sterile tip to add 100 μl of cell suspension from the + tube to the TSA+ plate, and another 100 μl to the TSA-AMP+ plate.
7. Repeat step 4 to spread the cell suspensions on the TSA+ and the TSA-AMP+ plates.
8. *Reflame the spreader one last time.* Let it cool in the air for a minute, and then put it down.
9. Place the four plates upside down in a 37°C incubator. Incubate them for 12 to 24 h. If an incubator is not available, incubate upside down at room temperature for 2 days. After this incubation, move plates to a refrigerator to preserve the culture.
10. Clean up responsibly. Put all waste that has come in contact with bacterial cells in the designated biological waste container. Wipe down your work area, and wash your hands before leaving the laboratory.
11. On the last day, sketch the four plates.

Detection of transformants by plating on solid media: expected results

Plate	Source of cells	Expected results (indicate many, few, or no colonies)	Observed growth	
			24 h	48 h
TSA-AMP	+ Tube			
TSA-AMP	– Tube			
TSA	+ Tube			
TSA	– Tube			

Questions

1. Compare the observed growth with what you expected. Account for the similarity or dissimilarity in the results.

2. What was the purpose of the TSA-AMP– plate? Of the TSA– plate? Of the TSA+ plate? Of the TSA-AMP+ plate?

3. Explain why growth occurred or did not occur on each of the plates or in the tubes.

4. On which plate(s) did you specifically detect transformants? Justify your answer.

5. Using the results of the laboratory, discuss the relationship of genotype to phenotype.

Conjugative Transfer of Antibiotic Resistance in *Escherichia coli*

Introduction

Conjugation refers to the transfer of genetic information from one bacterial cell to another in a process that requires contact between the cells. In conjugation, one cell always donates genetic information to the other; the donor cell is called the male cell or fertile cell, and the recipient is the female cell.

What makes a bacterial cell fertile? Fertile (male) cells contain special plasmids called conjugative plasmids. The first conjugative plasmid discovered was called F (for fertility) plasmid. Other conjugative plasmids work in a manner similar to F.

The F plasmid encodes proteins that enable the host cell to donate DNA. In fact, the DNA that fertile cells donate is the F DNA. F encodes proteins that form a special structure on the outside of the bacterial cell called a pilus. The pilus is a long tubelike appendage.

It binds to a recipient cell (a cell without the F plasmid) and draws the mating pair together. The fertile cell then copies the F plasmid and simultaneously transfers the copy to the recipient cell. The recipient becomes fertile, too (Figure 24.1).

Conjugative plasmids can contain extra genes, such as genes for antibiotic resistance. If you think conjugative transfer of antibiotic resistance can't possibly have any importance in your life, think again. Public health officials are quite concerned about the rise in antibiotic resistance among disease-causing organisms. For example, campylobacter is a bacterium that causes stomach cramps, diarrhea, and fever. In 1992, 1.3% of cases of campylobacter studied in Minnesota were resistant to a fairly new class of antibiotics called quinolones (Cipro is an example of a quinolone). By 1998, 10.2% of cases were resistant. In Spain, about 80% of campylobacter specimens isolated from humans are resistant to quinolones. Resistance to multiple antibiotics is

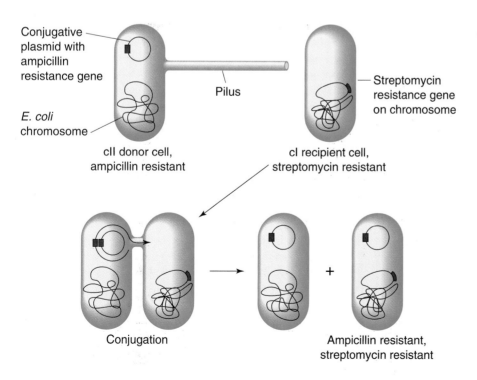

Figure 24.1 Conjugative transfer of an ampicillin resistance plasmid.

now common in many organisms. Anywhere you look for information on antibiotic resistance, you will find alarming statistics.

Virtually every clinically important antibiotic resistance gene is carried on a plasmid. Conjugation allows the spread of plasmids not only between different individuals of the same bacterial species but also between species and even between genera. Conjugation has been observed in the soil, on plant surfaces, in lakes, rivers, oceans, sediments, and sewage treatment plants, and inside plants, insects, chickens, mice, and humans. Conjugation is believed to be the most important route of transmission of antibiotic resistance in most disease-causing bacteria. A Reading on the spread of antibiotic resistance follows this chapter.

In the experiment you will conduct today, the fertile cell (cII) will transfer a gene for ampicillin resistance on a conjugative plasmid to the recipient cell (cI). The recipient is already resistant to streptomycin but cannot transmit this characteristic to the donor. How might you detect recipient cells that received the ampicillin resistance gene from cII and distinguish them from their cII parents?

Procedure

Obtain the following materials from your instructor:

- Agar media plate (no antibiotics)
- One ampicillin agar plate (label this plate Amp)
- One streptomycin agar plate (label this plate Strep)
- One ampicillin-streptomycin agar plate (label this plate Amp-Strep)
- Three small sterile containers
- Micropipette and sterile tips or other sterile measurement device
- Marker
- Optional: sterile inoculation tools

1. Label the three sterile containers cI, cII, and cI + cII.
2. Following your teacher's instructions, add 100 or 200 μl of cI culture to the tubes marked cI and cI + cII. Use sterile technique. Change micropipette tips between additions. Dispose of used tips in the containers provided by your teacher.
3. Add the same volume of cII to the tubes marked cII and cI + cII. Change tips between additions.
4. Let the tubes stand at room temperature for approximately 20 min. During this time, the cI and cII cells will conjugate.
5. While you wait, mark the back of each of your agar plates with a large Y, so that the plate is di-

vided into three approximately equal areas. Label one area cI, another cII, and the third cI + cII.
6. When the 20 min are up, use a sterile micropipette tip to pipette 5 μl of the cells in tube cI to the area of each agar plate marked cI. Change tips between plates. Dispose of used tips properly.
7. Repeat step 6, adding 5 μl of the cells in tube cII to the four plate areas marked cII.
8. Repeat step 6 once more, using the cells in tube cI + cII and the plate areas marked cI + cII.

Note: Your instructor may want you to use an alternative method to inoculate your plates. Listen for directions.

9. Incubate the plates overnight at room temperature or 37°C.

Day 2

Record your results by sketching any growth on the plates in Figure 24.2.

Figure 24.2 Diagram for recording results of conjugation experiment.

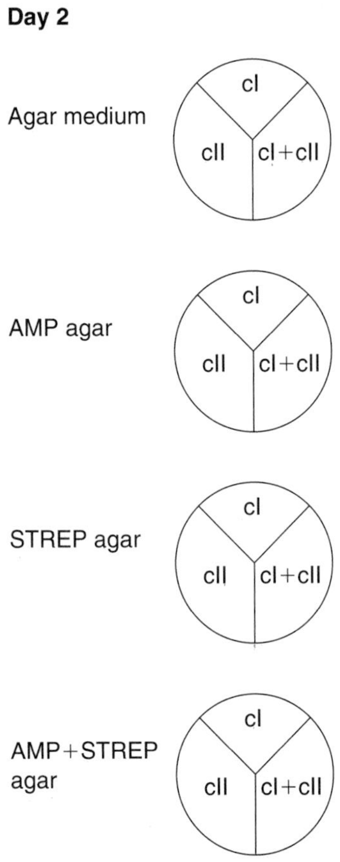

Day 2

Questions

1. In Table 24.1, indicate where you expect to see growth on your plates. Write why you expect to see growth or no growth. For example, on the Amp plate, you expect to see growth of the cells from tube cII because strain cII is supposed to be ampicillin resistant. On the other hand, you do not expect to see growth of cII on the strep plate because cII is streptomycin sensitive.

2. In this experiment, which cells grow on the Amp-Strep plate?

3. In this experiment, how does having the recipient cell be streptomycin resistant make it possible to detect conjugation products? What would happen in this experiment if the recipient cells were not streptomycin resistant?

Table 24.1 Expected results

Medium	Inoculated with:		
	cI	cII	cI + cII
Plain agar Do you expect growth? Reason?			
Amp agar Do you expect growth? Reason?			
Strep agar Do you expect growth? Reason?			
Amp-Strep agar Do you expect growth? Reason?			

Classroom Activities

Gene Transfer in Nature

The focus in this section of activities is on the transfer of genetic information between organisms other than parent and offspring. Although it's easy to think of transferring genes between organisms as a modern development in biotechnology, microorganisms routinely transfer genes among themselves in nature.

Natural gene transfer has important medical consequences. The spread of antibiotic resistance among disease-causing microbes is an example, but not the only one, of natural gene transfer that has profound effects on human health. The spread of antibiotic resistance is also an example of evolution in action. Because of the importance of this issue, we are taking time here to provide information about it.

The Rise of Antibiotic Resistance

For about 100 years after Darwin published his theory of evolution in 1858, scientists believed that favorable new traits, caused by rare mutations, would spread slowly through the population as the new mutant and its progeny reproduced. Over the past 40 years, the spread of antibiotic resistance among microbes has turned this notion on its head, at least where microbes are concerned.

In the 1940s, scientists discovered a microbial enzyme that neutralized the new antibiotic penicillin. They noted that this enzyme might interfere with antibiotic therapy. At the time, it was thought that such resistance would not present much of a problem because of the slow rate of mutation to antibiotic resistance observed in the laboratory. The scientists making these predictions could not have been more wrong.

Scientists began to see how wrong they were in the late 1950s and early 1960s. In the 1950s, an outbreak of dysentery that resisted antibiotic treatment occurred in Japan. The culprit, a *Shigella* species, was found to be resistant to four different antibiotics. The Japanese clinicians immediately knew something strange was happening. In the laboratory, mutants resistant to the antibiotic streptomycin arose at a rate of about 1 in 10 million (10^7) to 1 in 1 billion (10^9). The odds against four such mutations arising in the same cell would be at least one in 10^{28} [$(10^7)^4$]! Those odds were astronomical, but to make matters even more astounding, *Escherichia coli* cells isolated from the same dysentery patients were resistant to the same four antibiotics. How could resistance to the same four antibiotics arise simultaneously in two different genera of bacteria? The way antibiotic resistance was arising in nature must be different from what had been observed in the earlier laboratory experiments.

Researchers began to study the phenomenon, and they found that the antibiotic resistance could be transferred among *Enterobacteriaceae* (a family of gut-dwelling organisms that includes *Escherichia, Shigella, Salmonella,* and other genera). The researchers found that organisms isolated by virtue of their resistance to one antibiotic were frequently resistant to multiple antibiotics, like the *Shigella* described above. They found identical clusters of resistance genes in identical arrangements in different organisms. They also found genes for resistance to veterinary antibiotics in microbes isolated from humans.

To make a long story short, the microbes were sharing genes. We now know that microorganisms acquire new DNA and new traits by several methods other than transmission from mother cell to daughter cell and spontaneous mutation: transformation, conjugation, and transduction. Transmission of traits through any of these means is often referred to as horizontal gene transfer, to distinguish it from parent-to-offspring, or vertical, transmission. Horizontal gene transfer has been shown to take place between microorganisms of many different genera.

Examples of the Spread of Antibiotic Resistance

The bacterium *Staphylococcus aureus* can be found in certain places on the bodies of about one-third of all humans, many mammalian pets, and in the environment.

In healthy individuals, *S. aureus* is usually not a problem. However, it can infect wounds, burned skin areas, and immune-compromised individuals, and, untreated, it can kill. Hospitals provide a prime area for transmission of *S. aureus* infection. The organism can be spread by the hands of hospital workers or by equipment, and hospitals are full of wounded, burned, and immune-compromised individuals. Day-care centers are also particularly susceptible to outbreaks of staph infection, presumably because of the poor hygiene habits of small children.

In 1952, almost 100% of *Staphylococcus* strains were susceptible to penicillin. Because penicillin-resistant infections began to show up with increasing frequency, in the late 1960s doctors began routinely treating staph infections with methicillin instead of penicillin. The switch from penicillin to methicillin increased the cost of a typical drug treatment for a staph infection 10-fold. By 1982, fewer than 10% of all clinical infections could be cured with penicillin. Methicillin-resistant staph infections increased in frequency during the 1980s, and by 1993 only one surefire treatment for *S. aureus* infections was left: vancomycin. In 1992, about 23 million Americans underwent surgery, and it is estimated that up to 920,000 of them developed postsurgical infections, the majority of which were caused by *Staphylococcus* strains.

Streptococcus is a genus of bacteria that causes infections ranging from strep throat to ear infections to pneumonia to rheumatic and scarlet fevers. In 1941, 10,000 units of penicillin per day for 4 days was enough to cure strep respiratory infections. In the early 1970s, penicillin and erythromycin were still effective against strep infections, and scarlet fever and rheumatic fever were considered diseases of the past.

In the 1980s a major shift occurred. A new, more dangerous strain of *Streptococcus* appeared that was resistant to penicillin. Recurrent outbreaks of rheumatic fever appeared around the country. Streptococcal pneumonia that was resistant to multiple antibiotics broke out in Oklahoma, killing more than 15% of those infected. Streptococcal ear infections in children became increasingly resistant to penicillins, with associated hearing loss an urgent problem. Multiply resistant streptococcal infections were turning up all over the world, and by the 1990s some strains were resistant to nearly every major class of antibiotics save vancomycin.

Around 1988, vancomycin-resistant strains of *Enterococcus faecalis* and *Enterococcus faecium* emerged.

Enterococci are normally harmless inhabitants of the intestinal tract. In severely ill, usually hospitalized patients, these organisms can cause severe or even fatal infections. A 1994 Centers for Disease Control and Prevention survey of major hospitals found that nearly 8% of all *Enterococcus* infections were vancomycin resistant. Rates of resistance among infected intensive care patients were higher. The first cases of *Staphylococcus* infections that were partially resistant to vancomycin were reported in Japan in May, 1996. The first partially resistant cases in the United States were reported in the summer of 1997 in Michigan and New Jersey. In both cases, the patients had been repeatedly treated with vancomycin for infections that were resistant to methicillin. Interestingly, the mechanism of resistance in the partially resistant *Staphylococcus* strains was different from the mechanism of resistance in *Enterococcus* strains.

These are only a few examples illustrating the spread of antibiotic resistance among microbial populations. The same patterns have been seen in many other organisms, such as *Neisseria gonorrhoeae*, the causative agent of gonorrhea.

Origins of Antibiotic Resistance Genes

The genes for antibiotic resistance work in many different ways to protect bacteria from the drugs. Some encode proteins that act like tiny pumps that actively pump a particular drug out of the bacterial cell. Others encode proteins that chemically modify the antibiotic, turning it into a different chemical that does not harm the bacteria. Yet others alter the antibiotic's target within bacteria, so that it is no longer affected by the drug. Table 24.2 lists some examples of specific antibiotics, how they act to kill bacteria, and how resistance genes protect bacteria from them.

Where did all of these different antibiotic resistance genes come from? To understand the answer to this question, you need to know that antibiotics themselves have been around for eons. Most families of antibiotics were not invented by humans but were discovered from natural sources such as soil bacteria. The soil bacteria themselves evolved the ability to produce an antibiotic, presumably to kill other nearby bacteria and thereby gain an advantage in their local environment.

These soil bacteria would be killed by the very antibiotic they produced unless they also evolved some means of protecting themselves. Indeed, this is precisely where at least some antibiotic resistance genes

Table 24.2 Examples of mechanisms of action of and mechanisms of resistance to antibiotics

Antibiotic	Mechanism of action	Most common mechanism of resistance
Penicillins (such as penicillin G, ampicillin, amoxicillin)	Inhibits enzymes involved in bacterial cell wall synthesis	Plasmid-encoded β–lactamase enzymes inactivate the drug by hydrolyzing the β-lactam portion of the molecule
Erythromycin	Binds to 50S ribosomal subunit and disrupts translation	Plasmid-encoded enzyme that methylates a specific A residue in the ribosomal RNA of the 50S subunit, reducing its ability to bind to the drug
Streptomycin	Binds to 30S ribosomal subunit and disrupts translation	1. Plasmid-encoded enzyme that modifies the streptomycin molecule so that it cannot enter the cell 2. Chromosomal mutation of a protein component of the 30S subunit that alters the streptomycin binding site
Tetracycline	Binds to ribosomes and disrupts translation	Transposon-encoded protein that pumps tetracycline out of cells; usually transmitted on plasmids
Sulfonamides	Block folic acid biosynthesis by competitively inhibiting the enzyme dihydropteroate synthetase	Plasmid-encoded form of dihydropteroate synthetase enzyme is resistant to sulfonamides
Chloramphenicol	Binds to 50S ribosomal subunit and disrupts translation	Plasmid-encoded chloramphenicol acetyltransferase enzyme modifies the drug and renders it inactive

come from: they are the self-protective genes of the organisms that produce the antibiotic. A major group of antibiotic-producing soil organisms, the streptomycetes, is known to exchange genetic material with other types of bacteria via conjugation. In this way, the streptomycetes' resistance genes could be introduced into a wider population of bacteria. Over evolutionary time, the resistance genes must have been picked up by plasmids in the creation of our modern-day multiple-drug-resistance plasmids. Our widespread use of antibiotics has simply applied intense selective pressure favoring the increase in frequency of these genes.

Antibiotics are now chemically synthesized, and researchers have created chemically modified versions of the naturally produced antibiotics to increase their effectiveness and overcome bacterial resistance. The β-lactam antibiotics, penicillins and cephalosporins, are a good example. Resistance to these antibiotics is conferred by β-lactamase, an enzyme that hydrolyzes (splits by inserting a water molecule) the β-lactam portion of the antibiotic molecule.

Many different penicillins and cephalosporins have been developed and introduced over the years, and each time, resistance has arisen. When the β-lactamase enzymes and genes from organisms resistant to the new drugs are compared to the "old" resistance enzymes and genes, it appears that the "new" enzymes are slightly different versions of the "old" ones. The base differences in the "new" versions of the gene result in amino acid differences in the β-lactamase protein. The amino acid differences in the

protein give the enzyme a different specificity, allowing it to work on the new form of the antibiotic. We put "old" and "new" in quotation marks because the so-called new gene may not be any newer than the so-called old gene. The use of the new antibiotic (which really is new) may simply have favored the spread of a different ("new") version of the resistance gene that was in the microbial population all along.

Evolution of Resistance Plasmids

Most antibiotic resistance genes are transferred on plasmids. These plasmids, often called R (for resistance) plasmids, usually contain several different antibiotic resistance genes close together. Clusters of resistance genes form because most resistance genes are found within transposable elements. Transposable elements, or "jumping genes," are segments of DNA that can move from one DNA molecule to another in a process called transposition (see *Transposable Elements* in chapter 3).

Two features of the transposition process make it an excellent vehicle for creating clusters of resistance genes. First, the ends of the element play a key role in the transposition process, acting as molecular handles for moving whatever DNA is between them. Second, when many transposable elements move, they leave a copy of themselves behind in the original location. These two features allow transposable elements to combine into new, larger elements.

For example, imagine one transposable element jumping into the middle of a different one. When the target transposable element transposes, it will move all the DNA between its ends, including the transposable element that jumped there. Even if the first transposable element jumps again, it will leave a copy of itself behind in the target element. Thus, the first transposon has effectively become part of the original target transposon, creating a new, larger transposon. In this way, transposable elements recombine with one another to form larger elements carrying multiple resistance genes. The clusters of resistance genes found on R plasmids are complexes of transposable elements that were presumably generated in this way.

Environmental Selection of Antibiotic-Resistant Organisms

The spread of antibiotic-resistant and multi-drug-resistant organisms requires more than the formation of clusters of resistance genes on R plasmids. It also requires selection for the drug-resistance abilities conferred by the plasmids. Maintenance of a plasmid within a bacterial cell requires energy. In the absence of environmental pressure favoring plasmid-bearing cells, the cell with a plasmid is at a disadvantage compared with plasmid-free cells.

If an antibiotic is added to the medium, however, only resistant cells can continue to divide. If the resistance gene is on the R plasmid, only plasmid-bearing cells will continue to grow. Imagine a culture containing one plasmid-bearing cell and 1,000 plasmid-free cells. If you grew this culture in antibiotic-free medium, you might eventually be unable to find any plasmid-bearing cells. However, if you added an antibiotic to which the plasmid-bearing cells were resistant and the plasmid-free cells were sensitive, the situation would reverse. After several generations of growth, you might be unable to find any plasmid-free cells in the culture. In evolutionary terms, we say that the antibiotic is creating selective pressure for drug resistance, which in this case is pressure for maintaining the plasmid.

Our widespread use of antibiotics has created an environment that heavily favors drug-resistant organisms. Imagine the microbial population in the intestinal tract of a human being who begins to take an antibiotic for an infection. The antibiotic will kill not just the intended target, but also all other susceptible microbes. Any resistant microbes will flourish in the face of reduced competition, creating a larger population with antibiotic resistance genes. In patients taking oral tetracycline, for example, the majority of *E. coli* fecal isolates carry tetracycline-resistant R plasmids within 1 week of starting drug treatment.

Unfortunately, modern hospitals are a prime environment for the spread of drug-resistant organisms and genes. In hospitals, many sick people carrying disease-causing organisms are brought together in one place and treated with a variety of antibiotics. Many patients have depressed immunity, whether from a specific disease or general poor health, and are susceptible to infection to organisms that do not cause disease in healthy people. In this environment, a resistant organism from one patient may be transferred to another patient, flourish, and then pass its resistance genes on to that patient's microbial population. Many organisms that are dangerous to severely ill hospital patients (such as *Enterococcus*) are not hazardous to people in good health, but their resistance genes can be transmitted to organisms that could infect healthy individuals.

Inappropriate uses of antibiotics have probably helped the spread of drug resistance. Antibiotics are prescribed frequently for bacterial infections, which is certainly an appropriate use. In many cases, however, antibiotics have been prescribed for viral illnesses such as the common cold, which cannot be helped by the drugs. Additionally, in many parts of the world antibiotics can be purchased without a prescription. In these regions, people often don't know whether the antibiotic is appropriate or of adequate dosage. Taking antibiotics provides selective pressure for increasing the numbers of any resistant organisms that are present, organisms that might disappear from the population in the absence of selection.

Perhaps even more significant, vast amounts of antibiotics are used in farming. It was discovered years ago that feeding antibiotics to animals improved their growth. Now antibiotics are added to animal feed as a routine matter, whether the animals are sick or healthy. Feeding antibiotics to farm animals has been shown to increase the spread of drug resistance plasmids, and these plasmids are transferred from microbes within animals to those within people. Some of the first multi-drug-resistant organisms ever isolated contained a gene for resistance to furazolidone, an antibiotic used primarily to treat calves!

Unfortunately, as new antibiotics came on the market, they were introduced into animal food and veterinary medicine, both in the United States and abroad. This practice promoted the spread of resistance to new antibiotics. For example, a Minnesota study linked the use of quinolone antibiotics in poultry in the United States to the appearance of quinolone-resistant campylobacter infections in people. Campylobacter infections are usually caused by handling contaminated poultry or eating undercooked, contaminated poultry.

The U.S. Food and Drug Administration (FDA) approved the use of quinolones in poultry in 1995. At that time, quinolones were the newest class of antibiotics available.

The problem is, if anything, worse abroad, where there is less regulation of antibiotic use. In Spain, for example, quinolones were introduced into poultry and livestock farming in 1989. At that time, virtually no campylobacter infections in humans were resistant to the drug. By the end of 1991, fully 30% of human campylobacter infections in Spain were resistant, and in 1999, 80% were resistant.

The increasing ease and speed of global travel have also promoted the spread of drug resistance among microorganisms. Genetic fingerprinting allows epidemiologists to trace the spread of particular microbial strains. For example, a drug-resistant strain of *Streptococcus pneumoniae,* dubbed 23F, emerged in a Spanish hospital in 1978. An infected person carried the microbe to Ohio, where it acquired the ability to resist another antibiotic. Other descendants of the original Spanish isolate turned up in South Africa, Hungary, and the United Kingdom, picking up additional resistances along the way. Eventually, the even more resistant microbial strain made its way back to the United States and to Spain. In 1992, it was possible to trace every known type of 23F *S. pneumoniae* strain back to the single mutant clone that arose in Spain—a clone that arose from a single cell.

New Antibiotics

As microbes become increasingly resistant to the antibiotics in common use, it is clear that we need to develop new antibiotics. It takes time and lots of money to develop new antibiotics, and the newer antibiotics are more expensive than the older ones. In 1999, the FDA approved the first member of a new class of antibiotics intended to treat hospital patients infected with organisms that are resistant to all other drugs. The drug, Synercid, represents an entirely new class of antibiotics, the first new class to appear in the U.S. market in 10 years.

Synercid can only be given intravenously, and its administration into a small vein is usually so painful that it must be delivered straight into a large vein in the chest or abdomen. The cost of Synercid treatment is estimated to be $250 to $300 per day, more than 100-fold more expensive than some oral penicillin treatments. Other new antibiotics are currently under development, but we can be certain that microbes will eventually develop resistance to them and that they will share the resistance genes.

Clearly, it would be beneficial if we would alter some of our practices that promote the spread of resistance. In response to recent studies on the effects of agricultural use of antibiotics, the FDA is writing regulations that may reserve some drugs exclusively for human use. Unfortunately, these rules will apply only in the United States. It is hoped that other countries will follow suit. Many of the problems of drug resistance can be linked to the absence of sanitary infrastructure—waste disposal and water purification systems—and lack of basic medical treatment in poorer countries. Solutions to those problems will require more resolve on the part of richer countries and more cooperation between countries than is currently in evidence.

Summary

Gene transfer is a natural phenomenon. The rapid spread of antibiotic resistance among disease-causing microorganisms is a consequence of their ability to share genes. The spread of resistance is also a clear example of evolution: genetic change followed by selection. In the case of antibiotic resistance, we humans provide the selection force through our widespread use of the drugs.

Selected Readings

To read current reports dealing with drug resistance, visit the website of the Centers for Disease Control and Prevention at http://www.cdc.gov. Use their WONDER browser to search their publication *Morbidity and Mortality Weekly Report* for articles on vancomycin resistance or other topics of interest.

Garrett, L. 1994. *The Coming Plague.* Penguin Books USA, Inc., New York, N.Y. This is a long, fascinating book about emerging diseases. Chapter 13 deals with antibiotic resistance.

Levy, S. 1998. The challenge of antibiotic resistance. *Scientific American* 278(3):46.

Miller, R. 1998. Bacterial gene swapping in nature. *Scientific American* 278(1):66.

Transduction of an Antibiotic Resistance Gene

Transduction is a natural method of gene transfer that occurs in bacteria. The key player in transduction is a bacterial virus, or bacteriophage (phage, for short). Many different bacteriophages infect many different bacteria. You may have already met one of the phages: lambda. Today you will meet a different phage, one that also infects *Escherichia coli*. It is called T4.

How does T4 transfer genetic material between *E. coli* cells? The answer is found in its life cycle. T4 infects *E. coli* by attaching to its outer membrane and injecting its DNA into the bacterial cell. Once inside the cell, the phage DNA takes over. The *E. coli* cell becomes a factory for producing many copies of the T4 genome and large amounts of viral proteins. Some of these proteins help replicate the T4 DNA; others are assembled into new T4 heads and tails. After many copies of the T4 genome have been made and many new heads and tails are floating around in the cytoplasm, still other T4 proteins begin to put together new virus particles. These proteins fill the empty phage heads with T4 DNA and then attach the tails. After many new viruses are assembled, the *E. coli* cell bursts, releasing the virus progeny.

What does this have to do with transferring *E. coli* genes? The critical step is the point at which the new virus particles are assembled. Once in a while, the T4 assembly proteins make a mistake. Instead of filling a phage head with T4 DNA, they fill it with a piece of host DNA. The filled head gets a tail and becomes a virus particle fully capable of injecting DNA into a new bacterial cell. However, when it does so, the new host cell receives that bacterial DNA instead of dangerous viral DNA. When the new host expresses the bacterial DNA it received, it is said to have been transduced. (Remember, no virus infection took place, since the virus particle was like a "dummy warhead" filled with harmless bacterial DNA.)

Does transduction occur in nature? Absolutely. It has been observed in soil, lakes, rivers, oceans, and sewage treatment plants, and inside organisms such as shellfish and mice. Studies have shown that natural waters literally teem with bacteriophage particles. And transduction plays a key role in many human diseases. One example is the usually fatal food poisoning called botulism. This disease is associated with the bacterium *Clostridium botulinum,* but the fatal syndrome is actually caused by a single protein produced by that organism. The protein is the botulism toxin. The gene for the botulism toxin is not really a *C. botulinum* gene! Instead, it is carried on a bacteriophage that infects *C. botulinum* and is thought to have been transduced from another type of bacterium. Other examples of human disease in which transduction plays a role are *Staphylococcus aureus* food poisoning, diphtheria, and cholera.

This Activity

In this activity, you will observe the transmission of an antibiotic resistance gene by phage T4. The T4 virus particles you will work with were grown on a plasmid-containing host strain, and some of the virus particles produced from that infection contain plasmid DNA. Your job is to detect some of these plasmid DNA-containing particles by their ability to transduce antibiotic-sensitive *E. coli*. Figure 25.1 summarizes the transduction process.

How do you think you could detect transductants (*E. coli* that have received plasmid DNA)?

Procedure

Part A: transduction of E. coli

Obtain the following materials from your instructor:

- Four sterile test tubes
- Two micropipettes, one for small volumes and the other for larger volumes
- Sterile micropipette tips for both
- Four ampicillin medium plates
- Marking pen

1. Infection of plasmid-containing host cell

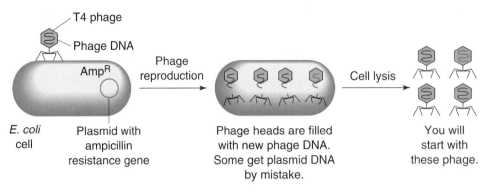

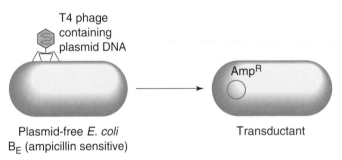

2. Second infection: transfer of plasmid DNA (today's activity)

Figure 25.1   Transduction of plasmid DNA by bacteriophage T4.

1. Label all the plates "Amp." In addition, label one plate "no phage," one plate "undilute," one plate "1 to 10," and the last one "1 to 100."
2. Label the tubes the same way (it is not necessary to write Amp on them).
3. When you have labeled the tubes, use the large micropipette to place 0.5 ml of *E. coli* B$_E$ cells in each tube. It is not necessary to change tips for each addition unless you touch the outside of the tube or some other nonsterile surface with the tip.

To avoid inhaling or splashing in your eyes any aerosol that might be created, keep your face away from the tip end of the pipette while you are pipetting the culture.

Your teacher is making 1-to-10 and 1-to-100 dilutions of the phage suspension for you to use. Why do you think the phage suspension is being diluted?

4. Take your tube of B$_E$ cells labeled "undilute" and add 10 µl of the undilute phage lysate to it, using a sterile tip. Thump the tube gently to mix the phage and cells. To the tube marked "1 to 10,"

add 10 µl of the 1-to-10 dilution. Mix gently. Finally, to the tube marked "1 to 100," add 10 µl of the 1-to-100 dilution, using a fresh tip. Mix gently.
5. Let all of the tubes stand at room temperature for about 15 min.

Two methods for spreading the phage-cell mixtures on the ampicillin plates are described below. Your instructor will tell you which option you will use.

Spreading Option 1
6. Obtain four wrapped sterile cotton swabs.
7. Take the tube labeled "no phage," and pour its contents on the appropriately labeled plate. Open one sterile cotton swab. Use the swab to spread the cells over the entire surface of the agar. Do the spreading gently so as not to tear up the agar. Place the used swab and the tube in the biological waste container provided for you.
8. Repeat for the other tubes and plates. Use a fresh swab each time.

Spreading Option 2
6. You will need a bent glass rod, a large beaker containing alcohol, and access to a flame.

7. Take the tube labeled "no phage," and pour its contents on the appropriately labeled plate. Replace the plate lid.

8. Dip the spreader in the alcohol, and let the excess alcohol drip off. Pass the spreader through the flame only long enough to ignite the alcohol (do not let the spreader heat up in the flame). Remove the spreader from the flame, and allow the alcohol to burn off completely. Repeat.

9. Remove the lid from the plate, and cool the spreader by touching it to the agar away from the cells. It is essential that the spreader not be hot, or it can kill the cells.

10. Spread the cells as evenly as possible over the surface of the agar. To do this, drag the spreader back and forth across the plate, then turn the plate a quarter turn, and repeat. Do this until you have turned the plate a full turn.

11. Repeat steps 7, 8, 9, and 10 with the other tubes and plates.

After the plates have sat a few minutes and the liquid is absorbed, invert the plates. Incubate the plates overnight at 37°C or at room temperature.

What do you expect to see on the no-phage plate?

Part B: detection of cell lysis by bacteriophage T4

In part A, you set up an experiment to detect transductants. The *E. coli* B$_E$ strain used in this experiment allows the transducing T4 strain to inject its genome but does not permit the virus to multiply inside it. Therefore, you will not see cell lysis (bursting) by the phage. Other *E. coli* strains such as CR63 will permit the transducing T4 strain to reproduce and will therefore be lysed by the phage. In part B of this activity, you will add T4 phage to CR63 cells to observe the lethal action of the bacteriophage.

To do this optional segment of the activity, you will need the following:

- Four more sterile test tubes
- Four agar plates *without* antibiotic

1. On each of these tubes, write "CR63." In addition, write "no phage" on the first tube, "undilute" on the second tube, "1 to 10" on the third tube, and "1 to 100" on the fourth.

2. Label the agar plates in the same manner, writing "CR63" and the rest of the information on each plate.

3. To each of the four test tubes, add 0.5 ml of the culture of *E. coli* strain CR63. It is not necessary

to change tips unless you contaminate one by touching some nonsterile surface.

4. Take your tube of CR63 cells labeled "undilute," and add 10 μl of the undilute phage lysate to it, using a sterile tip. Thump the tube gently to mix the phage and cells. To the tube marked "1 to 10," add 10 μl of the 1-to-10 dilution. Mix gently. Finally, to the tube marked "1 to 100," add 10 μl of the 1-to-100 dilution, using a fresh tip. Mix gently.

5. Let all of the tubes stand at room temperature for 5 to 15 min.

Spread the cells on the appropriately labeled plates as you did in part A. After the liquid is absorbed, invert the plates, and incubate them overnight.

Next day

Examine your plates. Are there colonies? If the plates were incubated at room temperature, you may need to let them grow another day. Record your results.

Results

Part A

How many colonies are on the no-phage plate?

How many colonies are on the undilute plate?

The 1-to-10 plate?

The 1-to-100 plate?

If there is a reasonable number on the plate, count them and record the number. If there are very many colonies on the plate, divide the plate evenly into four quadrants by making a large + on the back of the plate, and count the colonies in one quadrant. Estimate the total number on the plate by multiplying the number in the quadrant by four.

Part B

Record your results (drawing the plates may be helpful).

Describe the no-phage CR63 plate.

Describe the 1-to-10 plate.

Describe the 1-to-100 plate.

Questions

Part A

1. Why was the phage suspension diluted?

2. Did you see colonies on the no-phage plate? Was this what you expected? Why or why not? What was the purpose of this plate?

3. What are the colonies growing on the plates to which you added phage? Are there more of them on the undilute plate than on the 1-to-10 or 1-to-100 plate? Why?

4. Suppose you looked first at the plate to which you added the cells containing 10 μl of the 1-to-10 phage dilution and counted 50 colonies. How many would you expect to see on the plate with the cells to which you added 10 μl of the undilute lysate? On the plate with the cells to which you added 10 μl of the 1-to-100 dilution?

Part B

1. Do the plates to which phage were added look the same as the no-phage plate? If they are different from the no-phage plate, what is causing the difference? If they are the same, why do you think they are?

2. What was the purpose of the no-phage plate?

Classroom Activities

Agrobacterium tumefaciens: Nature's Plant Genetic Engineer

26

Would you believe that there exists in nature a genetic engineer that inserts new genes into plant cells by using a plasmid as its vector? Would you believe that this natural genetic engineer is a common inhabitant of the soil? It's all true, and the natural genetic engineer is the soil bacterium *Agrobacterium tumefaciens.*

A. tumefaciens infects certain types of plants (most dicots but not monocots) at wound sites. Once in the wound, the bacterium injects a segment of its plasmid, called Ti, for tumor-inducing, into the adjacent living plant cells. This piece of DNA, called T-DNA for transferred DNA, is only one region of the plasmid. The T-DNA inserts itself into the plant's genome, where it goes to work hijacking the plant's machinery to support the reproduction of *A. tumefaciens.*

Around the infected wound, living cells proliferate into a tumor, or gall, to house and protect the bacterium. The tumor cells synthesize new chemicals that provide critical nourishment to the bacterium but are useless to the plant. Both of these effects are driven by genes on the T-DNA.

In the late 1970s, plant scientists realized they might be able to take advantage of *A. tumefaciens'* natural genetic engineering abilities. They developed an important method for plant genetic engineering based on this organism and its Ti plasmid.

In this method, the T-DNA genes that induce tumor formation and nourish the bacterium are removed from the Ti plasmid and replaced with any gene of interest. This is accomplished through the use of restriction enzymes, DNA ligase, and other recombinant DNA techniques. The new plasmid is returned to *A. tumefaciens,* which is grown in culture so that many bacteria carrying the engineered plasmid are produced.

Plants to be engineered are then infected with the bacterium carrying the "designer" Ti plasmid. *A. tumefaciens* injects the engineered T-DNA into the plant. Instead of receiving genes for tumor formation, the plant gets the genes inserted into the Ti plasmid by the scientist. This method has been used to genetically transform more than 20 important agricultural plants. For example, tobacco has been genetically engineered to produce medically important proteins such as hemoglobin. In a more lighthearted experiment, scientists have also produced plants that synthesize the protein luciferase, the enzyme that causes the light of fireflies. These plants glow in the dark, though not as brightly as fireflies flash.

A. tumefaciens is an effective vector for tobacco, petunias, tomatoes, and other dicots (plants with two seed leaves), but *A. tumefaciens* can't penetrate the cells of most monocots. This is a major limitation, because monocots include most of the important cereal food crops, such as corn, wheat, barley, rice, and rye. This shortcoming forced scientists to look beyond *A. tumefaciens* for ways to deliver foreign DNA into monocots.

Microinjection

Microinjection is a new twist on an old idea. Biologists first used fine glass microtools in the late 1800s to dissect animal tissues. Today, scientists use compound microscopes and micromanipulators fitted with tiny glass pipettes to inject DNA directly into the nuclei of plant cells. Most plant cell walls are so tough, however, that they usually must be stripped with enzymes before DNA can be injected. These wall-less cells, called protoplasts, can then be cultured into whole plants in vitro, that is, in glass dishes.

Electroporation

When scientists use electroporation, they shock protoplasts with electricity until the cells become receptive to foreign DNA. A high-voltage electrical pulse temporarily opens small holes in the protoplast membranes, allowing the foreign DNA to slip through before the membranes reseal themselves. The protoplasts can then be cultured into whole plants.

Biolistics

When bacterial vectors, microinjection, electroporation, or other techniques aren't suitable for a particular plant, scientists increasingly turn to biolistics, a blending of ballistics and biology. With a .22-caliber gene gun nicknamed the "bioblaster," plant engineers bombard cells with metallic microprojectiles coated with DNA. Scientists have already used this gun to transform yeast, algal, higher-plant, animal, and human cells, and they predict it will become one of the more useful and versatile tools for plant genetic engineers.

Today's Activity

In this laboratory, you will infect a plant with *A. tumefaciens* to start the tumor formation process. Because gall formation is slow, you will need to observe the plant for several weeks to see development of the tumor.

Procedure

1. Observe your teacher's demonstration of the proper inoculation technique.
2. Wipe down the laboratory table with Lysol water or other disinfectant.
3. Use an alcohol swab to wipe the plant surface to be wounded (stem or upper leaf). Put the alcohol away.
4. Pick up a sterile toothpick by touching only one end. Pierce the stem of the plant with the other end; the toothpick should go all the way through the stem.
5. Remove the cap from the test tube of water, and flame the tube's mouth. Use a sterile inoculating needle to apply a small amount of sterile water to the wound on the plant. Flame the needle, and reflame the mouth of the test tube before replacing its cap.
6. Remove the cap from the slant tube of *A. tumefaciens,* and flame the tube's mouth. Flame an inoculating needle, and then cool the needle. Use the sterile needle to apply a small amount of *A. tumefaciens* from the slant culture to the plant's wound. Reflame the needle. Reflame the mouth of the test tube before replacing its cap.
7. Remove the paper from the sterile side of the Parafilm (don't touch that side). Apply the sterile side of the Parafilm to the plant's wound. It does not have to adhere tightly to the plant surface, but it should remain in place for a day or so.
8. With a second plant, repeat these steps but without *A. tumefaciens.*
9. Wipe down all laboratory surfaces with disinfectant.
10. Water the plants as you normally would.

Observe the plants regularly for gall formation. Evidence of gall formation may appear in 1 to 2 weeks. If this does not occur, however, do not discard the plants. Gall formation sometimes requires several months.

Record your observations on the log provided below. Supplement the log with sketches of the plant. Label each sketch with the date you make it.

Plant Observation Log

Date	Observations
Week 1	
Week 2	
Week 3	
Week 4	
Week 5	
Week 6	
Week 7	
Week 8	

Classroom Activities

D. Bioinformatics and Evolutionary Analysis of Proteins

In this section of activities, you will look at evolution at the level of a single protein: the enzyme amylase. First, you will test material from a variety of sources to determine whether amylase activity is present. You will then use protein electrophoresis to see the relative size of amylase from different sources. Next, you will compare a portion of the amino acid sequence of amylases from a variety of organisms. Using the degree of difference you find between the amylase protein sequences, you will make a hypothesis about the relatedness of the various organisms from which the amylases originated. Finally, you will use online bioinformatics resources (public databases of protein and DNA sequences) to identify the organisms, identify relatively constant regions of the protein, and find other closely related amylase sequences.

Testing for Amylase Activity

Introduction

In nature, plants use energy from sunlight to power the synthesis of the sugar glucose. The plants then break down the glucose to recover the energy and use it for growth and other life processes. Plants store excess glucose by linking the glucose molecules together in a long chain. These long chains of glucose molecules are what we call starch. When a plant needs to use the stored energy, enzymes break the starch molecule back down into its component glucose molecules. Animals also use starch for energy, but instead of making it, they obtain starch by eating plants. When an animal eats starch, enzymes in the animal's body break the starch into glucose molecules, which are used for energy.

All organisms that use starch for energy make enzymes that convert it back into glucose molecules. One of these enzymes is amylase, which breaks the chemical linkages between glucose molecules by inserting a water molecule into the bond, a process known as hydrolysis (Figure 27.1).

Like many other animals, humans use starch for energy: think of bread, potatoes, and rice. Humans secrete amylase in their saliva, so conversion of starch to glucose begins as soon as starch enters the mouth.

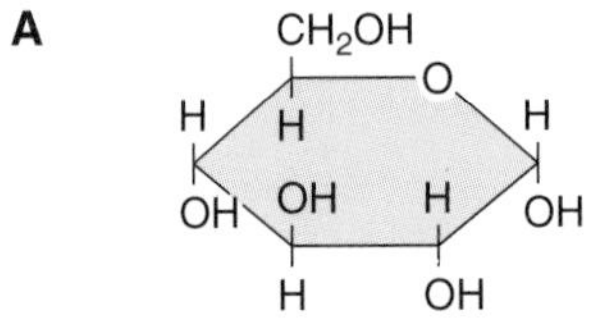

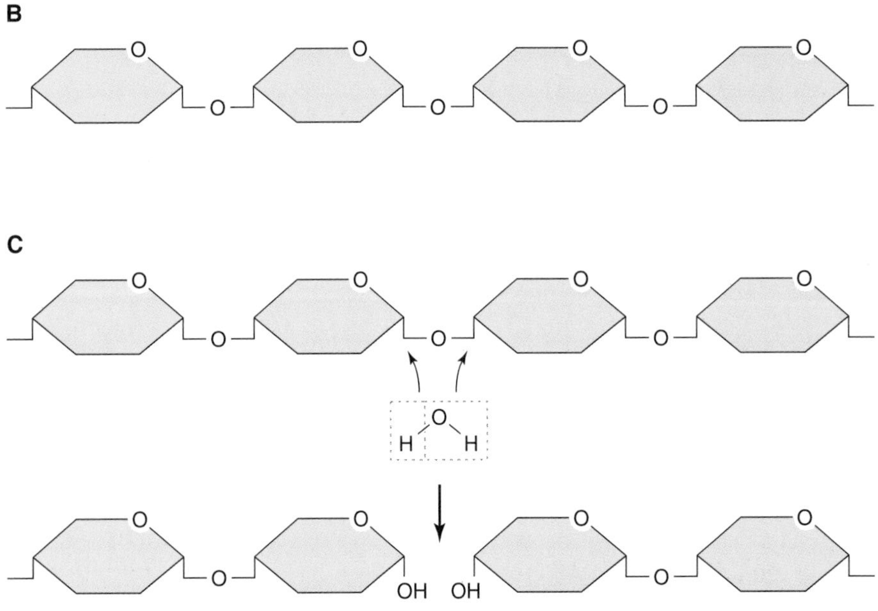

Figure 27.1 (A) The sugar glucose. The corners of the hexagon are carbon atoms. (B) Starch. Starch is a polymer of glucose. In this representation, only the hexagons and the linking oxygen atoms are shown. (C) Hydrolysis of starch. When starch is hydrolyzed, a water molecule splits the bond linking two glucose residues. Complete hydrolysis of starch yields glucose.

You can experience this by holding a saltine cracker in your mouth for several seconds and noticing that a slight sweet taste develops. The sweet taste is a result of amylase cleaving starch into sugars.

Humans also produce amylase in the pancreas, as do many other animals. Amylase is secreted from the pancreas into the small intestine, where digestion of starch continues. Salivary amylase and pancreatic amylase are encoded by separate genes, though the genes are very similar. The gene for salivary amylase has signals in the promoter region that cause it to be expressed in the salivary glands. Production of amylase in the salivary glands seems to be a fairly recent evolutionary development: many animals that use starch for energy produce amylase only in the pancreas, not in saliva.

Industrial uses of amylase

Besides being important for energy production in plants and animals, amylase is an important industrial enzyme. In fact, one of the most common industrial uses of enzymes today involves amylase. The process is the conversion of corn starch into high-fructose corn syrup.

In the 1970s, sugar prices soared. The high cost of cane sugar stimulated the development of an enzyme-based technology for producing a sweet syrup from corn, a crop often produced in surplus in the United States. Starch from corn is converted into high-fructose corn syrup by three enzymes. First, the enzymes amylase and amyloglucosidase (an enzyme similar to amylase) convert the corn starch to glucose. Glucose syrup does not taste as sweet as cane sugar, so a third enzyme, glucose isomerase, is used to convert some of the glucose to fructose. Fructose is very similar in chemical structure to glucose but tastes about twice as sweet. The result, high-fructose corn syrup, is the sweetener in virtually all soft drinks produced in the United States today, as well as in many other products. Look for it on food ingredient labels and think of amylase.

Amylase and its relatives also play an important role in the brewing industry. In traditional brewing, beer is produced by fermenting barley malt. Barley malt is germinated barley. During germination, the barley grains take in water and swell, activating their amylase enzymes. The barley amylases convert the starch in barley to "malt sugar," a sugar that consists of two glucose molecules linked together. The malt is crushed and steeped in water to extract the sugar. This liquid is boiled with hops for flavor, then cooled. Yeast is then added. Yeast cells cannot use starch for energy,

so they use the malt sugar, producing carbon dioxide bubbles and alcohol as products. We call the end result beer.

In modern large-scale brewing, manufacturers want to make large quantities of uniform product at an economical cost. Barley malt is expensive and can vary greatly in quality, so manufacturers use other sources of starch, e.g., corn and rice, and use purified amylase enzyme to convert the starch to sugars for the yeast. These and other industrial uses of amylase are listed in Table 27.1.

Industrial amylase

Where does purified amylase come from? The world's largest producer of industrial enzymes, a company called Novo Nordisk, uses the bacterium *Bacillus licheniformis* to produce amylase. *B. licheniformis* naturally produces an amylase enzyme that is very heat stable, a characteristic that is highly desirable in an industrial enzyme. Novo scientists genetically engineered *B. licheniformis* to produce more of the enzyme by giving it extra copies of its own gene.

In a Novo Nordisk plant, engineers and plant workers grow *B. licheniformis* in huge tanks, feeding it corn

Table 27.1 Industrial uses of amylases

Industry	Use of amylases
Baked goods	Increasing sugar content for yeast fermentation
	Ensuring quality of frozen doughs
	Retarding the staling process
Beverages	Liquefying gelatinized starch
	Increasing sugar content for yeast fermentation
	Clarifying juices by removing starch turbidities
	Producing low-calorie beer
Sugars and syrups	Producing high-fructose corn syrup
	Recovering sugar from candy scraps
	Refining sugar from sugar cane
Textiles	Removing the starch adhesive added to cotton and cotton blends to strengthen fibers during weaving (desizing)
Paper	Producing starch-based compounds to strengthen paper base or coat surface of paper
	Producing adhesive binding agents from starch
Environmental	Reducing insoluble carbohydrates in sewage
Household products	Improving laundry detergents by removing starch-based stains
Energy	Making ethanol, a fuel additive, from grains

and soy grits. As the bacteria reproduce, they secrete amylase. When growth is complete, the mixture is purified to remove the bacteria and other particles and then concentrated. Five hundred milliliters of Novo's industrial amylase can convert 1 ton of starch into sugars. Although you may think of enzymes, such as restriction enzymes, in little vials, Novo ships out its industrial amylase in tank trucks.

Today's Activity

Amylase activity can be easily detected by following the hydrolysis of starch. The presence of starch can be detected with iodine. Because of the bond angles between glucose residues in starch, starch molecules assume a helical form. Iodine molecules nestle inside the coils of the helix, forming a complex that has a dark blue color (Figure 27.2). It takes at least six turns of a starch helix (36 glucose residues) to form the colored complex with iodine. As amylase cleaves the starch molecule into smaller fragments, the colored complex can no longer form.

The purpose of this activity is for you to test different organisms for amylase activity. You will begin by detecting the activity in your own saliva and comparing it with the activity of an industrial preparation of amylase. You will then test a soil sample to see whether

Figure 27.2 The starch-iodine complex. A stylized sketch of the helical starch molecule is shown on the left. On the right, two molecules of iodine (I_2; represented by the dumbbell shapes) nestle inside the helix. Six turns of the helix are required to produce a gold color.

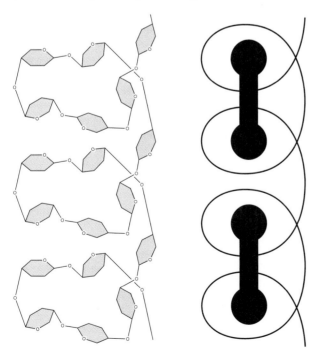

any soil organisms produce amylase, and you will collect and test a variety of other materials as well. As you prepare the materials for testing, think about the organisms you are testing and how they get energy. Based on your ideas, predict whether or not you will see amylase activity, and state why you think so.

Part 1: testing human saliva for amylase activity

Obtain a starch plate, two cotton-tipped applicators, iodine solution, and a sample of industrial amylase enzyme. You should have access to a container of 10 to 20% bleach.

Put the tip of the applicator in your mouth to make it moist. Use the applicator to write your name on one-half of the starch plate. Put the used applicator into the bleach.

Dip the second applicator into the industrial enzyme preparation. Try to get the tip moist but not drippy. Use the tip to write "amylase" on the other side of the starch plate. Put the used applicator into the bleach. (The industrial enzyme preparation is not dangerous, but the tip may have contacted saliva.)

Let the plate stand at room temperature for about 10 min.

Perform the iodine test for starch by flooding the plate with iodine. You should see your name and the word "amylase" appearing clear against a blue background. The clear area shows that amylase broke down the starch molecules where you touched the plate. The blue background shows the presence of starch in the rest of the plate.

Part 2: testing soil organisms for amylase activity

Soil contains a very large number of microorganisms. In this procedure you will dilute soil into water in a series of increasingly large dilutions and plate a sample from each dilution on starch agar. These plates contain starch and other nutrients for supporting the growth of microorganisms. After colonies of microbes grow on the plates, you will test for amylase secretion by flooding the plates with iodine. By plating a series of dilutions, you should have a plate or two on which you can see individual colonies.

This procedure is not written as a sterile procedure. Nevertheless, do not introduce additional contamination by touching the tips of the pipettes, the insides of containers, or the surfaces of the agar plates with your fingers. You will be plating the water as a control to see whether any amylase producers were present in the water (or inside the test tube or pipettes) to begin with.

Obtain five starch-agar plates, five disposable plastic pipettes, five cotton swabs, one container capable of comfortably holding 100 ml, and four 15-ml test tubes.

1. Label the large container "1:100." Label the four test tubes "water," "1:1,000," "10:10,000," and "1:100,000."
2. Label the five starch-agar plates "water," "1:100," "1:1,000," "1:10,000," and 1:100,000."
3. Put 99 ml of water into the large container.
4. Put 9 ml of water into each of the test tubes.
5. Use a pipette to remove 1 ml of water from the "water" tube and add it to the agar plate labeled "water." Discard the pipette.
6. Dip one cotton swab into the remaining water in the test tube and tap it on the edge of the tube to shake off excess. Use the cotton swab to spread the water over the surface of the agar plate. Replace the lid on the plate and discard the swab.
7. Weigh out 1 g of soil. Add it to the 99 ml of water.
8. Use a fresh pipette to stir soil in the water. Mix it well. Using the same pipette, add 1 ml of the suspension to the 1:100 plate. Still using the same pipette, add 1 ml of the suspension to the 1:1,000 test tube. Discard the pipette.
9. Dip a fresh cotton swab into the water tube and tap it to shake off the excess. Use the swab to spread the soil-water suspension on the 1:100 plate. Replace the lid on the plate and discard the swab.
10. Use a fresh pipette to stir the suspension in the 1:1,000 tube. Using the same pipette, add 1 ml of the suspension to the 1:1,000 plate. Still using the same pipette, add 1 ml of the 1:1,000 suspension to the 1:10,000 tube. Replace the lid on the plate and discard the pipette.
11. Dip a fresh cotton swab into the water tube. Tap off the excess and use it to spread the liquid on the 1:1,000 plate.
12. Use a fresh pipette to stir the mixture in the 1:10,000 tube. Use the same pipette to add 1 ml of suspension to the 1:10,000 plate. Still using the same pipette, add 1 ml of the 1:10,000 mixture to the 1:100,000 tube. Replace the lid on the plate and discard the pipette.
13. Dip a fresh cotton swab into the water tube and use it to spread the sample on the 1:10,000 plate.

14. Use a fresh pipette to stir the mixture in the 1:100,000 tube. Use the same pipette to add 1 ml of the mixture to the 1:100,000 plate. Discard the pipette.
15. Dip a fresh cotton swab into the water tube and use it to spread the sample on the 1:100,000 plate.
16. Incubate the plates at room temperature with the lid side up. It may take 2 or 3 days before many colonies appear.

When colonies have grown up on the plates, flood them with iodine solution and look for areas of starch hydrolysis.

Part 3: testing other materials for amylase activity

Amylase is a common enzyme, but not all organisms produce it, and multicellular organisms don't necessarily produce it in all their tissues. We are going to suggest a few additional materials for you to test for amylase activity, but the objective is for you and your classmates to come up with a variety of materials to test on your own.

Before you conduct any tests, use Worksheet 27.1 (see Appendix A) to predict whether each material will show activity and why you think so. After you conduct the test, record the results on the worksheet. If the result is different from what you expect, think about the sample organism again and see if you can come up with an explanation for your result. Ideally, the explanation should suggest an additional experiment that could verify it, even if the experiment is something you cannot do at present (such as check various tissues for amylase activity or do a Southern blot to look for an amylase gene).

Germinating Bean
Soak a dry bean such as a pinto bean or kidney bean in water overnight. Mash the bean with a little of the water and test the extract. (You will have extracted many substances from the bean into the water; therefore, it can be accurately referred to as an extract.)

Plant Leaves
Remove a leaf or leaves from a plant and grind them up with a little water. Test the leaf extract.

Plant Roots
Grind up or blend a fresh plant root with some water and test the extract.

Yeast

Suspend some baker's yeast in water with a little sugar, and let it sit until the mixture begins to bubble. Test a drop of the culture.

Dog Saliva

If you have a dog, take home two or three cotton swabs, a starch-agar plate, and iodine solution. Moisten one swab with dog saliva and test it on one-half of the plate. As a control, moisten the other swab with your own saliva and test it on the other half of the plate. Let the plate stand for 10 min and then flood it with iodine.

Additional Materials

Working with your classmates, identify additional materials that you can test for amylase activity. Record your predictions and the results on Worksheet 27.1.

Questions

1. Approximately what fraction of the soil organisms secreted amylase?

2. Starch is not a component of soil. Where might a soil organism encounter starch? Why do you think so many soil organisms make amylase?

3. Provide a reasonable explanation for the results of amylase testing on the germinating bean, the leaf extract, and the root extract.

Electrophoresis of Amylase Samples 28

Introduction

In the previous activity, you tested a number of substances for amylase activity. In this activity, you will use electrophoresis to separate the proteins in amylase samples such as saliva and will identify the amylase bands.

Electrophoresis of DNA is a little different from electrophoresis of proteins. The differences are due to the fact that proteins are quite different from DNA. DNA has a uniform backbone that is negatively charged at the pH of electrophoresis buffers. In addition, restriction fragments of DNA are usually the same shape (linear). Therefore, the major difference in DNA fragments is usually their size. During standard electrophoresis, they all migrate toward the positive pole at rates that depend on their size.

Proteins, in contrast, are made of chains of amino acids that can be positively charged, negatively charged, or neutral. Proteins themselves thus have quite different charges depending on their amino acid composition (as well as the pH of the environment, which we won't consider here). Proteins also have different shapes, and shape affects how easily a molecule migrates through a gel.

If a protein sample is mixed with the strong ionic detergent sodium dodecyl sulfate (SDS), the detergent molecules will bind to the proteins' backbones and impart a negative charge to all the proteins in the mixture. At the same time, SDS partially denatures the proteins by interfering with hydrophobic interactions and hydrogen bonds, causing the proteins to unfold into more similar shapes. The net effect is that the proteins will now all migrate toward the positive pole during electrophoresis (as do DNA molecules). You will be adding SDS to your amylase samples.

Samples such as saliva and bean extract contain a number of proteins. After you have separated the proteins on a gel, the next task is to determine which band is amylase. You are going to use the same ap-

proach to identify the amylase band that you used to detect its presence in different samples: detection of its activity. You (or your teacher) will add starch to the electrophoresis gel. The starch will make the gel cloudy. You will be able to identify the amylase bands because the enzyme will digest the starch in its vicinity, producing an area of clearing.

There is just one problem with this approach. Remember the SDS you will be adding to get the proteins to run in the same direction in the gel? It also partially denatures proteins, which can decrease or abolish their activity. (Do you know why?) To restore the amylase activity, it will be necessary to remove the SDS from the protein.

To do this, you will soak the gel (after electrophoresis is complete) in a solution of a nonionic detergent called Triton X-100. Triton X-100 will replace the SDS. After letting it soak for 15 min to 2 h, you will move the gel into water and let it incubate for a while and then look for starch hydrolysis.

The entire approach is as follows. You will prepare amylase samples with SDS and then load and run them in duplicate on a starch agarose gel. After electrophoresis, you will cut the gel in half so that you have two halves, each with a complete sample set. You will stain one-half of the gel so that you can see the protein bands. You'll soak the other half in Triton X-100 and then in water to restore amylase activity. Then you'll compare the stained half of the gel with the "activity" half of the gel and identify which stained protein bands are amylase.

Procedure

Sample preparation

Purified Fungal Amylase

Dissolve 0.5 g of the powdered amylase in 500 ml of distilled water, and mix well. Combine 36 µl of this solution with 12 µl of loading dye, and mix well.

Human Saliva

The preparation of saliva depends on the viscosity of the sample. Usually, a 1:2 dilution in water will dramatically ease the loading and running of the sample. In some cases, you will need to add even more water to the saliva. For a 1:2 dilution, mix 0.5 ml of saliva with 0.5 ml of water. Remove 36 μl of this mixture, and combine with 12 μl of loading dye.

If the sample is still too viscous, remove 0.5 ml of it to a fresh tube, add 0.5 ml of water, mix, and use 36 μl of this mixture with 12 μl of loading dye. You may need to heat the human sample to 65 to 70°C for 2 to 3 min to make it fluid enough to load. This heating decreases amylase activity but does not eliminate it. If there are several students in your work group, at least one is likely to have saliva that is not too viscous to load. It is normal for people's saliva to vary in viscosity.

Bean

Soak beans overnight in distilled water, 5 ml per g of beans. The next morning, crush or grind the beans in the water to make a paste. (More water may need to be added, depending on the absorption by the beans.) Combine 36 μl of bean paste with 12 μl of loading dye, and mix well.

Electrophoresis

1. Load 20 μl each of the samples into the gel in a repeating pattern. This is best accomplished by visually dividing the gel in half and loading each half in the same order with identical samples (see Figure 28.1). After electrophoresis, half the gel will be stained for total protein, and the other half will be assayed for amylase activity.
2. Electrophorese for 45 min to 1 h at 130 V.
3. Remove the gel from the electrophoresis chamber, and cut it in half vertically with a razor blade, ruler edge, or scissor blade. Make the cut between gel lanes.
4. Place one-half of the gel in 5% Triton X-100 solution. The gel should incubate in this solution for at least 15 min but can be left there for up to 2 h.
5. Place the other half of the gel in Coomassie protein stain solution. Incubate for exactly 1 min. After 1 min has passed, rinse the gel briefly with

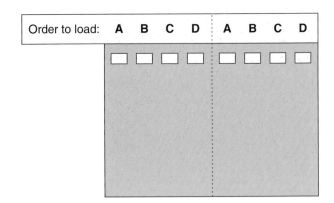

Figure 28.1 Suggested order for loading the samples into the gel.

water, and transfer the gel into the destain solution overnight. If possible, change the destain solution once, after it has turned blue.
6. After at least 15 min of incubation in Triton X-100, transfer the remaining half of the gel from 5% Triton X-100 to water, and incubate overnight as well, either at 37°C or at room temperature. The amylase activity is higher at 37°C.

The next day, the Coomassie-stained gel should contain fairly discrete bands showing all of the abundant proteins isolated from your samples. Many of the complex samples, such as bean extract or human saliva, will show multiple bands.

To determine which of these bands is amylase, carefully examine the unstained half of the gel for areas of visible clearing. Much of the gel will be opaque because of the starch present in the gel. In areas where amylase activity was present, the enzyme degraded the starch into disaccharides, which then left the gel. The result is a clear spot. Compare the pattern of clearing from this gel with the banding pattern in the other half of the gel to determine which of the bands present in the complex samples corresponds to the enzyme amylase.

Draw the stained half of the gel, labeling your sample lanes. Draw the unstained half, showing areas of clearing. On your drawing of the stained half of the gel, indicate the amylase bands.

Questions

1. The introduction to this chapter states that SDS disrupts hydrophobic interactions and hydrogen bonds. What other type of bond is a major contributor to a protein's three-dimensional structure?

2. Can proteins always be renatured after denaturation? If your answer is no, give an illustrative example.

3. Assume that you saw protein bands in the lane containing one of your samples, knew that the sample initially contained amylase activity, but couldn't detect clearing in the unstained half of the gel. Other than an error in procedure, what could be an explanation for this result?

 Classroom Activities

Constructing an Amylase Evolutionary Tree

Introduction

Our planet's life forms share the same basic chemistry. For example, we all use DNA as our hereditary material and express the information it contains through transcription and translation. We all make and use DNA polymerase, RNA polymerase, and ribosomes. We all metabolize glucose in fairly similar ways. Our genes encode and our cells produce many similar enzymes.

The theory of evolution holds that existing life forms evolved from earlier life forms through genetic change and natural selection. You are probably familiar with the idea that all organisms with, say, skeletons, are descended from an ancestral creature that introduced the innovation, survived, and prospered. You may be less accustomed to thinking of evolution at the level of proteins. Yet proteins are the "stuff" of which we are made and which carry out our functions (see chapter 4, *An Overview of Molecular Biology,* for a review). For a genetic change to affect the form or function of an organism, it has to affect a protein: the shape and function of the protein, the amount of it present, the cells in which it is expressed, the stage in development at which it is expressed, and so on. So proteins are a very reasonable level at which to consider evolution.

Protein evolution

You can think about the evolution of proteins in the same way you think of the evolution of skeletons: at some point, an ancestral form of the protein appeared. Modern versions of a particular protein are descended from an ancestral form of the same protein, just as modern versions of skeletons are descended from an ancestral form. Likewise, the gene for that protein is descended from an ancestral gene. In general, protein evolution tracks organismal evolution. Proteins that share a common ancestor are likely to come from organisms that share a common ancestor.

Protein evolution isn't always a simple matter of an enzyme accumulating amino acid differences. For ex-

ample, proteins can acquire entirely new functions, which is presumably why there are families of similar proteins that do different things. But for our purposes, we will focus on proteins that retain the same function even as they undergo changes to their amino acid sequences.

Recall what you learned about protein structure in chapter 4. Enzymes have sites that are directly involved in the activity of the protein, and they have other regions that are not directly involved. It seems plausible that a random change in the protein sequence would be more likely to harm the protein's function if the change were in an active site than if it were not. Scientists comparing the sequence of homologous proteins from different organisms often find regions in which the amino acid sequences are particularly similar. They call these regions *conserved regions*. Conserved regions are assumed to be particularly important to the structure and function of the protein.

You have now observed for yourself that amylase activity is present in a wide variety of organisms. If these amylases are descendants of an earlier, ancestral amylase, then you might expect that there would be some similarity in their amino acid sequences. You might also expect that the more closely related two organisms are, the more similar their amylase proteins will be. These are very reasonable expectations, and they can be used to construct an evolutionary tree.

Constructing evolutionary trees

There is more than one way to approach the construction of an evolutionary tree for a protein. Many evolutionary trees are quite sophisticated and require computers to keep track of all the data. You are going to use a very simple method that can be done by hand. You will count amino acid differences between protein sequences and use the number of differences as a basis for making your tree. For this kind of analysis to be meaningful, you must compare similar proteins.

Here is a simple example using three organisms and an imaginary protein called evolutionase. The evolutionase amino acid sequences from organisms 1 and 2 differ by 30 amino acids. The evolutionase sequences from organisms 2 and 3 differ by 29 amino acids. The evolutionase sequences from organisms 1 and 3 differ by 13 amino acids.

First we'll set up a scale of amino acid changes.

30   20   10   0
Amino acid differences

Organisms 1 and 2 differ by 30 amino acid changes, so we will draw their lineages diverging at the 30-amino-acid point.

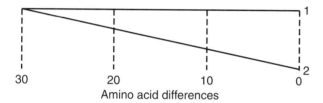

30   20   10   0
Amino acid differences

Organism 3's evolutionase differs from organism 2's by 29 amino acids, but differs from organism 1's by only 13. Organism 3's evolutionase is clearly more similar to organism 1's, so we make the assumption that organisms 1 and 3 are more closely related than organisms 2 and 3. We draw the organism 3 lineage diverging from the organism 1 lineage by 13 amino acids.

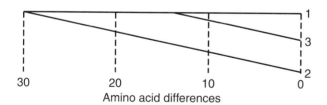

30   20   10   0
Amino acid differences

Does this tree take care of the 29-amino-acid difference between organisms 2 and 3? Yes. We will assume, since this is not an exact science, that 29 and 30 are not significantly different. Organism 3's lineage diverged from organism 2's 30 amino acids ago. At that point, organism 3 and organism 1 were still in the same lineage line. This picture is the simplest explanation of the data that we can construct.

Now let's try an example with a larger data set.

Comparison of amino acid sequences		Number of differences
Organism	1 vs. 1	0
	1 vs. 2	25
	1 vs. 3	17
	1 vs. 4	27
	1 vs. 5	8
	1 vs. 6	6
	1 vs. 7	26

30   20   10   0
Amino acid differences

First, draw a scale of amino acid changes.

There is a group of three organisms, 2, 4, and 7, that differ most from organism 1. Start by drawing the divergence of the most different one, organism 4.

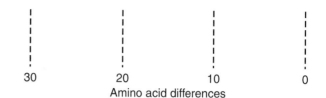

30   20   10   0
Amino acid differences

Now we need to determine whether organisms 2 and 7 are more similar to organism 4 than they are to organism 1. If they are, then they diverged from the organism 1 lineage when organism 4 did. If they are more similar to organism 1 than to organism 4, they diverged from organism 1 after organism 4 did.

Comparison of amino acid sequences		Number of differences
Organism	2 vs. 4	26
	2 vs. 7	24
	4 vs. 7	10

The amino acid sequence from organism 7 differs from the sequence of organism 4 by only 10 amino acids, so organism 7 branches from the organism 4 lineage. We will add organism 7 as shown.

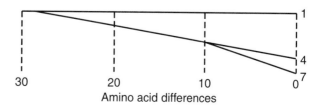

30   20   10   0
Amino acid differences

Classroom Activities

Organism 2's evolutionase, however, differs from the evolutionases of both organisms 4 and 7 by essentially the same number of amino acids as it differs from that of organism 1. So organism 2's lineage diverges from both organism 1's and organism 4/7's lineages at about the same point on the drawing. The numbers of differences are so close that the diagram could be drawn with the organism 2 lineage diverging from the organism 4/7 lineage at its base, or from the organism 1 lineage at its base. We'll draw it from the organism 1 lineage, realizing that the case is not compelling.

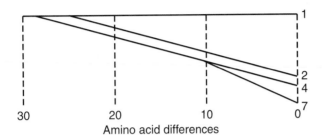

Organism 3's evolutionase differs from organism 1's by 17 amino acids, so we diverge its lineage from the organism 1 lineage at that point.

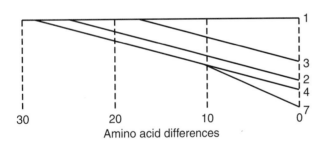

Organism 5's and 6's evolutionases differ from organism 1's by about the same number of amino acids. We need to compare them to see how many differences there are between the two of them.

Comparison of amino acid sequences	Number of differences
Organism 5 vs. 6	2

The evolutionases from organisms 5 and 6 have fewer differences from each other than from organism 1's, so their lineages were still together when they diverged from that of organism 1. We will add them to the diagram accordingly, and we are finished. Does the diagram account for all of our data? Check and see, comparing it with the data on amino acid differences.

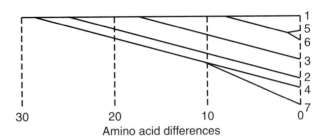

Why did we use organism 1 as a reference and put it on the horizontal line? Would it have been wrong if we had used another organism? No. Using organism 1 was entirely arbitrary. You could do the same analysis using any of the organisms as the horizontal line, and you'd get the same pattern of divergence. It would simply be rearranged on the page. Here's an example of the same diagram, using organism 4 as the reference point.

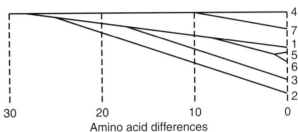

Amino acid differences and the passage of time

Can amino acid sequence changes be equated with the passage of time? In other words, can we use amino acid changes sort of like a clock, in which 10 amino acid changes might equal 5 million years of evolution? In a word, no.

When scientists first started accumulating enough protein and DNA sequence data to do this kind of analysis, they were very excited about this possibility. Further study showed that amino acid sequence changes do not occur at regular time intervals, or even close to them. For example, one study of a single common enzyme found a rate of 6.1 amino acid substitutions per 100 million years in the human-insect comparison, 23.7 changes per 100 million years in the human-mouse comparison, and 69.0 changes per 100 million years in the mouse-rat comparison. Although it's not possible to determine when two organisms diverged on the basis of sequences of a single protein, it is still generally true that the more closely related two organisms are, the more similar their protein sequences will be.

Some scientists still believe that a molecular clock may exist within noncoding regions of DNA. In other words, they believe that nucleotide changes may accumulate at fairly regular rates in these regions. Some of these scientists study differences in the noncoding regions of human mitochondrial DNA in an attempt to determine when various groups of humans diverged from one another. Other scientists are not so enamored of this approach. The data are interesting, but the jury is still out as to what it means.

Today's Activity

In this activity, you will construct an amylase evolutionary tree. We have provided a portion of the amylase amino acid sequence from seven different organisms in Worksheet 29.1 (see Appendix A). The sequences use a standard one-letter abbreviation system for amino acids as follows:

A alanine	I isoleucine	S serine
B asparagine or	K lysine	T threonine
aspartate	L leucine	V valine
C cysteine	M methionine	W tryptophan
D aspartate	N asparagine	X unknown or
E glutamate	P proline	nonstandard
F phenylalanine	Q glutamine	Y tyrosine
G glycine	R arginine	Z glutamine or
H histidine		glutamate

The sequences all contain an ellipsis (. . .) to signify where we omitted a portion of it. Before you begin constructing the tree, take a moment to look at the sequences. Do you see any conserved regions? Conserved regions don't have to match exactly; they need only be somewhat similar.

As a starting point, use organism A as a reference. Cut out the protein sequences on strips of paper to facilitate comparison. Compare the other six amino acid sequences to sequence A and record the number of differences. Some of the sequences have a hyphen in them. The hyphen was inserted because some of the organisms have an additional amino acid at that point. Count the hyphen and the corresponding amino acid (if the other organism has one) as one difference. If the other organism's sequence also has a hyphen at that point, count it as a match. Make a data table like the one below and fill it in.

	Comparison of amino acid sequences	Number of differences
Organism	A vs. A	0
	A vs. B	
	A vs. C	
	A vs. D	
	A vs. E	
	A vs. F	
	A vs. G	

Construct an evolutionary tree, using the examples above if you need a guide. Draw your evolutionary tree neatly, labeling each branch to match the corresponding organism's amino acid sequence. In the next activity, you will perform online searches of protein sequence databases to identify the organisms whose amylase sequences you have just compared.

Questions

1. Based on your tree, how many differences should there be between the amylase sequences of the following pairs of organisms? Fill in the table appropriately.

2. Compare the amylase sequences of the pairs of organisms and fill in the data. How well do the data match the prediction?

Organism pair	Predicted number of differences	Actual number of differences
B vs. C		
C vs. D		
C vs. E		

Bioinformatics

Introduction

One of the consequences of the development of the biotechnology techniques you've been learning about was an explosion in the amount of gene and protein sequence data being collected. For this information to provide the greatest benefit, it needed to be readily accessible and easy to compare with other sequence data.

In 1988, the U.S. government established the National Center for Biotechnology Information (NCBI) to be a national resource for molecular biology information. NCBI creates public databases of information, develops software for data analysis, and conducts research in computational biology. NCBI shares data with other international resources, updating its databases daily.

The information resources include DNA sequence databases, protein sequence databases, and protein structure databases. Powerful search-and-compare programs allow comparison of sequences. The availability of all this information has revolutionized certain aspects of biomedical research.

When a scientist determines a sequence of interest, the first thing she does is to compare it with the sequences in the databases. This comparison could tell her what the sequence is, or what her protein is similar to, or what certain parts of her protein are similar to. Comparison of protein sequences and structures led to the identification of protein structure motifs described in chapter 4 (*An Overview of Molecular Biology*). You can use the sequence comparison programs to search for a particular protein sequence for various motifs, which might provide clues about the function of the protein or about the mechanism of its function.

In fact, the availability of all this molecular information and relatively easy means of analyzing it have led to the coining of a new word: bioinformatics. The activities in this chapter will provide you with an in-troduction to it. This chapter is meant to open a door for you. Once you learn how to use the resources, you will be able to go back to them whenever you wish.

Part 1: performing a BLAST search to identify amylase source organisms

In this activity, you will use the search tool BLAST to identify the source organisms of the amylase sequences from chapter 29. This is not necessarily a cut-and-dried exercise. The data you are searching for was put into online databases by the scientists who collected it. The databases are constantly growing through the addition of new information. In addition, the people who maintain the NCBI website sometimes change the appearance of Web pages. We have included pictures of the Web pages as they appeared in January 2000. You may need to use your good sense to figure out what to do if the appearance has changed since then.

Procedure

1. Go to the NCBI home page at http://www.ncbi.nlm.nih.gov (no period after gov). Click on BLAST on the dark bar near the top of the page (Figure 30.1).
2. From the next screen (Figure 30.2), click on Basic BLAST search.
3. On the next screen (Figure 30.3), locate the word Program with a box next to it. Usually, this box contains the letters "blastn." Click and hold on the box to get a drop-down menu. Select "blastp" from the menu. The letters "blastp" should now appear in the box. This selects the protein database for searching.
4. A box next to the blastp box should have the letters "nr" in it. If it doesn't, click on the box and hold down. Select "nr" from the menu. This tells the search engine to search all available non-redundant data.
5. Further down the page is a box with the phrase "Enter here your input data as" in front of it. The box can say either "Sequence in FASTA format" or

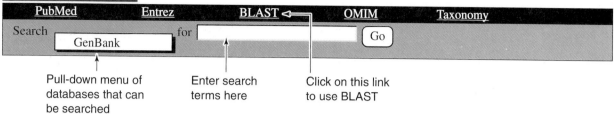

National Center for Biotechnology Information
National Library of Medicine   National Institutes of Health

| PubMed | Entrez | BLAST ← | OMIM | Taxonomy |

Search [GenBank] for [   ] [Go]

Pull-down menu of
databases that can
be searched

Enter search
terms here

Click on this link
to use BLAST

Figure 30.1

"Accession or GI number." Make sure the box says "Sequence in FASTA format." If it doesn't, click on the box and select that phrase from the menu.

6. In the data box, enter the first half of the protein sequence from organism A (nnngvikevtinpdttcgnd) from Worksheet 29.1, using the single-letter codes for amino acids on the worksheet; do not enter the ellipsis (. . .), and skip over any hyphens in the sequence. Verify that you have entered the sequence correctly and that you have not spanned the ellipsis. The program is not case sensitive. Ignore the formatting information beneath the input box. Click on Search.

7. The next screen (Figure 30.4) should tell you that your request has been submitted. If "blastp" and "Sequence in FASTA format" were not se-

lected, you will get an error message. If this is the case, go back and fix the problem. This screen will also tell you approximately how much time will elapse before your results are ready.

8. Click on Format results.

9. The next screen may be a waiting screen that tells you how much time is still needed, or it may be your results screen. The results screen will come up automatically when results are ready.

10. The results screen (an edited version is shown in Figure 30.5) presents results first in graphical form with a series of horizontal colored bars. The color of the bar corresponds to the alignment score. The alignment score is influenced by how long the query sequence was: the longer the query sequence, the higher the score can be. The color groups show the quality of the match in descending order, with the best-match group first.

Figure 30.2

BLAST

NCBI's sequence similarity search tool designed to support analysis of nucleotide and protein databases.

Please see the BLAST Frequently Asked Questions for tips on running BLAST searches.

Overview

Frequently Asked Questions

Web BLAST tutorial NEW

BLAST 2.0

- Basic BLAST search
- Advanced BLAST search

Click here to carry
out a basic search

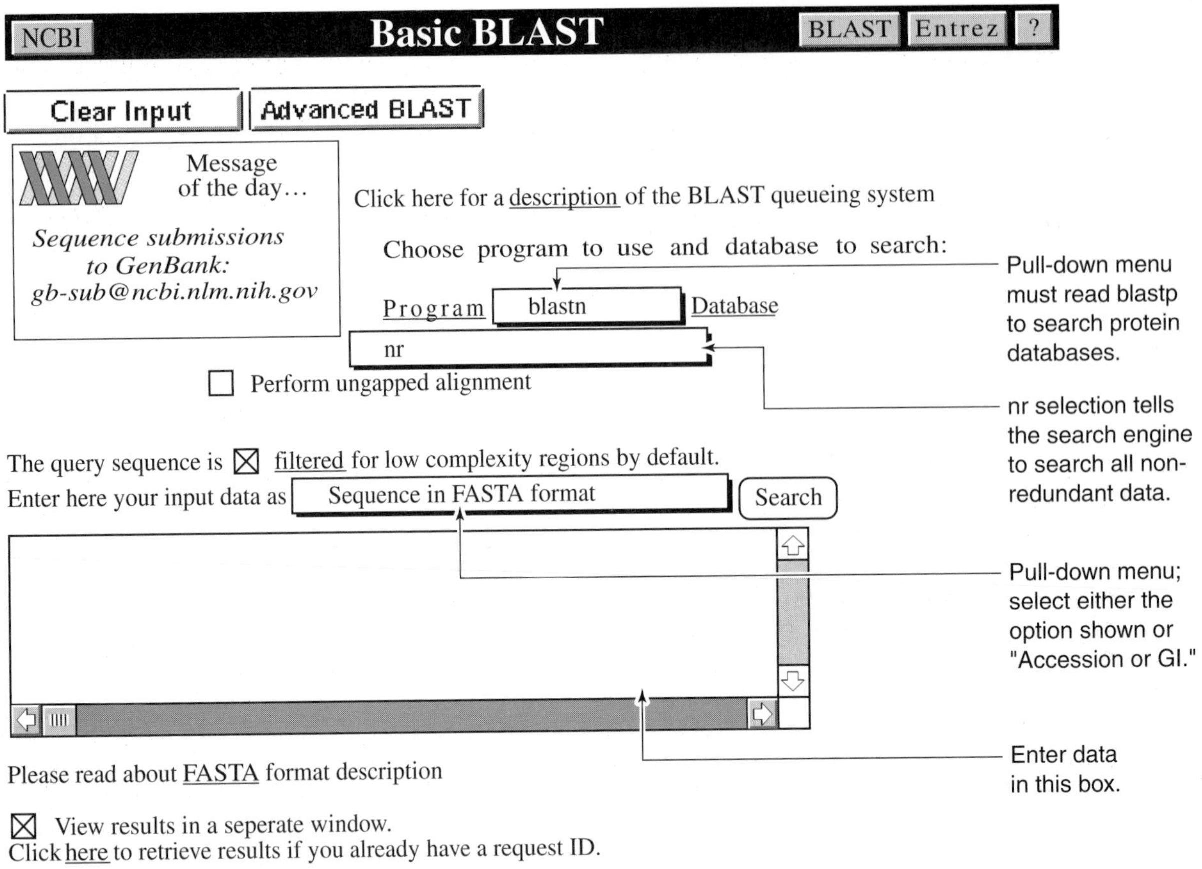

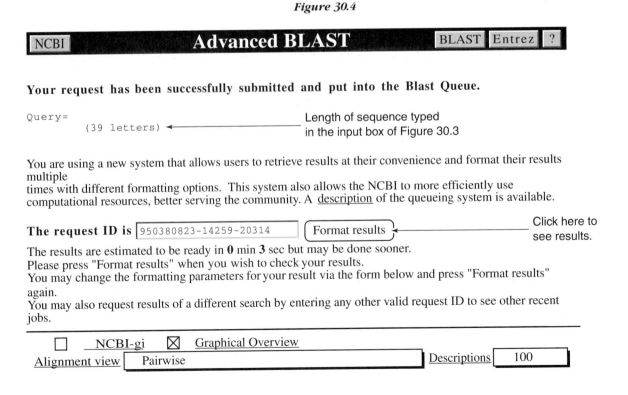

Clear Input   Advanced BLAST

Message
of the day...

*Sequence submissions
to GenBank:
gb-sub@ncbi.nlm.nih.gov*

Click here for a <u>description</u> of the BLAST queueing system

Choose program to use and database to search:

<u>Program</u>   blastn   <u>Database</u>

nr

☐ Perform ungapped alignment

The query sequence is ☒ <u>filtered</u> for low complexity regions by default.

Enter here your input data as | Sequence in FASTA format | (Search)

Please read about <u>FASTA</u> format description

☒ View results in a seperate window.
Click <u>here</u> to retrieve results if you already have a request ID.

— Pull-down menu
must read blastp
to search protein
databases.

— nr selection tells
the search engine
to search all non-
redundant data.

— Pull-down menu;
select either the
option shown or
"Accession or GI."

— Enter data
in this box.

Figure 30.3

Figure 30.4

Your request has been successfully submitted and put into the Blast Queue.

Query=
    (39 letters) ◀──────────

Length of sequence typed
in the input box of Figure 30.3

You are using a new system that allows users to retrieve results at their convenience and format their results multiple
times with different formatting options. This system also allows the NCBI to more efficiently use
computational resources, better serving the community. A <u>description</u> of the queueing system is available.

The request ID is | 950380823-14259-20314 | (Format results)

Click here to
see results.

The results are estimated to be ready in **0** min **3** sec but may be done sooner.
Please press "Format results" when you wish to check your results.
You may change the formatting parameters for your result via the form below and press "Format results"
again.
You may also request results of a different search by entering any other valid request ID to see other recent
jobs.

☐   NCBI-gi   ☒   <u>Graphical Overview</u>
<u>Alignment view</u> | Pairwise | <u>Descriptions</u> | 100 |

Classroom Activities (vertical side text)

Color plate 1 Midnight (left) and Cocoa (right). (Midnight's photo courtesy of the American Kennel Club. Cocoa's photo courtesy of Thomas A. Martin.)

Color plate 2 Two yellow Lab puppies. What color would they have been if their cells could produce functional MSH-R protein? (A) Babs lives with her human family the D'Aubins and black Lab buddy Ike in Scottsdale, Arizona. As you can tell from her black nose, Babs's melanocytes make TRP-1. (B) Angel (the puppy in front) lives with the Morgan family in Cedar Grove, North Carolina. Angel's nose is brown because her melanocytes don't make TRP-1. (Babs's photo courtesy of Mary Lynn D'Aubin. Angel's photo courtesy of Donna Morgan.)

Color plate 3 Kathleen, who has albinism, lives in Florence, Kentucky. She is active in Girl Scouts and school choir, takes tap, jazz, and ballet lessons, and competes at the state level in the Kentucky High School Speech League in acting and storytelling. Kathleen plans a career in the performing arts. (Photo courtesy of Rick Guidotti.)

Color plate 4 Humans express a wide range of hair and eye color. (Photo A courtesy of Hannah Vaughan. Photos B to I courtesy of Thomas A. Martin.)

Color plate 5 A few examples of dog coat colors and patterns. (A) Labrador retriever. (B) Kerry blue terrier. (C) Basset hound. (D) Maltese. (E) Australian shepherd. (F) Old English sheepdog. (G) Dachshund. (H) Dalmatian. (I) Doberman pinscher. (J) Beagle. (K) English setter. (L) Chow chow. (Photos A, B, D, E, F, H, I, J, K, and L courtesy of the American Kennel Club [photos by Mary Bloom]. Photos C and G courtesy of Helen Kreuzer.)

BLASTP   2.0.11   [Jan-20-2000]

Scroll over the bars with the mouse. As you pass over a bar, its identity will appear in
the box. Identities are listed below the graphic, in order of match quality.

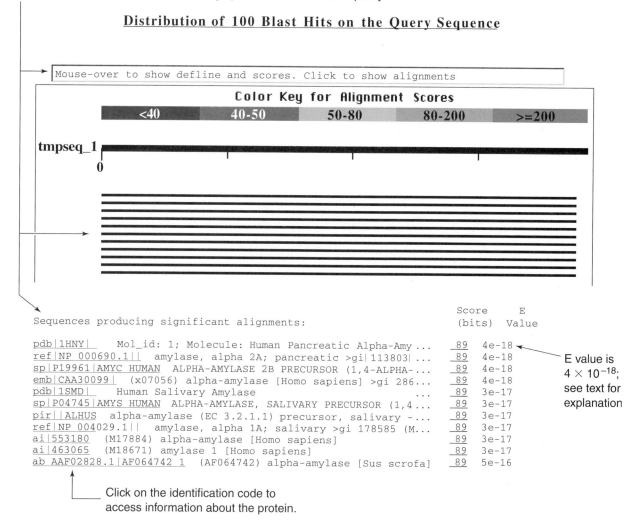

Distribution of 100 Blast Hits on the Query Sequence

Mouse-over to show defline and scores. Click to show alignments

Color Key for Alignment Scores

| <40 | 40-50 | 50-80 | 80-200 | >=200 |

tmpseq_1

0

Sequences producing significant alignments:

	Score (bits)	E Value
pdb|1HNY| Mol_id: 1; Molecule: Human Pancreatic Alpha-Amy ...	89	4e-18
ref|NP 000690.1|| amylase, alpha 2A; pancreatic >gi|113803| ...	89	4e-18
sp|P19961|AMYC HUMAN ALPHA-AMYLASE 2B PRECURSOR (1,4-ALPHA-...	89	4e-18
emb|CAA30099| (x07056) alpha-amylase [Homo sapiens] >gi 286...	89	4e-18
pdb|1SMD| Human Salivary Amylase ...	89	3e-17
sp|P04745|AMYS HUMAN ALPHA-AMYLASE, SALIVARY PRECURSOR (1,4...	89	3e-17
pir||ALHUS alpha-amylase (EC 3.2.1.1) precursor, salivary -...	89	3e-17
ref|NP 004029.1|| amylase, alpha 1A; salivary >gi 178585 (M...	89	3e-17
ai|553180 (M17884) alpha-amylase [Homo sapiens]	89	3e-17
ai|463065 (M18671) amylase 1 [Homo sapiens]	89	3e-17
ab AAF02828.1|AF064742 1 (AF064742) alpha-amylase [Sus scrofa]	89	5e-16

E value is
4×10^{-18};
see text for
explanation

Click on the identification code to
access information about the protein.

Figure 30.5

Below the series of horizontal bars is a list of the
match sequences corresponding to the bars. The
first entry in the list corresponds to the first bar,
and so on. At the right end of each entry is an E
score. This is the number of matches of this quality
that could be expected by chance in a database the
size of the one that was searched. The number in
Figure 30.5 is 4e-18, or 4×10^{-18}, over 100,000 times
smaller than 1 in 1 trillion (1×10^{-12}).

11. What is the best match to the first half of the
protein sequence of organism A? What is the
second-best match? What are the E values?

12. Go back to the input screen and enter the second half of the protein sequence of organism A
(grgnrgfivfnnddwsfsltlqtglpagtycdvisgdkingnctgi). Verify that the top menu still reads
"blastp" and that you have entered the sequence
correctly. Click on Search.

13. What is the best match for the second part of
the sequence of organism A? What is the second-best match? What are the E values? What is
organism A?

14. Repeat the above procedures to identify the
source organisms of the other amylase sequences.

Classroom Activities

Questions

1. What is organism A?

2. What were the second-best matches to the two different portions of the sequence from organism A?

3. Evidence strongly suggests that salivary and pancreatic amylase genes are descendants of one original pancreatic amylase gene. Propose two evolutionary explanations for the second-best match result, one in which the genes diverged before the organisms and the other in which the organisms diverged before the genes. Using each hypothesis, predict which entire protein sequence would be a better match to the entire pancreatic sequence from organism A.

4. Why is the E value of the best match to the longer portion of the sequence of organism A lower than the E value of the best match to the shorter portion of the sequence, even though the two portions of the sequence match the same thing?

5. What are the identities of the other organisms from chapter 29?

6. Discuss your evolutionary tree in the light of the identities of the organisms. How does it compare with what you know about the evolution of the organisms?

Whole-Protein Comparison

The BLAST searches described here used only portions of the amylase sequences. You saw that using the first and second portions of the sequence of pancreatic amylase turned up different second-best matches. What is the closest match to the overall structure of pancreatic amylase?

To avoid typing in the entire amino acid sequence, use the GI number for human pancreatic amylase, 1421331. Go to the Basic BLAST Search page (Figure 30.3). Select "blastp." Rather than entering the sequence in FASTA format, click and hold on that box until the pull-down menu appears. Select the box that says Accession or GI number. Enter the GI number in the screen and click on Search.

What are the closest matches to human pancreatic amylase?

Part 2: exploring the structure of human pancreatic amylase

In this portion of the activity, you will explore protein bioinformatics resources in more depth, using human pancreatic amylase as the example protein.

Procedure

1. Go back to the BLAST search screen and type in a portion of the human pancreatic amylase sequence as you did before. Click on Search.

2. On the results screen, find the human pancreatic amylase match with the identifier 1HNY and click on the link.

3. This takes you to a GenPept display screen with more information about 1HNY. Near the top of the entry is the GI number, an important general identification number. Human pancreatic amylase entry 1HNY has the GI number 1421331. DBSOURCE indicates that the sequence was retrieved from the Brookhaven Protein Data Base, a structure database. It indicates that the structure was determined by X-ray diffraction. The source organism (*Homo sapiens*) is given along with its Linnean classification.

4. Two journal articles are cited, with MEDLINE links. Click on these links to see abstracts of the published articles that describe the determination of the structure of human pancreatic amylase. Note that the entries are linked to the protein structure and to related articles. Explore the links and then go back to the GenPept screen.

Below the identification information and literature references are FEATURES. This gives structural data for the protein, which is logical since the sequence comes from a structure database. (For a review of protein structure, see chapter 4, *An Overview of Molecular Biology*). The phrase to the left indicates the type of structure feature being denoted. Source gives the source of the protein and the amino acids that were analyzed (1 to 496).

Region denotes domains. The first Region entry states that amino acids 1 to 117 together with 163 to 386 constitute domain 1. The second Region entry states that amino acids 118 to 142 constitute domain 2. The third and last region entry identifies amino acids 387 to 496 as domain 3.

SecStr denotes secondary structure. These entries identify regions of alpha helix and beta sheets. The beta sheets and helices are numbered in order. Bond indicates a bond; the bonds indicated are all disulfide bonds. Het denotes bonding to an ion or other cofactor. Human pancreatic amylase binds to a calcium ion.

The GenPept entry ends with the protein sequence.

Questions

1. Who published the structure of human pancreatic amylase? When did they publish the articles, and in what journal?

2. What is a disulfide bond? Is a disulfide bond a strong or a weak stabilizer of protein structure?

3. How many disulfide bonds are in human pancreatic amylase?

4. If possible, print the protein sequence of human pancreatic amylase. Using information from the GenPept entry, find the amino acids involved in the disulfide bonds, and draw a connection between the positions that are bonded together.

Viewing the Three-Dimensional Structure of Human Pancreatic Amylase

1. At the top of the GenPept screen for human pancreatic amylase 1HNY, you may have a box with the word Structure on it among other links. If so, click on it. If you don't have links in boxes, you may have two menu boxes reading Display and GenPept. Click on the GenPept box and hold down until you see the drop-down menu. Select Structure links from the menu and click on Display.

2. On the next screen, click on Structure. This takes you to a summary page for the protein structure. Click on View/Save Structure and save the structure to the data folder of Cn3D. Use Cn3D to view the structure.

3. Compare the structure with the structural information from GenPept. Play with the viewing options to get the most useful labels (none, every amino acid, every fifth amino acid, and so on) and the most useful display format (space-filling, wire, and so on). Rotate the structure and zoom in on various areas. See if you can locate the calcium ion. See if you can locate the disulfide bridges.

Part 3: exploring online biomedical literature references

Part of the responsibility of a scientist is publishing your results so that other scientists can use your data

and check your work. There are literally thousands of different journals in which scientists can publish their work. Most of these journals focus on specific areas or types of research.

Before the existence of the Internet, it could be quite a challenge to find published biomedical information. You would have to go to a science library, for one thing. Once you got there, it could take hours of work to find some good references, and even then you were not sure you'd found all the important ones.

After the Internet was established, the National Library of Medicine revolutionized the lives of researchers by creating an online database of the biomedical literature. Called MEDLINE, it allowed anyone with a computer and Internet access to search thousands of journals with the click of a mouse.

MEDLINE gives you reference information about an article and usually includes a short English-language summary of its contents. These summaries are called abstracts. Many biomedical journals are now published electronically, and subscribers can access the complete articles directly.

This activity is an introduction to the world of biomedical literature. You will learn how to search the

literature, and you can go back and look for information at any time in the future. If you become interested in a particular topic, you can search occasionally to see what new information has been published.

Procedure

1. Go to the NCBI home page (http://www.ncbi.nlm.nih.gov) and use the pull-down menu to select PubMed from the search menu. Enter "human amylase" in the search window. Note that PubMed appears in the dark bar above the search menu. If you click on this instead, you will get a slightly different screen from the one described below.

2. The next screen shows that over 7,000 documents were identified. We need to make the search more specific to get down to a manageable number of documents. Use the Add Term(s) to Search box to add the term "salivary" to the search. Click on Search.

3. Now the number is down to over 1,000. Use the Add Term(s) to Search box to add the term "gene" to the search. Click on Search.

4. Now over 70 documents are identified. Use the Add Term(s) to Search box to add the term "evolution." Click on Search.

5. Addition of the last term should reduce the number of hits to nine (this was in January 2000; there may be additional references now). Click on Retrieve Documents. You will get a list of the references.

6. Click on "The remarkable evolutionary history of the human amylase genes" by Meisler et al. and read the abstract.

Questions

1. Where did Meisler and Ting work at the time they published this article? When did they publish it?

2. According to the abstract, how many tandem copies of the human amylase gene are there?

3. What is responsible for the change in tissue specificity from pancreas to salivary gland?

4. What do the authors cite as evidence that there is strong evolutionary selection for amylase to be expressed in saliva?

Challenges

1. Remember that in chapter 27 you tested several substances for the presence of amylase. One substance we asked you to test was dog saliva, which tested negative. By searching the biomedical literature, find out whether dogs have an amylase gene. Can you find why their saliva tested negative for amylase activity?

2. Has anyone ever published an article about using amylase sequences as molecular clocks? If so, give the citation. What did they do, and what did they find?

E. Molecular Biology and Genetics

When you took biology, you probably had a unit in genetics, in which you learned about dominant and recessive traits. You may have wondered why certain traits (like tallness in Mendel's peas) were dominant. This section ties our growing knowledge of molecular biology to classical genetics. Through the readings and exercises, you will see how classical genetic observations can be explained and understood through molecular biology.

Humans typically are more interested in humans than in any other subject. Unfortunately, humans are a poor system for describing genetics. Compared to experimental systems with purebred strains and a wealth of genetic data, our understanding of human molecular genetics is slight. In fact, most of our knowledge of our own molecular biology comes from comparisons with animal systems.

We will present the connection between molecular biology and genetics by using an organism with whom most of you are familiar: the dog. This section begins with a discussion of simple cases of dog coat color variations, connecting the genetic observations with the molecular explanation. After establishing the system in the dog in chapters 31 and 32, we provide examples from human genetics that are based on the same cellular biochemistry. Chapter 33 includes a discussion of human genetic diseases that focuses on the molecular basis of these diseases. The section ends with a Reading on the molecular basis of cancer.

Classroom Activities

An Adventure in Dog Hair, Part I

Introduction

Cocoa and Midnight are Labrador retrievers. Cocoa is a brown female; Midnight is a black male. Cocoa and Midnight have puppies, and all of them are black, like Midnight. Can we use our understanding of genes and chromosomes to explain this?

On your worksheet, there are pictures of Cocoa and Midnight, with representations of their chromosomes bearing the coat color gene. Actually, dogs have many pairs of chromosomes, but the gene that determines black or brown color is on only one of them. Since we are currently interested in the black or brown color, we will only look at that chromosome pair.

Draw the gametes that Cocoa and Midnight could produce, showing the coat color chromosome. Review the steps of meiosis if you need to. Look at the gametes the parent dogs could produce. Fertilization is a random event, so to model it we would select one of Cocoa's gametes at random and one of Midnight's gametes at random and combine them.

What chromosome combination could the puppies have? Will all the puppies have the same combination?

As you know, it is not the entire chromosome that affects coat color; it is a specific gene on that chromosome. From now on, we will talk about the gene when we mean the gene and the chromosome when we mean the chromosome.

Fill in Table 1 on Worksheet 31.1 (see Appendix A) for Midnight, Cocoa, and the puppies, showing their coat color genes. Color the picture of the canine family.

Talk like a geneticist

You know all the puppies must have one of their mother's brown coat color genes and one of their father's black coat color genes, yet all the puppies are black, just like their father. Geneticists have a way to communicate about situations like this one. They say that the black coat color is dominant over the brown coat color. Similarly, they say that the brown coat color is recessive to the black coat color.

Geneticists also have a standard way of abbreviating names of genes when they know which is dominant and which is recessive. They usually use a capital letter to refer to the dominant gene, such as B for black, and the same letter in lowercase to refer to the recessive gene, i.e., b for nonblack. It is just a coincidence that the small letter b stands for brown in this case. Geneticists would use the small b no matter what the recessive color was—red, purple, or green. The name of the dominant trait determines which letter will be used.

Here's another handy term. The dominant black (B) and the recessive brown (b) are two different forms of the same gene. Geneticists call alternative forms of the same gene alleles. So B (black) is an allele of the coat color gene, as is b (brown).

Here's an important point. The terms dominant and recessive do not *explain* anything; they simply *describe* an observation like the one we made with Cocoa and Midnight's puppies. To understand why one version of a gene (black coat color) is expressed while another (brown coat color) is hidden, you have to understand what the proteins encoded by these versions of the gene are doing.

An Adventure in Dog Hair

Let's go on an adventure in dog hair color. We will use the scientific terminology to describe the pigments and the cells that produce them. Don't worry about the terms; instead, think about the logic of what is going on.

Pigmentation in dogs and other mammals (including you) is caused by the relative amounts and types of two classes of pigment: eumelanin and phaeomelanin. The eumelanins are the black and brown pigments, and the phaeomelanins are the red and yellow.

Both eumelanins and phaeomelanins are synthesized in pigment-producing cells called melanocytes.

Since we are interested in Cocoa and Midnight, who are brown and black, let's focus on the eumelanins. A schematic of the synthesis pathways is shown in Figure 31.1. First, the enzyme tyrosinase converts the amino acid tyrosine to a chemical called dopaquinone. If the enzyme called tyrosinase-related protein 2 (TRP-2) is present, it converts the dopaquinone to a version of eumelanin that has a brown color, Cocoa's pigment. If the enzyme called tyrosinase-related protein 1 (TRP-1) is present, it converts the brown version of eumelanin into the final, black pigment.

Remember that enzymes are proteins. How are proteins made? What tells the cell how to put a specific protein together? If you don't remember, go back to chapter 4 and refresh your memory, because you need to know where proteins come from in order to understand the rest of this story.

What would happen to pigment production if the gene encoding the instructions for making TRP-1 was defective, so TRP-1 couldn't be made? As you think about this question, assume that tyrosinase and TRP-2 are still present and functioning normally. Would tyrosine still be converted into dopaquinone? Would dopaquinone still be converted into the brown pigment? Would the brown pigment still be converted into the black pigment?

Chocolate Labs like Cocoa cannot make TRP-1. Consequently, their melanocytes produce brown pigment instead of black pigment (Figure 31.2). Look at the photo of Cocoa in Color plate 1. What color is her nose? Her lips? Because chocolate Labs do not make TRP-1, they cannot make black pigment, period. Their noses and lips are brown instead of black, too.

The b allele is actually a nonfunctional allele of the gene for TRP-1. Because Cocoa cannot make TRP-1, she is brown. Midnight's B allele is the functional allele of the gene for TRP-1. Because Midnight's melanocytes make TRP-1, they can convert the brown pigment to the black form, and Midnight is black.

Now let's think about the situation with Cocoa and Midnight's puppies. They have one recessive allele (b) from Cocoa, the defective gene for TRP-1. They have one dominant allele (B) from Midnight, the functional gene for TRP-1. What is going to happen in the puppies' melanocytes?

Let's think about what we know. We know that all the puppies' cells, including their melanocytes, have one functional allele for TRP-1 (the B allele) and one nonfunctional allele for TRP-1 (the b allele). Will the melanocytes be able to make TRP-1 from the instructions on the chromosome that came from Cocoa? What about the instructions on the chromosome that came from Midnight? What color are the puppies? Do the puppies' melanocytes make TRP-1?

Fill in Table 2 on Worksheet 31.1 (see Appendix A). Use the geneticist's abbreviations B and b when showing the genetic makeup of the dogs.

Figure 31.1 Synthesis of black and brown pigments (eumelanins) in melanocytes. First, tyrosinase converts tyrosine to dopaquinone. Next, TRP-2 converts dopaquinone to brown pigment. Finally, TRP-1 converts the brown pigment into black pigment.

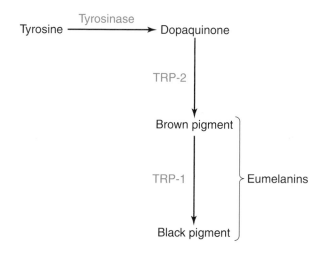

Figure 31.2 Pigment synthesis in Cocoa's melanocytes. First, tyrosinase converts tyrosine to dopaquinone. Next, TRP-2 converts dopaquinone to brown pigment. There is no TRP-1 to convert the brown pigment into black pigment. What effect does this lack of TRP-1 have on Cocoa's coloration?

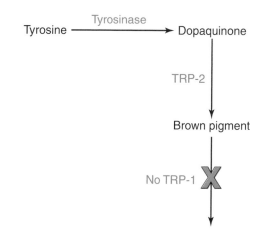

Talk like a geneticist

Now it's time to introduce two useful terms you probably already know: *genotype* and *phenotype*. The genotype is the nature of an individual's genes. Cocoa's genotype is bb, or brown/brown. The phenotype is the outward expression of the genotype. Cocoa's phenotype is brown. Midnight's genotype is BB, or black/black. What is his phenotype? The puppies' genotype is Bb, or black/brown. What is their phenotype?

Questions

1. A gardener has two strains of peas. One always produces yellow seeds, and the other always produces green seeds. The gardener crosses the two strains of seeds. All of the new plants have yellow seeds.

1a. What is the genotype of the green-seed parent plant? Its phenotype?

1b. What is the genotype of the yellow-seed parent plant? Its phenotype?

1c. What is the genotype of their offspring? The phenotype?

1d. Is one of the seed color genes dominant over the other? How do you know?

1e. What would be the geneticist's abbreviation for the genotype of the parent yellow plant? The parent green plant? The offspring yellow plants?

2. In the imaginary geneflower, a starting substance is converted by enzyme X to a blue pigment, which is converted by enzyme Y to a purple pigment (Figure 31.3). Most geneflowers have purple flowers, but there is a variety that always produces blue flowers.

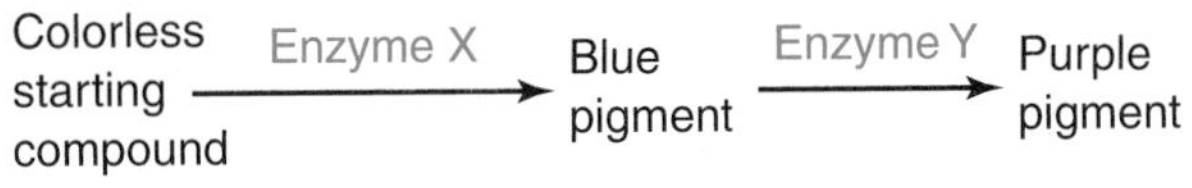

2a. Using what you know about the synthesis pathway for blue and purple pigments in geneflowers, give an explanation for the blue variety.

2b. When gardeners cross the purple-flowered geneflower with the blue-flowered geneflower, they always get purple-flowered geneflowers. Which flower color is dominant? Which is recessive?

2c. Use your knowledge of pigment synthesis in geneflowers to explain why one color is dominant over the other.

3. In a second imaginary flower, the roundbud, red pigment is synthesized from a white precursor by enzyme Q (Figure 31.4). It was believed that all roundbuds were red until a knowledgeable, observant teenager discovered a meadow full of white-flowered roundbuds while hiking with the Genetics Club.

3a. Use your knowledge of the synthesis pathway for red pigment to explain the phenotype of the white-flowered variety.

3b. Would you predict that red color would be dominant or recessive to white?

3c. If you crossed the white roundbud with the red roundbud, what would you expect to see, and why?

Figure 31.3 Pigment biosynthesis in the geneflower. Enzyme X converts a colorless precursor into a blue pigment. Next, enzyme Y converts the blue pigment into a purple pigment.

Colorless starting compound —Enzyme X→ Blue pigment —Enzyme Y→ Purple pigment

Figure 31.4 Pigment biosynthesis in the roundbud. Enzyme Q converts a white precursor into a red pigment.

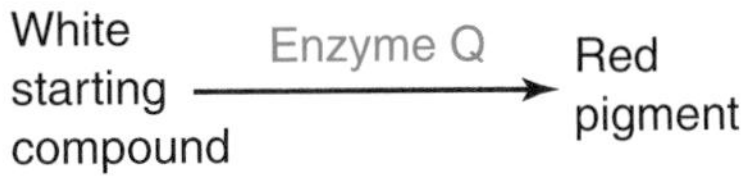

White starting compound —Enzyme Q→ Red pigment

3d. Fill in Table 1 on Worksheet 31.2 (see Appendix A) for the red and white roundbuds.

The teenaged discoverer of the white roundbud took some of the plants back for the Genetics Club to experiment with. They found that when they crossed white-flowered plants with white-flowered plants, they always got white-flowered plants. When they crossed the white-flowered plants with the red-flowered plants, however, they got a surprise. The offspring plants had neither red nor white flowers, but instead had pink flowers.

3e. Assume your predictions about the genotype for making enzyme Q in red and white roundbuds and in their offspring (which you probably expected to be red) are all correct, and nothing is going on that you don't know about. Using your knowledge of enzyme Q and its function, can you propose an explanation for why the offspring have pink flowers?

Talk like a geneticist

The descriptive term for what the Genetics Club observed with the roundbuds is *codominance*. When the red- and white-flowered varieties were crossed, neither color was dominant over the other. Instead, the phenotype of the offspring was intermediate between the two parents. In cases of codominance, geneticists use capital letters for both traits. In this case, the red-flowered roundbud is RR, the white-flowered roundbud is WW, and the pink-flowered roundbud is RW.

Remember, codominant is a term that merely describes an observation. It does not explain why the observed phenotype occurs. In roundbuds, the explanation involves the amount of enzyme Q synthe- sized in cells with two functional copies of the gene, versus the amount synthesized in cells with only one functional copy of the gene. In other cases, such as human blood types, there is a different explanation for the observation of codominance.

Blood groups A and B are caused by the presence of specific molecules on the surface of red blood cells. If an individual has the "A" allele, he has the A molecule on his cells. If an individual has the "B" allele, he has the B molecule on his cells. If an individual has both an A and a B allele, that individual has both molecules on the surface of his blood cells and belongs to blood group AB. In this case, the observed codominance is caused by the expression of two different forms of the same gene.

An Adventure in Dog Hair, Part II: Yellow Labs

32

Many of you have probably already wondered about the third possible coat color of Labrador retrievers, the yellow Lab. The yellow color is determined by an altogether different pair of genes on a different chromosome from the tyrosinase-related protein 1 (TRP-1) gene. To understand how the yellow color is produced, we must resume our adventure in dog hair.

When we last left our dog hair adventure, we had learned that pigments are synthesized in cells called melanocytes. In melanocytes, tyrosine is converted to a chemical called dopaquinone by tyrosinase. Then TRP-2 converts the dopaquinone to brown pigment, and TRP-1 converts the brown pigment to black. Chocolate Labs are brown because they cannot make TRP-1.

The story with yellow Labs starts here. In melanocytes, the synthesis of TRP-2 and TRP-1 is regulated by hormones. In particular, a hormone called melanocyte-stimulating hormone (MSH, sometimes referred to as melanocortin) travels through the bloodstream and binds to the surface of melanocytes found in hair follicles. They bind to these melanocytes because the melanocytes have a receptor protein on their surface that exactly fits the shape of the MSH molecule.

This receptor protein's job is to receive the hormone's signal and transmit it to the inside of the cell. When MSH binds to the receptor on the surface of a hair follicle melanocyte, the receptor protein changes shape. The change in shape generates a signal to the inside of the cell, telling it to produce TRP-2 and TRP-1 (Figure 32.1).

What happens if the melanocytes do not get the signal to produce TRP-2 and TRP-1? Tyrosinase is still there, converting tyrosine into dopaquinone. If TRP-2 and TRP-1 are not present in the melanocyte, the dopaquinone is converted into something else.

If TRP-2 and TRP-1 are lacking, all the dopaquinone will be converted to phaeomelanin (Figure 32.2). The phaeomelanins are the family of red and yellow pigments we mentioned briefly at the beginning of the dog hair adventure and haven't mentioned since. The exact color of the phaeomelanin depends on the enzymes available for its synthesis, as is the case with the color of the eumelanin (black or brown). In Labrador retrievers, the phaeomelanin is yellow.

Yellow Labs are yellow because the receptor for MSH doesn't work. So even though the dogs' bodies produce

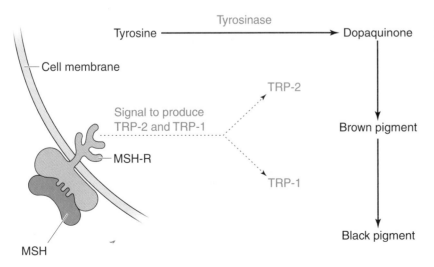

Figure 32.1 In black and chocolate Labs, MSH binds to its receptor, MSH-R, signaling the melanocyte to produce TRP-2 and TRP-1. The melanocyte can then synthesize black or brown pigment.

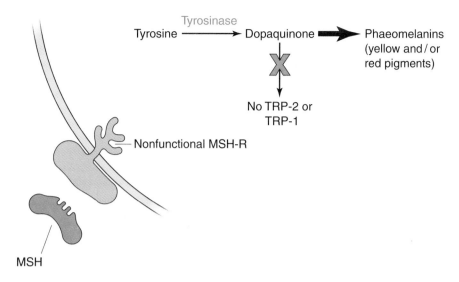

Figure 32.2 In yellow Labs, the MSH-R is nonfunctional. MSH cannot bind to it, and the hair follicle melanocytes never receive the signal to synthesize TRP-2 and TRP-1. The melanocytes therefore do not synthesize black or brown pigments. Instead, dopaquinone is converted into yellow phaeomelanin.

MSH and have functional genes for TRP-2 and TRP-1, the signal to produce the enzymes is never transmitted to the melanocyte. So the melanocyte never makes TRP-2 and TRP-1. All the dopaquinone is converted into yellow phaeomelanin instead of eumelanin.

The gene that causes yellow coat color in Labrador retrievers is actually a nonfunctional allele of the gene for the MSH receptor (MSH-R). We will call the functional gene R for receptor, and we will call the nonfunctional allele r. For a Lab to be yellow, it cannot make any functional receptor protein, so its genotype must be rr. Black and brown Labs have at least one functional allele, so their melanocytes receive the hormone signal to make TRP-2 and TRP-1. They have genotype RR or Rr.

If you crossed a Labrador of genotype RR with one of genotype rr, all the puppies would be of genotype Rr. Would these puppies be yellow or not?

Interestingly, the r allele affects *only* melanocytes in hair follicles. Melanocytes responsible for pigmentation everywhere else function normally. Now it gets fun. You can tell whether a yellow Lab would have been black or brown if its r gene were functional by the color of its nose and lips! Black Labs, who make TRP-1 and produce black pigment, have black noses and lips, in addition to black fur. Chocolate Labs, who do not make TRP-1 and produce no black pigment anywhere, have brown noses and lips. Look at the photos of Cocoa and Midnight (Color plate 1).

A Lab with genotype RRBB has black fur, black lips, and a black nose. Its melanocytes receive signals normally and can make enzyme 3. A Lab with genotype rrBB has yellow fur, but its nose and lips are black. A Lab with genotype RRbb has brown fur, brown lips, and a brown nose. A Lab with genotype rrbb has yellow fur, but its nose and lips are brown (Color plate 2).

Questions

1. Is the allele that results in yellow fur in Labrador retrievers dominant or recessive? Explain your answer.

2. What genotype must a Lab have to be yellow?

3. What are all possible genotypes of a yellow Lab with black lips?

4. What are all possible genotypes of a yellow Lab with brown lips?

Genetics with Yellow Labs

To think about the genetics of the B/b gene and the R/r gene at the same time, we can use the same techniques we used when we were thinking only about B/b. Let's start with a cross between a male Lab of genotype BBRR and a female Lab of genotype bbrr. What would the phenotypes of these dogs be? Color them in the picture on Worksheet 32.1 (see Appendix A).

To predict the genotypes and phenotypes of the puppies, we first have to figure out what gametes each parent could produce. Remember that in meiosis, each gamete receives a copy of only one member of each chromosome pair. (Assume the B and R genes are on different chromosomes.) What are the possible gametes the male parent could produce? The female parent? What will be the genotype(s) of the puppies? Fill in the blanks and color the puppies in the picture on Worksheet 32.1.

That was a simple example, because the parents were homozygous. Now let's have some fun and look at what would happen if two dogs of genotype BbRr were crossed. First, what is the phenotype of the parent dogs? Color them on Worksheet 32.2.

To predict offspring genotypes and phenotypes in a cross like this one, Punnett squares are very helpful. Use the same steps to construct one.

1. Assign symbols to the genotype of each parent. We already know in this case: BbRr.
2. Write out the cross, using the genetic symbols: BbRr × BbRr.
3. Determine the gamete possibilities for each parent, and fill in the outline of a square. This step is more complicated than it was in the previous example. One way to figure out the possible gametes is to make a table like this one:

Possible gametes for parents with genotype BbRr

Black/brown allele	Receptor allele	Gamete
B	R	BR
B	r	Br
b	R	bR
b	r	br

Since both parents have the same genotype, this table shows the gametes each one could produce. If the parents had different genotypes, it would be necessary to make a table for each one (unless

one was very easy, such as the homozygous parents above).

4. Fill in the outline of a square with the parents' gametes. This time each parent can make four different types of gamete, so the square has to have four slots down each side:

	BR	Br	bR	br
BR				
Br				
bR				
br				

5. Fill in the offspring genotypes produced by combining the different gametes.

	BR	Br	bR	br
BR	BBRR	BBRr	BbRR	BbRr
Br	BBRr	Bbrr	BbRr	Bbrr
bR	BbRR	BbRr	bbRR	bbRr
br	BbRr	Bbrr	bbRr	bbrr

6. Determine the genotype ratios by counting the contents of the boxes. Do this and fill in the table on Worksheet 32.2 (see Appendix A). What are all the different genotypes present in the offspring? How many times does each one appear in the Punnett square?
7. Interpret the genotypes in molecular terms. Fill in the chart, indicating whether each genotype will produce functional MSH receptor protein and functional TRP-1.
8. Determine the phenotype of each genotype. If the dog is a yellow Lab, indicate what color its lips will be. How many of each phenotype are predicted from the Punnett square?

The two dominant alleles in this cross are R and B. What are the associated phenotypes? Notice that you get the largest number of offspring (nine) showing these dominant phenotypes. The recessive alleles in the cross are b and r. What are the phenotypes associated with bb and rr? Notice that you get the intermediate number of offspring (three) for both combinations of one dominant and one recessive phenotype. What are these combinations in our cross? The smallest number of offspring show both recessive phenotypes. What is this phenotype in our cross?

Questions

1. A friend of yours has a pair of black Labs. These two dogs have a litter of puppies. To your friend's astonishment, about one-fourth of the puppies are chocolate Labs. Use genetics to explain this result to your friend, including why approximately one-fourth of the puppies show the chocolate color.

2. Another friend of yours also has a pair of black Labs. These two dogs have a litter of puppies. To this friend's astonishment, about one-fourth of the puppies are yellow (all have black noses and lips). Use genetics to explain this result to your friend, including why approximately one-fourth of the puppies show the yellow color.

3. A black Lab of genotype BbRr is crossed with a yellow Lab that has brown lips. Predict the genotype and phenotype ratio of the offspring. Hint: try using the Punnett square method outlined above.

Human Molecular Genetics

Introduction

Human genetics works the same way as dog genetics in terms of chromosomes, genes, enzymes, and other proteins. The big difference is that obviously we do not experiment with humans.

There are two basic approaches to learning about human genetics. In the classic approach, geneticists do what can only be described as detective work. First, a condition that seems to be inherited must be recognized. This usually happens through the medical profession: a patient comes in with an unusual condition, and the doctor learns that other family members have it. Geneticists would follow up on this report by visiting the family, interviewing relatives, and gathering as much information as possible. They would construct and analyze a family pedigree diagram to determine the inheritance pattern. This kind of detective work still occurs today, usually through medical centers associated with universities.

The second approach is based on animal studies and molecular biology techniques. The biochemistry of animals is pretty similar from species to species. So scientists do detailed studies in organisms that are small and easy to manipulate, such as mice and fruit flies. When an interesting gene is identified, such as the obesity gene first discovered in mice, scientists look for a homologous gene in humans through hybridization analysis or polymerase chain reaction. They use the gene sequence information from the animal to make probes and/or primers to try on human DNA. In this manner, the human counterpart to the mouse obesity gene was identified. Keep in mind, though, that a homologous gene doesn't necessarily have an identical function. Scientists must find additional evidence for the role of the homologous gene.

Here are some human genetics examples. See if you can apply what you learned from the Labrador retrievers to help you understand these stories. In these stories, the people are imaginary, but the diseases and their genetics are real.

John and Mary

John and Mary have brown hair, brown eyes, and medium-toned skin, as do their two children. They are astonished when their third child is born. He has white hair, very light, pinkish skin, and blue eyes. The doctor explains to them that their youngest son has *albinism*. His body cannot produce pigments (Color plate 3).

The doctor explains to John and Mary that pigments are produced from an amino acid called tyrosine. An enzyme called tyrosinase converts tyrosine to a chemical called dopaquinone, which is converted by other enzymes into all the body's pigments (Figure 33.1). Their young son's cells cannot make tyrosinase. Therefore, his cells cannot convert tyrosine to dopaquinone, and they cannot make pigment. The doctor reassures John and Mary that persons with albinism are otherwise normal and live normal lives, except for an extreme sensitivity to the sun. Why would that be?

This story should be reminding you of Cocoa and Midnight. Why was Cocoa brown and not black? What was happening in her cells?

Now think about the boy with albinism and his parents. Both his parents have normal pigmentation. How could they have a child with albinism? How could that child have siblings with normal pigmentation? Use what you have learned about genetics and the cellular chemistry of pigment synthesis to figure this out. Label the pedigree on Worksheet 33.1 (see Appendix A) with the genotypes of the family members (write all possibilities if there are more than one). Answer the questions on the worksheet.

Figure 33.1 Pigment production from tyrosine.

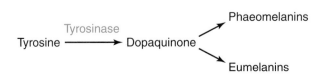

Bob and Susan

Bob and Susan have just had their first child. The baby underwent all the routine medical tests given to newborns. Afterward, the doctor and a genetic counselor come to talk with them.

The doctor explains that their baby lacks the ability to make an enzyme called phenylalanine hydroxylase (PH). This enzyme normally converts an amino acid called phenylalanine into the amino acid tyrosine. Tyrosine can then be converted to other things (Figure 33.2). The baby's cells cannot convert phenylalanine to tyrosine because they cannot make the PH enzyme. This condition is called phenylketonuria, or PKU.

The danger of this condition is that phenylalanine will build up in the baby's cells, since it cannot be converted to tyrosine. High levels of phenylalanine interfere with normal brain development in a child and therefore cause mental retardation.

The doctor quickly reassures Bob and Susan that their baby could develop normally, given proper treatment. Phenylalanine and other amino acids enter the body when protein-rich foods are digested (why would that be?). If Bob and Susan give their baby a strictly controlled diet that is quite low in phenylalanine, the buildup will not occur, and the baby's brain will develop properly.

Bob and Susan are upset about their baby's problem but relieved that the condition is treatable. They have several questions. Susan wants to know how she and Bob, who are both normal, could have a child with PKU. Bob asks the genetic counselor what would happen if he and Susan had another child. Would it also have PKU? Use your knowledge of genetics to answer Bob's and Susan's questions. Label the family pedigree on Worksheet 33.2 (see Appendix A) with the genotypes of Bob, Susan, and their child.

Here's a separate question you can answer based on what you know. People with PKU are very lightly pigmented. Use your knowledge of the biochemistry of pigment synthesis to explain this fact.

People with PKU must avoid phenylalanine-containing foods when they are children and when they are pregnant. Adults are less vulnerable than children because their brains have completed development. The natural source of phenylalanine in the diet is protein, but several years ago a new source was introduced into our food supply. The artificial sweetener aspartame contains phenylalanine. The next time you go to the grocery store, look at cans of diet soda and find the warning message on them.

Discovering PKU

Albinism is very obvious at birth, but PKU is not. The noticeable symptom of PKU is mental retardation, which has many other causes. Although children with PKU are lightly pigmented, they are not so light as to call attention to themselves.

PKU was first identified as a specific condition in 1934. An observant mother of two mentally retarded children noticed that her children's diapers had an abnormal, musty smell. She remarked on this observation to a relative named Folling, who was a physician and a biochemist.

Folling was curious and analyzed the urine of the children. He found that it contained a large amount of a substance normally present in only tiny amounts. The substance, phenylketone, was the direct result of the buildup of phenylalanine in the children's bodies. Folling did not know why so much phenylketone was present, but he published his findings about the two children. The condition was called phenylketonuria, which essentially means "lots of phenylketone in the urine."

In the years after Folling's discovery, many institutionalized mentally retarded people were tested for phenylketone in the urine. About 1% of the people tested were found to have PKU.

It was 1954 before the biochemistry of PKU was figured out, and 1961 before a reliable way of routinely testing newborns for the condition was developed.

Figure 33.2 The PH enzyme converts the amino acid phenylalanine into the amino acid tyrosine.

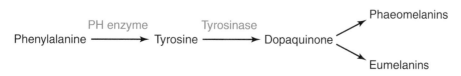

This test allowed affected infants to be identified before the damage to their brains began.

At that time, researchers knew that PKU was caused by a buildup of phenylalanine, so they reasoned (and hoped) that a diet very low in phenylalanine might protect affected babies' brains from damage. A diet was devised that was very low in phenylalanine but contained enough of the other 19 amino acids to allow the baby to make the proteins it needed to grow. The diet proved to be a miracle cure, but one that was based on scientific reasoning.

A problem with blood cells

Among African Americans whose families originated in West Africa, a specific type of blood disorder occurs more frequently than in other American subgroups. In this disorder, the red blood cells have an unusual shape. Instead of being round, they collapse into flat, curved shapes that resemble an ancient harvesting tool called a sickle. Because of this characteristic shape, the disorder is called sickle-cell anemia.

Sickle-cell anemia is caused by an abnormal form of hemoglobin, the blood protein that carries oxygen. Normally, hemoglobin molecules are distributed throughout the red blood cells. In sickle-cell patients, the abnormal hemoglobin molecules stick together, causing the red blood cell to collapse and assume the sickle shape. Because of their shape, sickled cells do not flow smoothly through capillaries. Instead, they form blockages that prevent adequate circulation and oxygen distribution, causing severe pain and tissue damage.

In the early part of the 20th century, the genetics of sickle-cell anemia was not understood. There appeared to be two forms of the disease: the severe form, which usually led to early death, and a mild form that had little effect on the individual. The severe form was referred to as sickle-cell anemia, and the mild form was called sickle-cell trait. The shape of blood cells in individuals with sickle-cell trait was not as distorted as it was in individuals with sickle-cell anemia, but the blood cells were abnormal (Figure 33.3).

Figure 33.3 (A) Red blood cells, ×4,600 (photograph courtesy of VU/©M. Bessir-D. Fawcett). (B) Sickle-cell anemia (photograph courtesy of Science VU). (C) Normal and sickled red blood cells, ×943 (photograph courtesy of VU/©George J. Wilder).

A

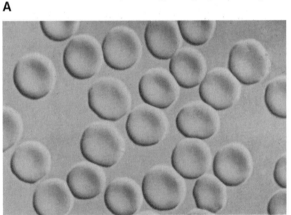

C

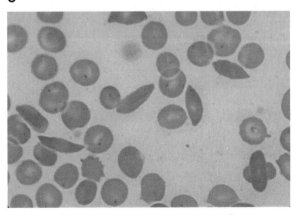

B

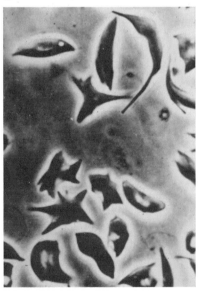

Children with sickle-cell anemia usually appeared in emergency rooms at an early age because of their disease. Their badly impaired blood circulation caused a number of severe problems. Individuals with sickle-cell trait were not affected by their blood cells' unusual shape and so were less commonly identified. The prevailing belief about sickle-cell disease in the 1940s was that it was caused by a gene that did different things in different people: sometimes causing the anemia, sometimes causing the trait.

In 1948, a physician and geneticist named James Neel decided to test an idea he had about sickle-cell anemia and sickle-cell trait. At that time, he was just beginning a job in Detroit, Michigan, which had a large African American population. Dr. Neel began to contact the families of sickle-cell patients. With their permission, he tested family members for the presence of sickled red cells.

Dr. Neel found, as he suspected, that both parents of children with sickle-cell anemia had sickle-cell trait, though they usually did not know it. Furthermore, many of the siblings of the anemia patients had sickle-cell trait. Dr. Neel drew his conclusions and published his data and interpretation in the journal *Science* in 1949. (*Science* is still published today, and important findings are often first described there.)

A few months later, Neel read an article written by scientists who were studying the hemoglobin protein. These scientists had found a way to characterize hemoglobin isolated from blood. They had found that hemoglobin from sickle-cell patients and hemoglobin from normal individuals migrated differently during electrophoresis under certain conditions. Consequently, they could identify which type of hemoglobin was present in a blood sample (Figure 33.4). These scientists reported that when they examined blood samples from individuals with sickle-cell trait, they found both forms of hemoglobin. When Dr. Neel read this report, he concluded that his hypothesis about the inheritance of sickle-cell anemia was correct.

Here is a challenge for you. Put together an explanation that takes all these facts into account: the inheritance of sickle-cell anemia, sickle-cell trait, and the hemoglobin test results. Use Worksheet 33.3 (see Appendix A) as a guide.

It's usually not that simple

You've been learning about the genetics of fairly rare but medically important diseases. But what about the inheritance of more normal traits such as height, general build, and hair color? We wish we knew! We are

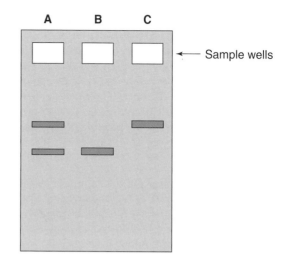

Figure 33.4   Electrophoresis of hemoglobin from a normal individual (lane A), from an individual with sickle-cell anemia (lane B), and from an individual with sickle-cell trait (lane C).

not being sarcastic when we say that. We say it because the inheritance of most human traits cannot be explained by the action of one or two genes. Inheritance of most traits is complex and poorly understood, if it is understood at all.

Most characteristics and the development of diseases such as heart disease and high blood pressure seem to be influenced by many genes and also by the environment in which an individual lives. Here are some simple examples of environmental influence. Suppose an individual had genes that could make him grow very tall, but he grew up impoverished and malnourished. His growth might be quite stunted. Despite his genes, he might turn out to be a short or average-sized adult. Suppose an individual had genes for normal intelligence but was exposed to high concentrations of alcohol while a developing fetus. Despite his genes, this person might be born with the mental retardation called fetal alcohol syndrome. Environmental conditions—what you eat, what you are exposed to, what diseases you have, and so on—can and do contribute to an individual's traits and fates. By the way, fetal alcohol syndrome is the most common cause of mental retardation in newborns today.

Most traits are influenced by many genes

If you think about most human characteristics, e.g., hair color, height, and eye color, you realize that there is no specific set of categories. We do not come in a specific set of heights, nor do we have a specific set of eye colors. Instead, our heights range from very short to very tall, with everything in between. Our eyes range from colors like light blue, amber, gray, and green to dark brown and everything in between (Color plate 4). Incidentally, blue eyes are not the result of blue pigment. Instead, they are the result of lack of pigment in the iris. Because of the way light is absorbed in the human eye, the iris appears blue if no pigments are present. That is why people with albinism have blue eyes.

Hair color, which is the equivalent of coat color in mammals, is the same way. Humans show every shade from fire-red to platinum blond to black. What if you were to try to study hair color inheritance in humans? You could easily observe families like one of ours: a man with black hair marries a woman with medium-brown hair. They have four children: a boy with medium brown hair, a girl with golden blonde hair, a girl with white-blond hair that slowly darkens, and a girl with reddish-brown hair. What can you make of that? Probably only a lot of confusion.

Why does it look so simple in dogs?

Some of you may be wondering why we can explain coat colors in Labrador retrievers if hair color inheritance in humans is so complicated. The answer to that question involves an important difference between the study of human genetics and the study of animal genetics. While we have genetically pure strains of mice, dogs, fruit flies, etc., there are no "pure breeds" of humans.

It may sound comical, but there's an important point here. We can figure out many of the genes that work in dogs and experimental animals like mice and fruit flies because people have bred them for generations and created pure breeds that differ very little in their genetic makeup from one individual to another.

Think about the Labrador retrievers. Labs come in three colors: black, brown, and yellow. Does that mean they only have two genes (the TRP-1 gene, which governs black or brown, and the MSH-R gene, which governs yellow) that can affect coat color? No, what it means is that Labradors have been bred so purely that all the other coat color genes are the same in every Lab. The only differences are in the TRP-1 gene and the MSH-R gene. So in Labrador retrievers, coat color really is determined by only two genes.

If you think about other breeds of dogs, you realize that coat color inheritance in dogs is not so simple, either. Think of examples of different coat colors, such as black-and-white spotted Dalmatians, red Irish setters, golden retrievers, the typical beagle color pattern, the large splotches of color found in some basset hounds, the tiny spots and blue-gray color of blue tick hounds, the black-and-tan pattern of some Dobermans and dachshunds, and so on. Dogs have many, many genes that affect coat color. However, in pure breeds almost all the genes are the same, so only a very few genes control the coat color in that particular breed (Color plate 5). Coat color inheritance has been extensively studied in mice, and over 100 different genes have been found to affect it! Some genes are more influential than others. The same is probably true of humans.

Don't give up on human genetics, though. Pedigree and population analyses, studies in animal breeding, and molecular biology point the way to understanding many human genes, and molecular biology now allows us to identify, clone, and study them. The Human Genome Project, by providing more and more DNA sequence information, will make it easier and easier to use data gleaned from animal studies to understand human molecular biology. If you were a researcher and identified a gene in an animal model (for example, the gene for an enzyme that converts yellow phaeomelanin to red in dogs), you would be able to compare the animal gene sequence with more and more of the human genome sequence to look for counterpart genes. The foreseeable future will be a very exciting time for human genetics.

Questions for Thought and Discussion

1. Sickle-cell anemia is caused by a specific change in the sixth amino acid of the β-globin protein that forms part of hemoglobin. Shown below are a DNA sequence encoding the first seven amino acids of the normal β-globin protein and a sequence encoding the first seven amino acids of the sickle-cell protein.

Normal amino acid sequence:

5′ GTT CAT CTA ACC CCT GAG GAG . . . 3′

Sickle-cell sequence:

5′ GTT CAT CTA ACC CCT GTG GAG . . . 3′

1a. Translate these sequences. What is the amino acid change that causes sickle-cell disease?

1b. The restriction enzyme *Mst*II cleaves the DNA sequence CCTGAGG. In human DNA, there is a *Mst*II site 1,150 base pairs 5′ (to the left as shown) of the β-globin coding region and 200 base pairs 3′ of the region shown. Design a DNA-based test for sickle-cell anemia using the enzyme *Mst*II. Draw a gel outline and show the expected results of your test on a homozygous normal individual, a sickle-cell anemia patient, and a person with sickle-cell trait (a heterozygous carrier). Provide a scale of base-pair sizes to the left of the gel outline, based on the expected results from your test.

2. A chromosomal translocation occurs when a segment of one chromosome breaks off and attaches to the end of a different chromosome. These events can occur during formation of the gametes, or they can happen in individual cells during the life of the person. If the gamete carries a translocation, all the cells in the resulting person will have it.

2a. Occasionally, healthy adults who carry a translocation in all their cells will be identified. The assumption we make about these people is that they did not gain or lose genetic information during the translocation, and the particular rearrangement (shifting of a segment of one chromosome to another) caused no harm. The translocations in these individuals are called balanced translocations. Would you expect the offspring of a person with a balanced translocation to be healthy?

2b. Translocations that happen in individual cells during adulthood can be harmless, or they can lead to cancer. For example, translocation of the end of chromosome 22 to chromosome 9 is associated with leukemia. Why do you think some translocations can be harmful and some can cause cancer or other problems?

3. Gaucher's disease is a recessive genetic disease caused by the lack of the enzyme glucocerebrosidase. By looking at the gene for that enzyme in Gaucher's patients, scientists have identified many different mutations that cause the disease. Interestingly, some mutations lead to a mild form of the disease (in homozygotes), and other mutations lead to a very severe form (again in homozygotes). Heterozygotes with one normal copy of the gene are healthy no matter what form of the mutant gene they have on the other chromosome.

3a. Why do you think some mutations in the gene can cause mild disease and others can cause severe disease?

3b. Some patients with Gaucher's disease have two different mutant forms of the gene. If a person has one copy of the gene with a mutation leading to mild disease and the other copy of the gene with a mutation leading to severe disease, that patient will have mild disease. Propose an explanation for why the disease would be mild.

4. In the following cases, would the associated disease be observed to be dominant, recessive, or in between?

4a. The "disease gene" is a nonfunctional form of the gene. No protein product (or a harmless nonfunctional one) is produced from this gene. A cell with one functional copy of the gene can synthesize enough normal protein to take care of its function.

4b. The "disease gene" is a nonfunctional form of the gene. No protein product (or a harmless nonfunctional one) is produced from this gene. A cell with one functional copy of the gene cannot synthesize enough normal protein to achieve full function, but it can survive. If none of the protein is present, the cell cannot live.

4c. The "disease gene" is a nonfunctional form of the gene. This nonfunctional form encodes the production of an altered form of its protein. The altered form interferes with the functioning of the normal version of the protein, so the cells of a person with one nonfunctional allele and one normal allele cannot carry out the function of the protein any better than the cells of a person with two nonfunctional alleles.

Classroom Activities

33. **Human Molecular Genetics** • 261

5. Huntington's disease is a dominant genetic disease. Symptoms of this disease do not appear until middle age, typically when the patient is in his or her 40s. However, when symptoms appear, they are devastating. The patient gradually loses the ability to think. At the same time, he begins to experience increasingly uncontrolled body movements, usually twitching and shaking. There is no cure. Since the disease is dominant, there is a 50% chance that the child of an affected parent will also develop it. However, since the disease doesn't show until middle age, an individual can reproduce before he or she knows that the disease gene is present.

Suppose that when you are a teenager, one of your parents develops Huntington's disease. You watch your strong, intelligent parent gradually lose mental and physical function. At first, muscle coordination is slightly impaired, and forgetfulness, confusion, and personality changes develop. As Huntington's disease progresses, these conditions worsen steadily until both voluntary and involuntary movement is uncontrolled, with jerking and writhing. Speech is slurred, thought processes diminish, and psychiatric conditions such as depression and uncontrolled rage appear. In the final stages, the Huntington's patient is mute, cognitively nonfunctional, and frozen in a contorted position.

You know that there is a 50-50 chance that you will suffer the same fate and could transmit it to your children. There is now a DNA test that will tell you if you carry the Huntington's disease gene. Would you choose to take the test? What factors would enter into your decision?

The Human Genome Project: Its Science, Applications, and Issues

Many of the DNA-based technologies described in this book are being applied to the largest project ever tackled by biologists in a systematic, coordinated, interactive way: the Human Genome Project. The development of recombinant DNA technology encouraged scientists to dream about the possibility of identifying and locating all of the genes in the human genome. As recently as the mid-1980s, however, such an undertaking seemed like an impossible dream. Using the technologies available at that time, it would have taken 1,000 researchers 200 years to complete the task, even if they devoted all their time to that one project. Excluding salaries, this project would cost between $15 billion and $20 billion. Such an investment of money and labor in one project seemed foolhardy. But a few visionary scientists began to consider strategies that would make the job feasible, and the initiative began to take shape.

Goals and Objectives

In 1990, Congress formally initiated the Human Genome Project by providing funding for 5 years at less than $200 million per year. The project's initial goal was twofold:

• Develop detailed maps of the human genome
• Determine the complete nucleotide sequence of human DNA by the year 2005

To accomplish these two goals, a number of objectives had to be met. The first involved developing the software and database management systems to support the mapping and sequencing goals. The Human Genome Project generates massive amounts of data, and researchers needed consistent methods for storing, managing, accessing, and integrating the data before they began to collect significant amounts of data.

Organizers of the project also recognized early on that they did not have the technology for such large-scale sequencing, so one of the objectives of the first 5 years was to invent the methods and tools for large-scale sequencing. The chain termination method of

DNA sequencing described previously remained a fundamental strategy, but only with automation could this become cost-effective. Automated sequencing replaced manual sequencing, and laboratory robots were developed to automate other parts of the process, such as preparation of the DNA templates for sequencing.

While some scientists were developing better sequencing methods, others focused on the goal of developing detailed maps. To achieve this goal, they needed better mapping techniques and better methods for cloning large amounts of DNA that could be assembled into overlapping fragments.

Because of technological innovations, the development of new laboratory techniques, and increasing experience, the Human Genome Project is progressing ahead of the schedule originally proposed in 1990. Project leaders publicly updated their goals and shortened the schedule in 1993 and 1998. As of now, the Human Genome Project is scheduled for completion in 2003, the 50th anniversary of Watson's and Crick's discovery of the structure of DNA.

Model Organisms

Since the human genome is so large and complex, much of the early work is being done with the simpler genomes of a set of "model" organisms.

• *Escherichia coli:* common human gut bacterium
• *Saccharomyces cerevisiae:* a yeast, the simplest unicellular eukaryote
• *Drosophila melanogaster:* fruit fly
• *Caenorhabditis elegans:* nematode, or roundworm
• *Mus musculus:* laboratory version of the house mouse
• *Arabidopsis thaliana:* a small plant in the mustard family

We are mapping and sequencing genomes of other organisms for a number of reasons. First, their genomes are much smaller and therefore easier to analyze. Their genomes do not contain as much "junk" DNA as the human genome, so we obtain much more useful information for each base pair that is sequenced. As a result, we can locate specific genes and identify their functions much more quickly. Second, because many genes are conserved as a species evolves, as we learn about the nucleotide sequence and function of one of their genes, we are learning about our own.

Why have we selected these organisms as models and not others? Because we already know a great deal about their genetics, biochemistry, development, and physiology. Most have been intensively studied for decades. This means that it will be easier to interpret and give meaning to the DNA sequence information we obtain from them.

Whose genome is *the* human genome? No one's. There is no "model human." According to information provided by the National Center for Genome Research, DNA for the Human Genome Project was collected from "a number of volunteers whose identity will remain secret to protect their privacy."

Making the Maps

Geneticists use two types of maps to characterize the human genome: physical maps and genetic linkage maps.

Physical maps

A physical map shows the actual physical distance between two points on a chromosome. In most cases, the points are restriction enzyme recognition sites, and distances between points are deduced from the sizes of restriction fragments as measured by how fast they migrate in electrophoretic gels. The actual points may have no biological significance. Distances between points are expressed in numbers of base pairs of DNA. Physical maps can be more or less detailed, depending on how much attention a chromosome has received. Less detailed maps may show only markers on the chromosome, while other maps may show some of the DNA base sequences between the markers.

Genetic linkage maps

Genetic linkage maps, in contrast, are not derived from direct physical measurements. Rather, they are biological maps, obtained by analyzing the frequency with which genes are inherited together. Linkage maps are built from observing the effects of crossing over, or recombination. The closer two genes are on a chromosome, the less likely they will be separated by crossing over during meiosis, and the more likely they will be inherited together.

Linkage maps are important because they allow us to locate genes associated with particular diseases or other observable physical characteristics. Such genetic locations (or loci) are also called *markers*. But the number of genes whose presence we can deduce by observing the patterns of inheritance of particular characteristics is limited. To make maps that are detailed enough to be useful, we need more markers. To be useful for tracing inheritance, markers on a genetic linkage map must vary between individuals.

Mapping a chromosome requires both types of maps. The two different types of maps and their levels of detail serve various needs of researchers. Genetic linkage maps are most useful if researchers are interested in an overall view, while physical maps provide a more detailed view of a segment of the chromosome. Ultimately, the linkage maps and physical maps of a genome will be correlated and used as guides for obtaining the detailed nucleotide-by-nucleotide sequence and for finding out exactly where the genes are located.

Sequencing DNA

The final step in making a physical map is providing the actual DNA sequence between markers. To read a sentence, you begin at the beginning and read one word after another until you reach the end. But DNA sequencing doesn't work that way. At present we have no way to hold on to one end of the huge linear DNA molecule that forms a human chromosome and simply start reading.

Because DNA molecules are so large, we can only handle them in pieces. Using restriction enzymes, we cut large DNA molecules up into shorter pieces. But, in doing so, we lose the order of the pieces. By subjecting many copies of the same DNA molecule to different restriction enzymes, however, you have fragments where certain nucleotide sequences overlap randomly. Eventually, when you have sequenced enough short pieces, you will find that a bit of sequence on one end of one piece is the same as a bit on one end of another piece, for example:

Fragment 1
AATGCCGTAGCTGGGTA**CCGTATTGCTTG**

CCGTATTGCTTGATTGCGCCTTCGAAATTGGGCT
Fragment 2

If you continue sequencing more and more fragments cut with different restriction enzymes and match up the overlaps, you can order the pieces. A set of DNA pieces ordered by overlaps into one contiguous sequence is called a *contig*.

Once a stretch of DNA has been sequenced and positioned on a physical map, it can then serve as a marker for future work. Such a marker is called a sequenced tagged site (STS). These markers may or may not be within genes. Most are in noncoding regions, because so much of the human genome is not expressed. A second type of sequencing, based on the use of messenger RNA, provides nucleotide sequences for expressed genes. These sites are called expressed sequence tags (EST).

The Genetic Information Explosion

Since the Human Genome Project began, progress has been astonishingly rapid. The rate of acquisition of sequence has been so fast that a new technological problem has had to be solved: how to organize and make sense of all the data. Correlating and managing genomic data is now a new field in biology—bioinformatics (see chapter 1).

You can get an idea of the explosive pace of the Human Genome Project from the increase in the number of base pairs of known sequence in Gen-Bank, the major publicly accessible DNA sequence da-tabase (Figure 33.5). In 10 years, this database has increased from 2 million to 200 million base pairs of known sequence.

Although the Human Genome Project is funded by Congress, research on mapping and sequencing the human genome or genomes of other organisms is not strictly a government initiative, nor is it strictly a U.S. initiative. Private companies also have research groups dedicated to genome sequencing, and the sequencing of genomes is truly an international effort.

Mapping and Sequencing Are Not Enough

If the only output of the Human Genome Project con-sisted of huge stretches of nucleotide sequences and a map of markers on chromosomes, the information would not be very meaningful. Even though those may have been the explicit goals when the project began, the implicit, understood, and widely accepted goals then and now are to identify and fully understand our genes—where they are, what they do, how they are regulated, and what happens when they malfunction. This field of research is called **functional genomics.** With information derived from functional genomics, we will be able to diagnose disorders that have a ge-netic component even before symptoms appear, de-velop new pharmaceuticals for treating and prevent-ing diseases, and even correct faulty genes. We will better understand development, gene regulation, and variation in susceptibility to environmental factors.

Figure 33.5   The growth of information on the human genome from January 1995 to August 1998. Data provided by the National Human Genome Research Institute.

The best way to discover what a gene does and how it is regulated is to study not genes but proteins. A new discipline within molecular biology, **proteomics,** focuses on the structure and interactions of proteins. Similar in concept to functional genomics, proteomics is the attempt to identify all proteins in a cell type, determine their function, and map their interactions. The success of this field, like genomics, will depend on the ability of scientists to develop technologies to rapidly identify the type, function, and amount of thousands of proteins in a cell.

Researchers have developed a number of new techniques for accomplishing the difficult task of assessing the activity patterns of thousands of genes simultaneously. One technique, serial analysis of gene expression (SAGE), is based on sequencing gene fragments and builds on the information we have compiled on EST. Another technique relies on an old stand-by, electrophoresis, to separate proteins and fluorescent dyes to determine the amount of proteins. A mass spectrometer is then used to determine the amino acid sequences. As many as 5,000 proteins can be assayed in a few days.

Societal Issues

Some members of the scientific community have objected to the Human Genome Project since its original inception. They fear that this project, like all "big science" projects funded by the federal government, will drain limited funding dollars from other, smaller projects that are just as worthy. They also worry about the efficiency of projects administered by large bureaucratic agencies and are not at all convinced that spending millions of dollars to generate massive amounts of sequence information is an effective expenditure of federal monies. Proponents of the project agree with their critics that pages and pages of sequence information in and of itself are not useful but that the detailed linkage and physical maps will be extremely useful. They counter the critics' "big science administered by bloated bureaucracies" complaint with the argument that a coordinated, systematic, and coherent approach is a more efficient method for collecting these data because it minimizes duplication.

While scientists argue about the best approach to gathering genetic information, nonscientists worry about how the information will be used. No doubt, various individuals and institutions will use the information generated by the project in many ways that will affect us, only some of which we can predict beforehand.

The Human Genome Project is surely unique among big government science projects in that a portion of the funding is earmarked for the ethical, legal, and social issues (ELSI) that will arise as a result of our new-found knowledge. Some of the issues being addressed by the project are listed below.

- Fairness in the use of genetic information by insurers, employers, courts, schools, adoption agencies, the military, and other government institutions

  Who should have access to your genetic information? How will the information be used by those who have access to it?

- Privacy and confidentiality of genetic information

  Who owns it and controls access to it?

- Psychological impact and stigmatization due to an individual's genetic differences

  How does the information affect an individual's view of himself or herself and society's perceptions of that individual?

- Genetic testing of an individual for a specific condition due to family history (prenatal, carrier, and presymptomatic testing) and widespread population screening for inherited disorders (newborn, premarital, and occupational)

  Should testing be performed when no treatment is available?
  Should parents have the right to have their minor children tested for adult-onset diseases?
  Are genetic tests reliable and interpretable by the medical community?

- Reproductive issues, including informed consent for procedures, use of genetic information in decision making, and reproductive rights

- Effective use of genetic information in clinical settings, including education of health service providers, patients, and the general public, and implementation of standards and quality-control measures in testing procedures

- Commercialization of products, including property rights (patents, copyrights, and trade secrets) and accessibility of data and materials

- Conceptual and philosophical implications regarding human responsibility, free will versus genetic determinism, and the concepts of health and disease

The Human Genome Project ELSI 2003 goals include the following.

- Examine the issues surrounding the completion of the human DNA sequence and the study of human genetic variation;
- Examine issues raised by the integration of genetic technologies and information into health and public health activities;
- Examine issues raised by the integration of knowledge about genomics and gene-environment interactions in nonclinical settings;
- Explore how new genetic knowledge may interact with a variety of philosophical, theological, and ethical perspectives;
- Explore how racial, ethnic, and socioeconomic factors affect the use, understanding, and interpretation of genetic information, the use of genetic services, and the development of policy.

Addendum in Proof

On June 26, 2000, as this edition of *Recombinant DNA and Biotechnology* went to press, President Bill Clinton, accompanied by Dr. Francis Collins, Director of the National Human Genome Research Institute, and Dr. Craig Venter, President and Chief Scientific Officer of Celera Genomics Corporation, announced the completion of "the first survey of the entire human genome." This announcement, which was widely reported by media all over the world, could easily be misconstrued as the completion of the Human Genome Project. In reality, only 85% of the human genome has been sequenced by the federally funded Human Genome Project when the announcement was made. The private corporation Celera Genomics has completed, in the words of President Clinton, the "first assembly of the human genome." Determining precisely what milestone had been attained and by whom is difficult because we are forced to rely on transcripts of the White House announcement and subsequent press conferences and not on published scientific papers. We could speculate about the facts underlying the content and timing of the announcement, but we will not. For information on the current status of the Human Genome Project and the transcripts of the June 26 White House announcement, visit the Human Genome Project Information web site, http://www.ornl.gov.hgmis, and select "About the HGP."

Selected Readings

Collins, Francis S., Ari Patrinos, Ellke Jordan, et al. 1998. New goals for the U.S. Human Genome Project: 1998–2003. *Science* 282:682–689.

Cooper, Necia Grant (ed.). 1992. *The Human Genome Project*. Los Alamos Science no. 20. Los Alamos, N.Mex.

Cutter, Mary Ann G., et al. 1992. *Mapping and Sequencing the Human Genome: Science, Ethics, and Public Policy*. BSCS and the American Medical Association, Colorado, Springs, Colo.

Kevles, Daniel, and Leroy Hood (ed.). 1993. *The Code of Codes: Scientific and Social Issues in the Human Genome Project*. Harvard University Press, Cambridge, Mass.

Lander, Eric, and Robert Weinberg. 2000. Journey to the center of biology. *Science* 287:1777–1782.

Lawrence Berkeley National Laboratory. Human Genome Sequencing Department. http://www-hgc.lbl.gov/Genome Home.

Murphy, Timothy F., and Marc A. Lappe. 1994. *Justice and the Human Genome Project*. University of California Press, Berkeley, Calif.

National Center for Genome Research. http://www.ncgr.org.

National Human Genome Research Institute. http://www.nhgri.nih.gov.

Spengler, Sylvia. 2000. Bioinformatics in the information age. *Science* 287:1221–1223.

Strausberg, R. L., E. A. Feingold, R. D. Klausner, and F. S. Collins. 1999. The mammalian gene collection. *Science* 286:455–457.

Vaughan, Douglas. 1996. *To Know Ourselves*. U.S. Department of Energy and The Human Genome Project, Washington, D.C.

Classroom Activities

Molecular Genetics of Cancer

Few diseases cause as much dread as cancer. In the United States, approximately 500,000 people die from cancer each year. Sometimes a person's life choices contribute to the development of cancer, such as with cigarette smoking and lung cancer, but often cancer seems to randomly strike apparently healthy individuals. Over 200 types of cancer syndromes have been identified, affecting almost every type of cell, tissue, and organ system.

Our current cancer therapies, though more successful than earlier ones, are actually crude, blunt instruments. Cancer cells are dividing cells, so most anticancer therapies are strategies for killing rapidly dividing cells. Of course, many normal body cells are dividing, too, which is why current cancer treatments have so many difficult side effects. It is hoped that a thorough understanding of the molecular basis of cancer will lead to more precise treatments and even perhaps prevention.

A Genetic Disease in a Single Cell

Most cancers start from a single cell. To become a cancer cell, a normal cell must accumulate mutations in several different genes. These mutations are then transmitted to the cell's cancerous descendants. Cancer can therefore be thought of as a genetic disease at the level of the cell.

A mass of cancer cells descended from a single parent cell is called a *tumor.* If a tumor poses no danger to life or health, it is called benign; if it does pose a danger, it is called malignant. Tumors can sometimes be completely removed by surgery.

Some tumors continue to grow at their original sites, while others stop growing and shrink. In other cases, some of the tumor cells break free of the original tumor, migrate to new sites in the body, and establish many new tumors. The process of spreading throughout the body is called *metastasis*. Metastasis is what makes cancer so lethal. When cancer metastasizes, it may reach many sites at which tumors can cause se-

vere damage, and it lodges in so many places that treatment by surgery alone is impossible.

A cancer is usually classified according to the type of cell from which it originated. For example, 85% of all cancers originate from cells that line organs and form parts of the skin. These cancers are called carcinomas. Cancers originating from cells of connective tissue, bone, or muscle tissue are called sarcomas. Leukemias arise from white blood cells (leukocytes). Cancers of glandular tissue are called adenocarcinomas; cancers of the nonneuronal cells of the brain are called gliomas and astrocytomas.

A Laboratory Model of Cancer

What kind of genetic changes are involved in cancer? The first clues came from studies of cultured animal cells. In the 1970s, it was discovered that infection by certain viruses could transform cultured cells. Normal cultured cells behave in a preordained manner. They do not grow on top of one another but instead form a confluent layer in a dish. They usually live for a certain number of life cycles and then die.

When cultured cells are infected by the viruses, however, dramatic changes occur. In some cases the cells become immortal, losing their predetermined life span. They begin to grow over one another in an uncontrolled manner. Other changes are also observed. The combination of several of these changes is called transformation. (Do not confuse this use of the word with the usage referring to the addition of new DNA to bacteria; they both refer to change, but in different contexts.) In the body, cancer cells lose their normal growth inhibitions and continue to multiply in an uncontrolled and inappropriate manner. Because of the similarities, transformation of cells in culture is considered the equivalent of cancer in a dish.

The ability to produce the cancerlike changes was traced to certain genes the viruses carried. These genes were christened oncogenes ("onco" is a prefix meaning tumor). Quite a number of oncogenes were identified; they were given three-letter names such as *ras, src,* and *myc.* Shortly after the viral oncogenes were discovered, scientists found that copies of these genes, christened proto-oncogenes, resided in our chromosomes. The "viral" oncogenes were actually abnormal versions of these cellular proto-oncogenes that the viruses had acquired and transmitted by transduction.

Oncogenes and Proto-Oncogenes

What do oncogenes do? An interesting and logical picture is emerging from 2 decades of intense study in many laboratories. Proto-oncogenes are a normal, essential part of our genetic material that belong to the group of genes in charge of causing and regulating cell growth and division. It is actually changes in these genes that can lead to the development

of cancer. Oncogenes are abnormal forms of proto-oncogenes.

Cell growth and division is a complex system involving many proteins. Growth is usually initiated by a signal, such as a hormone, received by a receptor on the cell membrane. After the receptor receives the "grow" signal, it transmits the message across the cytoplasm to the nucleus. There the signal initiates the cell division cycle (Figure 1). Mutations affecting any step of the signaling pathway could cause a cell to "think" it is constantly being instructed to divide. For example, many receptors respond to outside signals by adding phosphate groups to themselves and/or other proteins. If a receptor were mutated in such a way that it added phosphates whether or not it was being signaled, the cell would "believe" it was constantly being instructed to divide. Such a genetic change would

Figure 1 Extracellular growth signals are relayed to the cell nucleus via a series of protein-protein interactions, beginning at the cell membrane. The change in shape triggered by a growth factor binding to its receptor initiates a cascade of changes in associated proteins, resulting in a signal being transmitted to the nucleus.

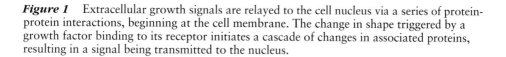
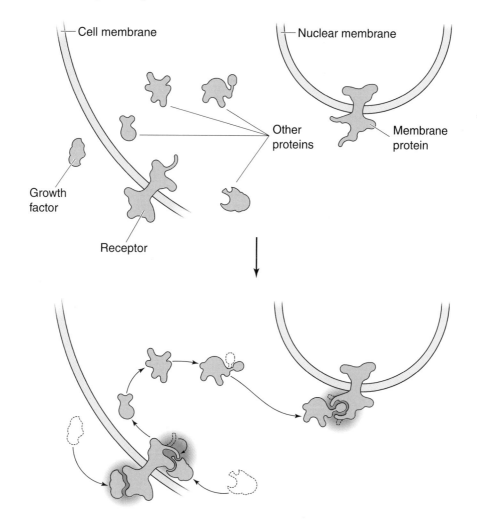

promote uncontrolled cell growth, rather like stepping on the accelerator in a vehicle. Likewise, the cell division cycle itself is heavily regulated. If any of the proteins regulating the cell division cycle were to malfunction, the division cycle could run amok. The oncogenes identified in the transforming viruses had undergone changes with this kind of effect.

Tumor Suppressor Genes

Evidence from genetic studies of cancer patients led scientists to believe that another type of gene might also be involved in the development of cancer. Subsequent research has proved their expectation to be correct. The product of this type of gene is like a brake on cell growth rather than an accelerator. Instead of being part of a signaling chain that tells a cell to divide, this type of protein represses cell growth and division. These repressor proteins apparently must be inactivated before full cancer can develop. The genes encoding these proteins are called tumor suppressor genes.

As might be expected, mutations in oncogenes that lead to cancer are usually different from mutations in tumor suppressor genes that lead to cancer. In oncogenes, the mutations must *activate* the protein to promote growth inappropriately. This kind of mutational change includes a change in a single amino acid that leads to an altered form of the protein, multiplication of the gene within the chromosome to provide greater activity, or alteration of the control regions of the gene to deregulate its expression or to cause it to be regulated inappropriately. These mutations also tend to be dominant; the presence of one normal copy of the oncogene cannot make up for the mutant, activated form. For example, the chromosome 9,22 translocation that brings about the inappropriate activation of the *abl* gene causes leukemia, even though a normal *abl* gene remains on the other copy of the chromosome. The cancer-causing viruses were able to transform cultured cells even though those cells had normal copies of the proto-oncogenes.

For tumor suppressor genes, however, *loss* of the protein function is required to promote cancer development. For this reason, mutations in suppressor genes tend to be recessive; one good copy of the gene can provide active protein. You might expect that individuals could inherit one copy of a recessive mutation in a tumor suppressor gene and that these individuals might be more likely to develop cancer during their lives. You would be right.

The first tumor suppressor gene to be identified was the retinoblastoma gene. Retinoblastoma is a tumor of the eye that can be hereditary. Analysis of patients with hereditary retinoblastoma revealed that one copy of a certain gene was inactivated in all of their cells and that both were inactivated in the tumors. A comparison with other patients whose retinoblastoma was not hereditary showed that the same gene, now called the retinoblastoma (*rb*) gene, was also inactivated in their tumor cells. Since loss of function of *rb* was apparently required for the tumor to develop, it was concluded that the *rb* gene product must suppress tumor development. Further study of the *rb* gene has shown that its protein product resides in the cell nucleus. One of its functions is to regulate transcription factors that initiate the expression of proteins required for cells to proceed from the G_1 stage of the cell cycle to the S phase (Figure 2). At present, we do not know why loss of function of the retinoblastoma protein affects retinal cells more than any other type of cell.

Children with hereditary retinoblastoma inherit one inactive copy of the *rb* gene from their parents. Every cell in their bodies has only one good copy of the *rb* gene. In these children, all that is needed is for one retinal cell to suffer a mutation to the other copy of *rb*, and the retinoblastoma tumor can develop. In genetically normal individuals, a single retinal cell would have to suffer two independent mutational events to knock out both copies of *rb*, an unlikely series of events. Retinoblastoma is a rare cancer. Most cases run in families, who apparently transmit the defective gene.

Mutation of another tumor suppressor gene, *p53*, is associated with a different inherited cancer syndrome. The Li-Fraumeni syndrome is a rare inherited syndrome of cancers. Members of Li-Fraumeni families develop cancers (often multiple cancers) of the

Figure 2 The cell division cycle. The stages were named for what could be seen under the light microscope. G stands for gap, because no visible activity was occurring. S signifies DNA synthesis, and M, mitosis. Most nondividing cells are in G_1. Together, G_1, S, and G_2 constitute interphase.

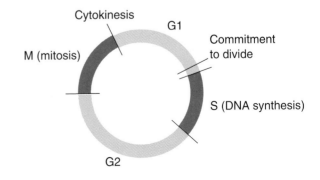

breast, brain, bone, and other tissues and leukemia, usually at early ages. Patients with Li-Fraumeni syndrome inherit one defective copy of the *p53* gene, so every cell in their bodies has only a single good copy. Apparently, inactivation of this good copy in any number of different cell types can lead to cancer.

The *p53* gene is also important in noninherited cancer. Scientists who analyze genetic alterations in cancerous cells have found that *p53* mutations are present in more different kinds of cancer than any other known cancer-related genetic alteration: fully 50% of all cancers. Needless to say, this finding has sparked enormous interest in the biological role of *p53*.

The p53 protein is a nuclear protein with several important regulatory roles. It is involved in regulating the cell cycle at both the G_1 to S phase and the G_2 to M phase transitions (Figure 2). If cells with normal p53 suffer DNA damage, the p53 protein prevents the cells from dividing, presumably until the damage is repaired. If the damage is too extensive, p53 initiates events leading to the death of the cell. Cells that lack normal p53 continue to divide. The replication of damaged DNA leads to changes (mutations) in the DNA sequences of daughter chromosomes. So a cell that lacks p53 activity apparently can accumulate mutations at a much higher rate than normal cells do. According to this line of thinking, if one of these mutations leads to activation of an oncogene, the result is cancer.

Genetic analysis of families with hereditary cancer and individual cancer patients is helping us identify additional genetic loci associated with cancer development. So far, the genetic picture of most common cancers looks very complicated, and no clear interpretation is available. It is clear that cancer is actually many diseases. It is also clear that different sets of mutations can cause similar cancers.

Breast cancer provides a good illustration of the genetic complexity of cancer. Approximately 2% of women who live to age 50 will develop breast cancer, and about 10% of women who live to age 80 will develop it as well. Most of these cases are sporadic, not associated with the inheritance or predisposing mutations. However, a tendency to develop breast cancer does run in some families. By studying such families, scientists identified two genes, *BRCA1* and *BRCA2*, that are associated with inherited breast cancer. Some of the families had the *BRCA1* mutation, and others had the *BRCA2* mutation. Screening of other families with inherited breast cancer showed that about 50%

of them carried the *BRCA1* mutation, and 30% of them carried the *BRCA2* mutation, indicating that still other genes associated with inherited breast cancer remain to be discovered. Male breast cancer is associated with *BRCA2* but not *BRCA1*.

Sporadic cases of breast cancer, about 90% of all cases, do not seem to be associated with *BRCA1* or *BRCA2* mutations. Currently, no obvious oncogenes or tumor suppressor genes are consistently associated with sporadic breast cancer. Alterations in a number of genes encoding growth factors, receptors, and cell division proteins have been reported.

Over 100 different oncogenes and tumor suppressor genes have been identified. As the Human Genome Project progresses toward its goal of understanding all human genes, more will likely be found. Table 1 lists a few of the known oncogenes and tumor suppressor genes and gives their role in the control of cell growth.

Does our increasing understanding of cancer genetics hold out any hope for improved therapies in the future? Many physicians and scientists believe so. Since cancer is the result of a malfunction of genes, the best way to fight it may be to restore proper genetic function. Possibly gene therapy could be used to introduce good copies of tumor suppressor genes into cells whose own copies are mutant. Antisense technology could be used to "turn off" the expression of tumor-promoting mutant genes. Scientists hope that the next few years will yield less toxic, more effective genetic therapies for one of our most dreaded genetic diseases.

Table 1 Some cancer genes and the physiological roles of their products

Gene	Role
Oncogenes	
sis	Growth factor
erbB, fms, neu	Cell surface growth factor receptor
ras, src, abl	Signal transmission within the cell
bcl2	Blocks cell-initiated death
myc, fos, myb	Regulators of transcription
Tumor suppressors	
rb	Regulation of replication and transcription
p53	Regulation of cell division cycle: stops cells from dividing if their DNA is damaged, allowing time for repair; initiates cell death if DNA is too damaged

Selected Readings

Cancer special issue. 1996. *Scientific American* 275(3). This issue contains articles on how cancer arises and spreads, new therapies, genetic testing for cancer genes, and more.

Cavenee, W. K., and R. L. White. 1995. Genes and cancer. *Scientific American* 272(3):72-79.

Greider, C., and E. Blackburn. 1996. Telomeres, telomerase, and cancer. *Scientific American* 274(2):92.

Pasternak, J. 1999. *An Introduction to Human Molecular Genetics*. Fitzgerald Scientific Press, Bethesda, Md.

Perera, F. 1996. Uncovering new clues to cancer risk. *Scientific American* 274(5):54.

Classroom Activities

Societal Issues 273

Societal Issues

PART III
Societal Issues

*S*ocietal issues raised by biotechnology, or any technology, are introduced by using background information and classroom activities designed to provide tools for analyzing difficult issues rationally. The complex relationship between scientific understanding, technological development, and societal structure is introduced and elaborated on in chapters on assessing and debating risks and conducting productive, thoughtful discussions of bioethical dilemmas. The final chapter provides information on careers in biotechnology.

Societal Issues: Introduction

Parts I and II provide information on science and technology, or, more specifically, on the science of genetics and the technologies we have developed by using our understanding of molecular genetics. In this part, we introduce a related body of knowledge that is finding increasing relevance and importance in today's world: the nature of science and technology—their history, interrelationship, and societal effects.

The reasons for studying the sciences that spawned biotechnology and for practicing the hands-on techniques used in biotechnology research and product development are clear. Scientific knowledge and laboratory skills spell jobs. But why should one bother to learn *about* science and technology?

In short, because our lives have repeatedly been and will continue to be transformed by the omnipresent influences of science and technology. These influences are both obvious and subtle, both simple and profound. Science and technology permeate all aspects of our lives and our relationship to the natural world. As a result, not only do they shape our values, goals, and world view, but they also reflect these parts of ourselves.

Technological innovations develop as solutions to specific problems and obvious needs, but they often provide doors to other innovations and novel applications. As a result, predicting all of the potential societal effects of any technology is impossible. Imagine if we had been so prescient as to realize that the introduction of automobiles would lead not only to the obvious risk of collisions with other autos and pedestrians, but also to polluted air and fragmented family units.

Recombinant DNA technology began to generate a great deal of discussion about its potential uses—from genetically engineered crops to recombinant microorganisms to gene therapy—at the earliest stages of research, before the introduction of any products. The debate about the societal effects of modern biotechnology has continued unabated throughout its development. This debate is healthy and appropriate, for recombinant DNA technology and other new biotechnologies will give us powers we have never had before.

The powers provided by biotechnology can be used for both good and evil. While this thought may be troubling, it is not new, nor is it unique to this technology. Societies have dealt with precisely the same issue in different forms ever since humans began to fashion crude tools from stone. Democratic societies have given their citizens the power to determine how to use these technologies. As is clear from this quote by Thucydides, author of *The Peloponnesian War*, the public in Greek democracies viewed itself as the best judge of matters that affect the public:

> *. . . our ordinary citizens, though occupied with the pursuits of industry, are still fair judges of public matters; for, we alone regard the man who takes no part in public affairs not as one who minds his own business but as good for nothing. We Athenians are able to judge all events, and instead of looking on discussion as a stumbling block in the way of action, we think it an indispensable preliminary to any wise action at all.*
>
> Thucydides, 405 B.C.

Do you think a historian would describe democracy in today's America in the same way?

But is it appropriate to expect the citizens of pure or representative democracies to be "able to judge all events" in our high-tech world? Can discussion actually become a stumbling block to wise action or even lead to unwise action if that discussion is ill informed? A founder of our democracy believed the solution to the potential problem of having a misinformed or uninformed public with the power to make decisions or, in the case of a republic, to exert its political will on decision makers was not to remove power but to provide information.

I know of no safe depository of the ultimate powers of the society but the people themselves; and if we think them not enlightened enough to exercise their control with a wholesome discretion, the remedy is not to take it from them, but to inform their discretion by education.

Thomas Jefferson, 1787

Like all technological applications, even the beneficial applications of recombinant DNA technology will likely be accompanied by some costs. To use the power of biotechnology wisely and to assess its costs and benefits correctly, we need to be conversant in the science, the application of the science, and the potential impacts of the science on society.

Dismissing decisions about the appropriate uses of recombinant DNA technology as someone else's concern could well lead to dissatisfaction with the way the technology develops. Citizens who abdicate their oversight of technological development to a handful of scientists, engineers, lawyers, and politicians are ignoring the power, privilege, and responsibility of a democratic way of life. They are also sacrificing a portion of their personal freedom.

The Activities in Part III

We begin this part with an overview of the interdependent relationship between science, technology, and society. Each of these ingredients influences the others and is in turn influenced by them. Analyzing this interdependence from a variety of perspectives will illustrate the extent to which they are interwoven.

Chapters 35 and 36 on risk assessment and weighing risks and benefits teach skills that will enable us to make a rational analysis and participate in a rational discussion of the risks and benefits of technological development in general. Many of the examples used are drawn from applications of recombinant DNA technology, but our experience with the products derived from recombinant DNA technology is relatively limited compared with our experience with earlier technologies. Because factual information and objective analysis of data play a large role in formal risk assessment and risk-benefit analysis, we have also included exercises focused on other technologies for which you can more easily locate information.

The factual and quantitative methodologies of formal risk assessment and risk-benefit analysis were developed to minimize the contribution of emotion to decision making. In chapters 37 to 39, we move from these relatively unemotional topics to the highly charged and contentious area of morals, values, and ethical dilemmas. To facilitate productive and informed discussion, we have provided a decision-making model for addressing the ethical dilemmas that arise in two scenarios relevant to gene-based biotechnology: gene therapy for the treatment of a genetic disease and genetic screening for inherited disorders. As will be evident, the model can be used to analyze thorny issues in areas other than biotechnology and biomedicine.

Finally, we end this part with a chapter on careers in biotechnology and the educational requirements for obtaining various jobs in this young and exciting field.

Societal Issues

Science, Technology, and Society

Introduction

Today's world has been profoundly and unalterably shaped by advances in science and technology. Some of the ways in which science and technology have affected our lives are immediately obvious: airplanes, automobiles, television, computers, and microwave ovens. Others are so subtle that we fail to recognize them as benefits of science and technology. Instead, we take them for granted as a "natural" part of life. The clothes you're wearing, the chair you're sitting in, the pen you may be holding, the paper this book is printed on, the food you had for lunch are all products of technology.

If you doubt the extent to which technology has penetrated every tiny facet of your life, pay attention to your actions during the rest of the day and take note of all the ways technology, both simple and sophisticated, has altered your world. As you are paying attention to the ways technology impinges on your life, also look at the flip side of the coin. How often do you come face to face with nature? In those encounters with nature, how many times do you experience a love for nature, and how many times do you wish it would go away and leave you alone? Thousands of years ago, we rejected the idea of living with nature. Our distaste for the adversities nature indifferently dispenses has driven our constant quest for new technologies. We use all of these technologies for a similar purpose: to change our environment so that the natural world suits us better.

Science and technology not only alter our environments—they also shape us. Our view of the world, the questions we ask, our sense of values, and the way we think are the products of life in a world shaped by scientific discoveries and technological innovations.

Has it always been this way? Yes and no. Technologies have always influenced human societies, our mental framework, and our relationship to the natural world. Today, however, technological advances are more pervasive, and their effects are more powerful. Also, the changes they bring are occurring at an increasingly rapid rate. We can trace the increased presence, power, and rate of technological change to the role scientific understanding now plays in propelling technological developments.

Which Came First, Science or Technology?

Technology predates science. People initially developed technologies empirically, using the manual skills required in crafts such as clock making, lens grinding, and metal forging, and independently of an understanding of the laws of nature, the domain of science. Early scientists used those technologies to observe and question natural phenomena and thus to increase their understanding of the world around them. As their understanding of the natural world broadened and deepened, science and technology began to converge.

Equipped with a more sophisticated understanding of the world they sought to control, people utilized scientific understanding to drive technological change. They replaced the empirical and relatively inefficient approach to technological advancement with the methodical, experimental approach that is characteristic of scientific investigation. Buttressed by sound scientific principles and procedures, the ensuing technological changes were more forceful, and the pace of technological change quickened. With ever more powerful technologies at their disposal, scientists were able to probe deeper and deeper into the causes and effects of natural phenomena.

Because the relationship between scientific understanding and technological innovation is circular and not linear, reciprocal and not hierarchical, we can expect the pace of science and technological change to accelerate even faster in the 21st century. A constantly accelerating rate of increase in scientific knowledge and technological achievement carries with it important implications for our species.

The Nature of Science and Technology

Many people equate science with technology, but the two differ in a number of fundamental ways. A few of the ways in which science differs from technology are provided in Table 34.1.

Despite their differences, science and technology converge in subtle ways that may not be apparent to people outside the scientific and engineering communities. Most people would readily agree that decisions to develop certain technologies and not others are often driven by societal factors such as politics, economics, and cultural values. Or, in other words, the development of certain technologies instead of others is value laden. Fundamental, value-based assumptions drive the quest for more and better technologies spawned by ever-increasing scientific knowledge. One assumption is that science and technology result in progress. A second is that we should dominate nature.

What most people are not aware of is that science is influenced by these same societal factors of politics, economics, and cultural values. People view science as an objective search for the truth, and in many ways it is. Adhering to the scientific method of systematic observation, hypothesis generation, and experimentation, today's researchers attempt to establish their hypotheses objectively and gather data through careful, methodical experimentation.

But scientists were raised in the society, and that society has shaped the way they view the world. The scientific method cannot shield scientists from the social context in which they conduct research. Society has molded their sense of values and goals since the day they were born. The narrower social context of the scientific world in which they function imposes pressures and erects powerful filters for selectively viewing the world. Only certain questions from the universe of available questions are asked; data are in-terpreted through these same filters (see the section on Dr. Barbara McClintock in chapter 3 for a real-world example of this phenomenon).

So, while the methodology scientists use to answer questions may be "objective," the questions asked and the interpretation of the results are not value free.

Science, Technology, and Responsibility

Science and technology have altered the characteristics of societies throughout history. Some of the forces they have exerted have been revolutionary, and some have been minor. Different aspects of life—social, cultural, ecological, economic—have been affected, and different groups have been affected in different ways. Often, one group has benefited at another's expense.

Because of the extraordinary degree to which our environment and we ourselves are shaped by science and technology, individuals must understand the interrelationship between scientific understanding and technological change and must respect the impact technologies can have on the quality of life. While scientific investigation and technological advance may be the province of a few members of society, all citizens in a democracy should be prepared to consider the societal issues spawned by a science and its derivative technologies. In the absence of public understanding and interest, the stage is set for a handful of knowledgeable individuals to direct the course of scientific investigation and technological change and therefore, over time, society's future.

The effects of science and technology can be difficult to predict. It often seems that with every intended benefit comes an unintended cost. Daily we come face to face with the conflicts and contradictions brought about by technological innovation: medical advances increase the average life span as well as the national debt; agricultural advances have provided an

Table 34.1 Fundamental differences in science and technology

Science	Technology
The search for knowledge	The practical application of knowledge
A way of understanding ourselves and the world	A way of adapting ourselves to the world
A process of asking questions and providing answers or broad, generalized explanations	A process of providing solutions to human problems to make our lives easier
Looks for order or patterns in nature	Looks for ways to control nature
Evaluated by how well the data support the answer, model, or theory	Evaluated by how well it works
Limited by our ability to collect relevant data	Limited by financial costs and risks
Discoveries give rise to technological advances	Advances give rise to scientific discoveries

Societal Issues

abundant food supply and increased leisure time, leading to obesity and heart disease. Some people question whether the benefits of technology are worth the price. They wonder whether the assumption that science and technology give rise to progress is valid.

Science and technology have given humans tremendous power over the natural world. With that power comes a responsibility that all of us must share: to make far-sighted, informed, and judicious choices that ensure we do not misuse our power.

Barry Commoner, a noted environmentalist, has applied the adage "There is no such thing as a free lunch" to the inevitable price we will pay if we continue to ignore the environmental costs of human activities. With every technological advance designed to control our environment, we attempt to remove ourselves from the natural order and live further outside the boundaries Mother Nature has established. We want to live by our own rules and not nature's. Whether we can manage to have it our way, and at what cost, remains to be seen.

Suggested Activities

Your instructor will request that you complete one or more of the following assignments.

1. Make a list of *all* of the technologies that influence your activities during a day. On the same day, list all the ways in which you interact directly with the natural world.

2. Rank several technologies according to their degree of importance in your life. If the need to conserve resources arose, which of these could you eliminate?

3. Pick a major technology—antibiotics, electricity, television—and thoroughly analyze its development and its predicted and unpredicted effects on society.

4. Pick an emerging technology or one application of an all-encompassing technology like biotechnology (see chapter 1) and describe what scientific knowledge, existing technology, or other resources are required for this technology to be fully developed. In addition, try to predict what some of the intended and unintended consequences of its development will be.

5. What factors should be used in evaluating technologies? Historically, technologies have been evaluated by three criteria:

   - Can it be done?
   - Will it sell?
   - Is it safe?

   Are these criteria sufficient? If not, what others would you add?

6. Do you think technologies increase or decrease our options? Explain.

7. Write a brief essay explaining the following quote by Edward Wenk: "Those who control technology control the future."

Weighing Technology's Risks and Benefits

Introduction

Day in and day out, consciously or unconsciously, we take risks. We ride in cars, cross streets, and play sports without thinking twice about the risks involved, because we have "decided," usually subconsciously, that the risks are outweighed by the benefits.

Sometimes our subconscious risk-benefit analysis makes good sense. At other times, our decisions seem to involve no rational thought. We accept huge risks and worry about trivial ones—cigarette smokers protest the building of nuclear power plants; people risk pregnancy and AIDS by having unprotected sex and then worry about pesticide residues on their food— yet many more people are harmed by smoking and unprotected sex than by accidents at nuclear power plants or pesticides on food. Our choices are irrational because we sometimes make decisions about acceptable and unacceptable risks based not on facts but on emotions.

Emotions and Unconscious Risk Assessment

What are some of the factors that trigger emotions, interfering with our ability to make rational assessments of true risks?

Voluntary versus involuntary

If the risk is one we consciously decide to take (a voluntary risk, like smoking), we accept a much higher level of risk than if we feel we have had no choice in the matter (an involuntary risk, such as the location of a nuclear power plant).

Control versus no control

We are more fearful of risky situations over which we have no control. Many people choose to ride in their cars rather than fly in airplanes because they fear flying, even though more people are hurt or killed in auto accidents than in plane crashes. Despite the evidence, we perceive the risks of flying as greater than the risks of driving because we are not in control of the plane.

Familiar versus unfamiliar

Our view of risk also varies with our degree of familiarity with the risk. We have come to accept the large risks involved in driving a car, sunbathing, and drinking alcohol. A new risk elicits our concerns and avoidance not because the risk is greater, but because it is unfamiliar.

Natural versus man-made

Another factor that affects our concept of the degree of risk is whether the risk is natural or man-made. We tend to view nature as more benevolent despite all the evidence to the contrary. Just because something is natural does not mean it is risk free. Microbes are natural, and yet millions of children in underdeveloped countries die every year from diarrhea resulting from microbial contamination of food and water. In developed countries, man-made products and processes prevent this same tragedy from occurring. How can we ignore this and other examples and continue to remain skeptical of man-made products as we unthinkingly embrace natural products as safe?

Our confusion on the issue of natural and man-made risks deserves additional attention because an objective analysis of biotechnology requires a clear understanding of the pitfalls in equating natural with risk-free and man-made with risk.

Food Safety: Natural and Man-Made Risks

Some of the most instructive examples of problems caused by an irrational preference for natural products and an irrational fear of man-made products come from issues surrounding food safety. Even though food safety is very important to today's consumer, little has been done to place the issue of the relative risks of man-made versus natural compounds in perspective.

Natural toxins

Some people are frightened by the use of agricultural chemicals on our food crops. They have heard that these chemicals harm our health because they are "toxic" and "cancer-causing." To avoid these chemicals, some consumers purchase "organically grown" food at much higher prices.

Some man-made chemicals, particularly those we first developed in the 1940s, are highly toxic. But are all man-made agricultural chemicals more toxic than substances we readily ingest like coffee and soft drinks? No. For example, in a study to assess the toxicity of various compounds, half the rats died when given 233 mg of caffeine per kg of body weight, but it took 4,500 mg of glyphosate—the chemical in the herbicide Roundup—per kg to cause the same percentage of deaths. Even more to the point, do some of our crop plants contain natural chemicals that are toxic? Yes. In fact, some of the naturally occurring compounds in the plant foods we eagerly consume as "healthful" are more toxic than some of the agricultural chemicals we use for pest control.

How can it be that the fruits and vegetables we think of as healthy make their own toxic chemicals? Take a moment and think about the biology of plants. Plants did not evolve to serve as food for humans or any other organism. For millions of years they have done their very best *not* to be eaten. Stuck in one place, how can a plant defend itself? Chemicals. Plants make hundreds of different chemicals to ward off would-be predators. Biologists call these defense chemicals "secondary plant compounds."

For many years, crop scientists have directed much of their plant breeding to decreasing or eliminating these natural toxins. Without such selective breeding, many of the plants Mother Nature cooked up would not be fit for human consumption. Despite years of work to strip our crop plants of secondary plant compounds, many of the plants we consume still contain toxins and carcinogens (cancer-causing agents), as defined by federal regulatory agencies such as the Food and Drug Administration and the Environmental Protection Agency. Does this mean we should stop consuming fruits and vegetables? Of course not. Most of these plant toxins occur in concentrations that are so low they have no effect on our health (see the box "Parts per Million and Parts per Billion"). The benefits of eating fruits and vegetables far outweigh the negligible to nonexistent risks posed by naturally occurring toxins. Does that mean all man-made chemicals are so safe they can be ignored as health risks? No. Our point is only that we must keep issues of chemical toxicology and food safety in perspective.

Natural carcinogens

According to Bruce Ames, the well-known biochemist who developed the most widely used test for measuring the cancer-causing potential of chemicals, in any given meal, an individual consumes about 100 to 150 natural carcinogens and 10,000 times more natural

Parts per Million and Parts per Billion

Part of the difficulty in accurately evaluating statements about the toxicity of substances and the safety of food stems from our extraordinary ability to measure chemical compounds. Our measuring techniques can detect quantities so small that they have no real significance for health and safety. We often hear the expressions "parts per million" and "parts per billion." We can now measure substances at the level of 1 part per trillion! What do these values really mean?

How much is a billion?
A billion seconds ago, World War II ended.
A billion minutes ago, St. Paul was writing the Epistles.

What is 1 part per million?
1 gram per ton
1 mouthful in a lifetime
1 drop in 1,000 quarts of water

How much is 1 part per billion?
1 inch in 16,000 miles
1 minute in 2,000 years
1 cent in 10 million dollars
1 drop in a million quarts of water

How much is 1 part per trillion?
1 inch in 16,000,000 miles

carcinogens than man-made carcinogens. Cabbage alone has over 40 natural carcinogens!

To shed more light on the relationship between food safety and natural versus man-made chemicals, here are the carcinogenic potentials of some familiar compounds. The higher the number, the greater the carcinogenic potential.

Substance	Carcinogenic potential
Wine	5.0
Beer	3.0
Mushrooms	0.1
Peanut butter	0.03
Chlorinated water	0.001
PCBs	0.0002

You may be surprised by these cancer-causing potentials. You probably did not expect peanut butter to be 150 times more carcinogenic than polychlorinated biphenyls (PCBs) because you have heard that PCBs are very dangerous. Yet no one has warned you about eating peanut butter. How could you have gotten such misleading information on something as important to you as the food you eat? The answer is probably found in the information source you use.

Most of us get our information about science and technology from newspapers, magazines, and television. Most representatives of the media want to convey accurate information to the public, but a reporter writing a story on science and technology may not know where to get solid scientific information, how to assess the scientific merit of research, or how to interpret scientific data. These understandable shortcomings are sometimes made worse by other constraints. Newspapers, magazines, and television stay in business by selling their products. Fear sells; solid science doesn't.

Fear may sell newspapers, but it also ends up costing consumers a lot. Unnecessary fears may drive consumers to spend more money on products with no additional benefits, avoid products that are perfectly safe, and fret over perceived but nonexistent problems. These same unfounded fears sometimes inspire Congress and regulatory agencies to pass certain regulations that do nothing to increase the protection of the public but much to increase the cost of goods and services (see the box "Thoughts on Public Perception of Risks"). The relationship between increased fear and increased cost is best illustrated by an example.

The case of decaffeinated coffee

In the mid-1980s, much attention was given to the use of methylene chloride to decaffeinate coffee. Members of the media reported that methylene chloride was a carcinogen in rats when inhaled. They neglected to mention that it has a much lower carcinogenic potential when ingested. This distinction is very relevant, because in the case of decaffeinated coffee, the only variable coffee drinkers need be concerned with is the carcinogenic potential when ingested. Nonetheless, in response to consumer fears, some manufacturers decaffeinated coffee by a different process, and the price increased.

Thoughts on Public Perception of Risk

"A lot of our priorities are set by public opinion, and the public quite often is more worried about things that they perceive to cause greater risks than things that really cause risks. Our priorities often times are set through Congress. Their decisions may or may not reflect real risk; they may reflect people's opinions of risk or the Congressmen's opinions of risk."

Linda Fisher,
former Assistant Administrator of the
Environmental Protection Agency

"For many years there has been an argument among some technical experts that the public is all mixed up, that they just don't have their risk priorities right. But experimental results in psychology in recent years have suggested that that is not true. If you give the public a list of hazards and say "Sort these in terms of how many people die each year from each of them," they can do it. If instead you give them the same list and they are asked to sort them by how risky they are, you get a very different order. The point is that to most people risk does not equal expected numbers of deaths, it involves a lot of other things, things like equity, whether you can control the hazard, whether you understand it. And so partly the arguments derive from this difference between making judgments just on the basis of expected numbers of deaths versus considering all these other factors."*

Granger Morgan,
Carnegie Mellon Institute

Societal Issues

Coffee itself has over 300 chemical compounds, many of which are more toxic and more carcinogenic than methylene chloride. To ingest enough methylene chloride to reach the value shown to cause cancer in rats, a person would have to drink 50,000 cups of coffee a day. In those 50,000 cups, natural coffee carcinogens would occur in far greater amounts than methylene chloride. More important, perhaps, a person would die from drinking 50,000 cups of *water* a day. Long before the 50,000th cup was consumed, a person's kidneys would shut down.

Again, the message is not that man-made is safe and natural is unsafe. The situation is not that simple. The moral of the story is that all of us work from misconceptions derived from inaccurate or misleading information. Unfortunately, misinformation made worse by emotionally driven evaluations of risk doesn't yield smart choices very often. How can we do a better job of making choices involving food safety risks or any other risks?

Facts and Conscious Risk Assessments

What drives your assessment of risks—emotions or facts—may or may not seem particularly important to you, yet it is. The risks you are willing to assume and the experiences or products you avoid because of faulty assumptions and misinformation affect the quality of your life and the lives of those around you. In today's world, people suffer from great anxiety about the risks involved in simply living. Some of the fears that grip us are real; others are not. Wouldn't it be useful if we had a mechanism for helping us determine the difference? We do.

Toxicity versus hazards

First, to better understand and evaluate discussions of risks, particularly those involving food safety, we should differentiate between toxicity and hazard. Toxicity can be defined as the capacity of a substance to do harm or injury of any kind under any conditions. Given the definition of toxicity, anything can be toxic. In the decaffeinated coffee example above, water is toxic.

Contrast this to the definition of hazard. A hazard is the relative probability that harm will result when a particular quantity of a substance is used in a particular manner. Given these two definitions, it becomes clear that meaningful assessments of food safety, or the safety of any substance, must be based on the hazard a substance poses and not its toxicity. If we use

this principle, we will be one step closer to assessing risks objectively.

Risk-benefit analysis

Using this concept of hazard, scientists have developed a process for assessing risks and benefits that attempts to maximize the contribution of factual information and minimize the contribution of emotion to the analysis: risk-benefit analysis. Risk assessments are often quantitative evaluations, but we will limit our discussion to the conceptual aspects of risk-benefit analysis.

Assessing Risks

First of all, to analyze risks as thoroughly and unemotionally as possible, scientists conduct a risk assessment. The first step is defining what a risk actually is.

Risk is the probability of loss or injury and is dependent on both the hazard and exposure to the hazard, or

$$\text{Risk} = \text{hazard} \times \text{exposure}$$

In other words, a hazard, which can be viewed as a source of danger, becomes a risk only if you are exposed to it. Swimming pools are hazardous but are not a risk to someone who has never been near one. Your probability of loss or injury increases in relationship to the increase in hazard and/or exposure.

Once you have clearly defined the concept of risk, you are ready to begin a risk assessment. Assess the amount of risk involved in a particular situation by answering these three questions:

1. What could go wrong? (Identify the specific risk.)
2. How likely is it to happen? (Assess the probability of the risk occurring.)
3. How harmful is it if it does occur? (Assess the consequences.)

A simple formula for approximating the amount of risk is

$$\text{amount of risk} = \frac{\text{hazard} \times \text{exposure}}{\text{safeguards}}$$

If you are able to establish effective safeguards, the amount of risk decreases dramatically.

These conceptual formulas illustrate the first steps in using rational risk assessment methodology to get a handle on real, not perceived, risks. By no means are these steps perfect or devoid of subjectivity, but they provide a starting point for rational discussion of risk.

Why worry with all of this? Why not just overestimate risk to make sure we're safe? As the example of decaffeinated coffee demonstrates, responding to risks that don't exist can limit our options, increase the money we spend, prohibit the development of products, and raise the costs of production of goods because of excessive regulatory burden. It is possible to be overly cautious. For a relative ranking of the risks of familiar activities and products, see Table 35.1.

Assessing Benefits

Assessing risks is a simple task compared with assessing benefits. Benefits are often very difficult to identify in the abstract. We need to experience them. Beyond the obvious and intended benefits of electric lights over gaslights, we would never have foretold all the benefits we have derived from electrification of our homes. We would never have imagined other unintended, but direct, benefits of electrification: electric washing machines, irons, hot-water heaters, air conditioners, refrigerators, electric guitars, home computers, and VCRs. Nor could we have predicted the secondary, indirect effects of some of the unintended effects: homes in the desert and frozen north, the Internet, home shopping, more free time, the disappearance of household servants, more women in the workforce, and, as a result, a need for day-care centers.

In addition to being difficult to predict, many benefits are very subjective. They are based on emotions, values, and ethical principles and they therefore vary from person to person. Some of the factors we use to measure benefits include the following:

- Saves lives
- Improves health
- Solves problems
- Improves the quality of life
- Increases emotional well-being
- Saves money

Putting Risks and Benefits Together

Defining and assessing the amount of risk and then comparing the risk with the perceived benefits and their probabilities of occurring, are the components of a thoughtful, methodical analysis for assessing risks. Using this methodology, we are in a much better position to make judicious choices on issues that involve real or perceived risk.

Once we have conducted our own risk-benefit analysis on a certain issue and arrived at a decision regarding the acceptability of a risk, we may find that the person next to us has arrived at a different conclusion. This difference of opinion is to be expected, given the difficulties in assessing the amount of risk and the subjective and emotional nature of benefit analysis.

You may think that if the decision involves an individual's acceptance or rejection of personal risk, differences of opinion on the costs and benefits of individual behavior should present no problem. In other words, if I want to accept the risks when I skydive or ride a motorcycle without a helmet, it's my business. Once again, it's not that simple. When does someone's personal decision about the acceptability of a personal risk infringe upon the right of others?

Table 35.1 Ranking of the relative amount of risk[a]

Risk	Source
0.2[b]	PCBs in diet
0.3[b]	DDT[c] in diet
1[b]	1 qt of city water a day
8[b]	Swimming 1 h/day in a chlorinated pool
18	Electricity—chance of dying of shock in a year
30[b]	2 tablespoons of peanut butter a day (risk from peanut mold)
60[b]	12 oz of diet cola a day
100[b]	3/4 teaspoon of basil a day
367	Accidents in the home
600[b]	Indoor air containing formaldehyde vapors from furniture
667	Contribution of air pollution in eastern United States to respiratory illness
800	Auto accident—chance of dying in 1 yr
2,800[b]	12 oz of beer a day
12,000[b]	One pack of cigarettes a day
16,000[b]	One tablet of phenobarbital a day

[a]The higher the number, the greater the risk. Drinking a quart of city water, which contains chloroform, a by-product of chlorination, serves as a reference point. For example, the risk in eating 3/4 teaspoon of basil a day is 100 times greater than the risk in drinking a quart of city water a day. The data are taken from articles by Bruce Ames and Richard Wilson in the April 17, 1987, issue of *Science*.

[b]These values indicate the risk of getting cancer.

[c]DDT, dichlorodiphenyltrichloroethane, a pesticide.

People who drive drunk may feel comfortable with their own risk-benefit analysis of their behavior, but other drivers on the road may not. Given that example, most people would probably come to a similar conclusion regarding individual rights versus the rights of other members of society. Driving on public roads is a public matter made possible with public money. An individual's right to accept the risk of driving drunk is secondary when compared with the public's right to use roads without fear of drunken drivers.

What about private matters like safe sex and AIDS? Surely the risks of having sex are relevant only to the people involved. If people risk contracting AIDS by having unprotected sex, isn't that their business and only their business? How can that choice infringe on the rights of others?

This may not be as simple and straightforward as it first appears or as we wish it were. As the cost of providing care for a growing population of AIDS victims increases, health insurance costs will increase. That cost increase will be borne by many people in ways that seem completely unrelated to AIDS and unprotected sex. For example, today many employers pay for the health insurance of their employees. As health insurance costs increase, employers must pay more money to insure their workers. The company must then compensate for this increase by decreasing costs in other areas or by increasing income. Cutting jobs is one way to cut costs. To increase income, companies usually raise the prices they charge for goods and services. So, an increase in health insurance costs could mean fewer jobs and/or higher prices. Should everyone pay the price for those who have chosen to have unprotected sex and, as a result, been infected with the human immunodeficiency virus?

Issues of individual rights are never as simple as we would like. As members of society, we have both rights and responsibilities to ourselves and one another. Finding the proper balance between the two can be very, very difficult.

A Case Study:
Monarch Butterflies and
Bacillus thuringiensis (Bt) Corn

In May 1999, a story hit the front pages of major newspapers and popular magazines and got the attention of nature lovers around the world. Researchers at Cornell University had shown that corn pollen from plants genetically engineered to contain a gene from *Bacillus thuringiensis* (Bt) would kill monarch butterflies (*Danaus plexippus*). Attention-getting headlines such as "Attack of the Killer Corn" and "Nature at Risk" triggered strong emotional responses as well as political actions. The European Union's regulatory commission, which decides which crops can be grown and imported, stopped the approval process for Bt corn varieties under review. Some activists called for a worldwide ban on agricultural biotechnology. More moderate groups demanded that the U.S. government remove Bt corn seeds from the market.

The story of the monarch butterfly and Bt corn provides an excellent example of the way in which emotional thinking can cloud objective analysis of risk, sometimes with negative consequences. The popular media's coverage of the Cornell study left the impression that scientists had discovered something that had been overlooked by the companies that developed Bt corn and by regulators who did not thoroughly review the potential ecological impact of the commercial use of Bt corn. In fact, the companies, agricultural scientists, and regulators were not surprised by the predictable findings of the Cornell study, which were published as correspondence and not as a research article about a new discovery. As described below, Bt corn was thoroughly reviewed and tested before its commercialization, and farmers around the world had used Bt as a biocontrol agent since 1938. Nonetheless, U.S. regulators were being pressured to reverse their decision and remove Bt corn from the market or limit the amount farmers could grow, and European Union regulators were reluctant to permit farmers in Europe to grow Bt corn.

Will banning Bt corn benefit monarch butterflies, other insects, farmers, or the public? An unemotional, fact-based risk-benefit analysis can help provide the answer. What are the facts behind the monarch-Bt story?

Facts about monarch butterflies

You may already be somewhat familiar with the famous monarch-viceroy mimicry relationship. Most introductory biology books contain pictures showing the striking similarity between the unpalatable monarch and the tasty (to birds, at least) viceroy butterfly (*Limenitis archippus*).

Monarch butterflies are members of the insect order Lepidoptera, which includes butterflies and moths. Like all lepidopterans, monarchs undergo complete metamorphosis in which there are four distinct stages: egg, larva (caterpillar), pupa (chrysalis), and adult (butterfly). The development from egg to adult takes about 1 month in monarchs. Like some other lepidopterans, monarchs are feeding specialists. The

Societal Issues

35. Weighing Technology's Risks and Benefits • **285**

larvae of monarchs eat only milkweed plants (*Asclepius* spp.) (milkweeds have compounds that make the monarchs so unpalatable). Females usually lay their eggs singly on the underside of the leaves of milkweed plants. Only the larval stage feeds on the leaves and flowers of milkweeds. The adults drink nectar from milkweed flowers or, if they emerge from cocoons before milkweeds bloom, from a variety of flowers.

In North America during the summer, there may be as many as four to six generations of monarchs in the south and one to three generations in the northern states and southern Canada. The adults from most generations live only 2 to 6 weeks during the summer. However, the last generation of adults migrate south for the winter and gather in huge roosts in trees. Millions of butterflies can live in a roost. They live in these large aggregations for 6 to 9 months before migrating north in the spring, starting in late February to early March. Monarchs migrating northward from Mexico usually arrive in the middle portion of Texas in mid- to late March and then are usually sighted in Kansas by mid-April.

There are two distinct populations of monarchs in North America: an eastern population that overwinters in Mexico and lives and breeds east of the Rocky Mountains, and a western population that overwinters in California and lives and breeds west of the Rocky Mountains. The two populations have little contact with each other, so there is probably little, if any, interbreeding between them. Under such conditions, scientists would expect to see genetic differences in the two populations. Surprisingly, mitochondrial DNA studies reveal virtually no genetic variation, either between the eastern and western populations or within either population.

Monarchs also live throughout Central America and most of South America, in Australia, and in Hawaii and other Pacific islands.

Facts about Bt

As described in chapter 1, Bt is a naturally occurring bacterium with a worldwide distribution that is com-

monly found in soils and enclosed insect-rich environments such as grain storage facilities. Bt produces large amounts of a crystalline protein that, when ingested by certain insects, slows development and ultimately kills the insect if enough is consumed. The protein must be ingested by the insect to exert its insecticidal effects. Bt is lethal to certain insects but nontoxic to other insects, birds, mammals, and other organisms.

Because of its selective toxicity, Bt has long been a favorite form of insect pest control in small-scale agriculture. First commercialized in France in 1938, Bt was registered as a biopesticide in the United States in 1961. Organic growers and home gardeners throughout the world have used a number of commercially available Bt sprays as biocontrol agents for decades.

Within the Bt species there are also strains or subspecies that exhibit remarkable specificity. Certain subspecies are specifically toxic to certain insect orders (Table 35.2).

Facts about Bt corn

Bt corn is a generic name for a number of corn varieties that have been genetically engineered to contain the gene that codes for the insecticidal crystalline protein produced by the Bt strain *B. thuringiensis kurstaki*. Bt corn was developed to control a specific lepidopteran pest, the European corn borer (ECB), which causes an estimated $1 billion worth of damage to U.S. farmers in lost corn yields every year. Illinois corn growers alone lose $50 million a year because of decreased yields resulting from ECB damage. The Bt gene can also provide protection against other lepidopteran corn pests such as the cornear worm, Southwestern corn borer, and fall army worm. Currently, there are nine varieties of Bt corn commercialized by four companies available to U.S. farmers. In 1999, 30% of the country's corn acreage was planted in a Bt corn variety.

The U.S. Regulatory Process for Bt Corn Approval

Before companies moved from laboratory research to field tests and then to commercial production of Bt corn seed in the United States, they had to receive approval from three federal regulatory agencies.

Table 35.2 Insect orders and species susceptible to certain subspecies of Bt[a]

Bt subspecies	Insect order	Target pests
B. thuringiensis subsp. *kurstaki*	Lepidoptera (butterflies and moths)	Gypsy moth, European corn borer
B. thuringiensis subsp. *israelensis*	Diptera (flies and mosquitoes)	Yellow mosquito, black fly
B. thuringiensis subsp. *tenebrionis*	Coleoptera (beetles)	Colorado potato beetle

[a]Different subspecies of *B. thuringiensis* are selectively toxic to certain species within a given insect order. Scientists have genetically engineered crop plants to contain the gene from the Bt subspecies that is toxic to the primary pests of that crop. Scientists are also developing Bt products to control insects that are disease vectors, such as mosquitoes.

- The U.S. Department of Agriculture (USDA), which regulates small-scale field tests and importation and interstate transport of genetically engineered plants and must provide clearance for large-scale production of genetically engineered plants;
- The Environmental Protection Agency (EPA), which regulates field trials larger than 10 acres, if the plant has been genetically engineered to contain a pesticidal compound such as the Bt protein, and also the sale and distribution of these plants;
- The Food and Drug Administration (FDA), which has jurisdiction over the nutritional and food safety aspects of Bt corn.

Companies also have to obtain permission from appropriate regulatory agencies in other countries before testing and selling Bt corn in those countries.

For the earliest varieties of Bt corn approved for sale in the United States, the step-by-step regulatory approval process, from submitting a request to the USDA for a permit for a small-scale field test to introducing seeds to the commercial market, took approximately 3 to 4 years. What type of information did the regulatory agencies require from the companies before they would approve Bt corn for field tests or commercial use? Because our case study focuses on the environmental effects of Bt corn, we will not discuss FDA's regulatory requirements. If you are interested in the type of nutritional and food safety information that companies must provide to the FDA before a genetically engineered food is approved for sale, visit the FDA website at http://www.fda.gov.

To ensure that environmental consequences of proposed field tests are reviewed before outdoor testing of a genetically engineered organism, the USDA requires the submission of large amounts of technical information so it can conduct an environmental assessment (EA), which is required under the National Environmental Policy Act. The USDA scientists who review the submitted information use a portion of the required data to assess ecological attributes such as potential for movement of genetic material to plants of the same species (outcrossing), distance required to prevent outcrossing, potential for movement of genetic material to related species, methods to prevent dispersal of genetic material from seeds and vegetative plant parts, potential of the crop to become a naturalized weed, potential to outcompete native flora and displace indigenous plant species, increased susceptibility to pathogens or herbivores, and *potential to harm animals near and around the test site.*

Once the USDA has reviewed the potential environmental impact of the field test, they issue a field test

permit to the product developer if there is a "finding of no significant impact" (FONSI). The USDA requires that certain experimental methods (such as required isolation distances and borders of unmodified crops), data collection, and monitoring procedures be followed. The USDA is required by law to

- Provide an opportunity for public input during the EA process;
- Make both the EA and FONSI statements available to interested or affected parties;
- Delay action if a member of the public challenges the sufficiency of the EA or FONSI. For information on the USDA regulatory requirements for genetically engineered organisms and other relevant public documents, see the website http://www.aphis.usda.gov.

If the results of small-scale field tests are positive, the product developer, which may be a company, research institute, or university, moves to large-scale tests and, ultimately, to commercial production. At this stage of product development, the EPA regulations become applicable if the genetically engineered crop contains a pesticidal compound.

All pesticides, including genetically engineered pesticidal plants, must be registered with the EPA and can be distributed and used only in accordance with the conditions specified in the registration. Before EPA can register a pesticide, it must have sufficient data to determine that the product, when used in accordance with commonly recognized practices, will not cause (or significantly increase the risk of) unreasonable adverse effects to humans or the environment.

Before product registration, researchers must obtain a permit to conduct large-scale tests (>10 acres). (EPA also has legal authority to require a permit for small-scale tests if it determines that EPA oversight is necessary.) To obtain a permit and product registration, scientists must submit extensive data on

- The characteristics of the product, including information on the biochemistry and bioactivity of the pesticidal substance, the molecular biology of trait introduction, the biology of the recipient plant and the source of the introduced genes, and the expression levels and environmental stability of the pesticidal protein.
- The environmental fate of the pesticidal gene product, including information on the amount found in soils when plants die; rate of degradation, dissipation, and transport; the degree to which the pesticidal protein is exuded or volatilized from the plant during the growing season; and persistence in soil, water, and air.

Societal Issues

- Human health effects, which will not be enumerated here but can be found under the Office of Pesticide Programs at the EPA website http://www.epa.gov.
- Ecological effects, such as potential for cross-pollination, gene flow to wild relatives, and weediness; increased or decreased disease and herbivore resistance; and *effects on nontarget organisms.* Because of the pertinence of the ecological risk analysis to the discussion of monarchs and Bt corn, more information on specific data requirements for this section of the permit application is provided in Table 35.3.

Scientific findings of the Cornell study

In controlled laboratory feeding experiments, researchers at Cornell University clearly demonstrated that 3-day-old monarch larvae that fed on milkweed leaves dusted with Bt corn pollen ate less, grew more slowly, and suffered a higher mortality rate than larvae that were fed milkweed coated with non-Bt corn pollen or milkweed with no corn pollen. Only 56% of the larvae that were fed leaves with Bt corn pollen survived after 4 days of feeding, compared with a survival rate of 100% for the other two treatment groups. The larvae feeding on pollen-free leaves ate significantly more leaf material than the larvae fed leaves dusted with non-Bt corn pollen (1.61 versus 1.12); in turn, the larvae fed leaves dusted with non-Bt corn pollen ate significantly more than the larvae fed leaves dusted with Bt corn pollen (1.12 versus 0.57). After 4 days, the average weight of the surviving larvae fed Bt corn pollen was less than half the average weight of the larvae fed leaves with no pollen (0.16 g versus 0.38 g).

Table 35.3   Data and tests required by the EPA for assessing potential environmental impacts[a]

For the hazard and risk analysis, the following questions must be answered adequately with data from experiments or information from the scientific literature:

- Can the plant become a weed/pest through dispersal and persistence?
- What plant parts contain the gene product? Do pollen and seeds?
- Is the gene product released from the plant or from flowering parts?
- Is the plant a copious producer of wind-borne pollen?
- Is the plant naturally self- or cross-pollinated, or both?
- Is the plant's pollen transmitted by wind, insects, and/or other vector?
- Are sexually compatible, nontarget plants nearby? If so, can the plant transmit the newly acquired trait to nontarget plants?
- If transmission to nontarget plants occurred, what would be the ecological consequences?
- If the new trait is pest resistance, would there be significant ecological consequences if it were transferred to related plants?
- Would natural control of the wild plant/weed populations by insects or disease be curtailed, and if so, what would be the ecological consequences?
- Will the gene product (pesticide) persist and move in the soil environment?
- What effect does the plant protein have on beneficial soil invertebrates?
- Will wild birds/mammals feed on the plants or moribund insects?
- Will nontarget animals/insects feeding on or exposed to the plant be adversely affected?
- What effect would the pollen have on pollinating insects?
- When the active principle is a metabolic product (e.g., chitinase, protease inhibitor, a lectin, hypersensitive response proteins), will it pose a hazard to insects or beneficial fungi?
- Can animals, birds, or bats distribute seed to locations containing weedy relatives?
- When in aquatic plants, will these be consumed by fish or aquatic invertebrates?
- When in aquatic plants, will these be released from plants into water?
- Is the gene product expected to reach the estuarine/marine environment in significant concentration?

To assess the potential effects on nontarget organisms and fate in the environment, the following studies must be conducted:

- Avian oral toxicity test on upland game bird (quail) and waterfowl (duck) species
- (Avian injection and inhalation tests may also be required)
- Wild mammal oral toxicity test (rodent species)
- Freshwater fish oral toxicity test (on both cold water and warm water species)
- Freshwater invertebrate testing (on *Daphnia* or another aquatic insect species)
- Estuarine and marine animal testing (grass shrimp and fathead minnow species)
- Nontarget plant studies (terrestrial, aquatic, and outcrossing issues)
- Nontarget insect testing—toxicity to predators and parasites (green lacewing larvae, ladybird beetle, and parasitic wasp species)
- Nontarget insect testing—toxicity to pollinators (honey bee larval and adult stages)
- Nontarget soil organism testing—toxicity to earthworm and springtail/*Collembola* species
- Terrestrial environmental expression testing (environmental fate/degradation rates of the proteins in soil)
- Plant tissue expression data (and degradation rates of the proteins)

[a]The EPA is one of three U.S. regulatory agencies responsible for evaluating the potential risks of genetically engineered organisms to the environment or human health. To obtain a permit to conduct a 10-acre (or larger) field test on a genetically engineered pesticidal plant, such as Bt corn, applicants must provide EPA scientists with data for assessing potential environmental impacts.

Societal Issues

Risks of Bt corn to monarch butterflies in nature

Is Bt Corn Pollen Toxic or Hazardous to Monarchs?

The laboratory study clearly shows that, as expected, Bt corn pollen can be toxic to monarch larvae that ingest it. Bt corn has been engineered to contain the gene from *B. thuringiensis* subsp. *kurstaki* to protect it from the ECB. This Bt strain produces a protein specifically toxic to lepidopteran insects, such as monarchs. Although monarch larvae are not the target organism for the pest control Bt provides, they and other nontarget lepidopterans could be harmed if they ingest any part of a Bt corn plant that is expressing the Bt gene.

The study does not provide information on the number of Bt pollen grains that must be consumed in order to observe the lethal or developmental effects of Bt pollen on 3-day-old larvae. Nor does it provide information on effects on older, larger larvae, which, like other lepidopteran larvae, are expected to have a higher tolerance to Bt toxicity.

Is Bt Corn Pollen a Risk to Monarchs Under Natural Conditions?

Recall that risk is dependent on both the hazard and the amount of exposure to the hazard. A number of questions need to be answered to determine the probability that monarchs will be exposed to Bt corn pollen in amounts large enough to present a significant risk.

- Do monarch larvae occur in areas where Bt corn is grown?

  Yes. A map showing where most corn is grown in the United States is provided in Figure 35.1A. These are also the areas with the worst infestations of ECB. Farmers in areas with regular outbreaks of ECB are much more likely to plant Bt corn than farmers in parts of the country where ECB infestations do not occur. Therefore, it appears that only certain monarchs in the eastern population would coexist with Bt corn varieties currently marketed. Approximately 25% of the U.S. monarch population is concentrated in the Midwest, the primary corn-growing region of the country (Figure 35.1B).

  The Bt corn varieties grown to date are all field corn varieties used as a source of grain for feeding livestock. If companies begin to sell varieties of sweet corn genetically engineered to contain the Bt gene, the geographic area over which Bt corn is grown will probably increase as farmers attempt to control other lepidopteran pests, such as the corn earworm (*Heliothis virescens*).

- What is the likelihood that milkweed plants occur near cornfields?

  Monarch larvae eat only milkweed plants, so the degree of exposure of monarchs to Bt corn is related to the distribution and location of milkweed plants. The most common milkweed species are usually found in open areas: along roadsides, in ditches, at the edge of forests, and in open fields such as pastures. So milkweed plants could occur in the vicinity of cornfields, but not all milkweed plants that are potential hosts for developing monarchs are located near cornfields. Farmers treat milkweed plants as weeds, so they do not usually occur within the cornfield.

- What is the likelihood that Bt corn pollen would be found on leaves of milkweed plants growing near cornfields?

  We can answer this question by looking at studies of corn pollen movement. Because corn pollen is relatively heavy, it does not tend to move very far outside the borders of the cornfield. Iowa State University researchers have shown that 70% of the pollen released by corn plants stays within the confines of the cornfield, and only 10% of the pollen travels farther than 3 meters from the edge of the field. Researchers at the University of Guelph in Ontario recorded similar distances of pollen movement for corn: within 5 yards of the field's edge, 90% of the pollen had been deposited.

- What is the likelihood that a female monarch will lay eggs on a milkweed plant with Bt corn pollen?

  To answer this question, we need more information about the distribution of milkweed plants. If most milkweed plants that a monarch female is likely to encounter do not occur next to cornfields planted with Bt corn, then the probability is small that a monarch female will lay eggs on milkweeds with Bt corn pollen.

- What is the likelihood that Bt pollen will be found on milkweed leaves when monarch larvae are feeding on milkweed?

  This question relates to the relative timing of monarch egg laying and corn pollen shedding. A given field of corn sheds pollen for 8 to 10 days sometime between June and August, the corn-growing season in the entire United States. The approximate date of pollen shedding within that 2-month period varies with the latitude at which the corn grows. To be harmed by Bt corn pollen, monarch

Societal Issues

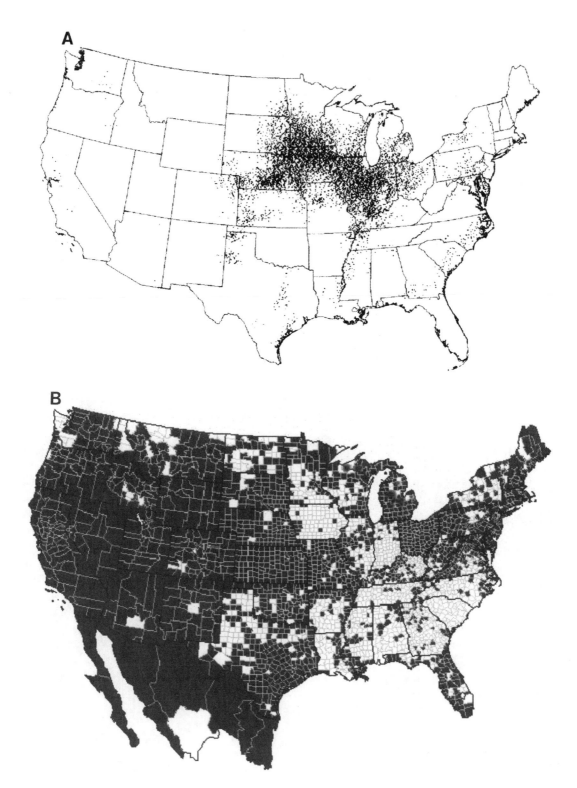

Figure 35.1 (A) In the United States, more than 85% of the field corn acreage lies in only 10 states—the Corn Belt—and more than 70% of the country's field corn grows in only seven of those states. This is also the section of the country where the greatest percentage of farmers grow Bt corn, because ECB infestations can cause major losses to corn growers in this area. (Data provided by the USDA. 1 dot equals 10,000 acres.) (B) Distribution of monarch butterflies in the United States. Darkened areas indicate areas in which they reproduce and, therefore, where larvae would occur. Approximately 50% of the monarch butterflies in the United States breed west of the Rocky Mountains. Approximately half of those breeding east of the Rockies breed in the Midwest. (Graphic provided by Dr. Paul Opler and the Biological Resources Division of the U.S. Geological Survey.)

larvae must be feeding on milkweed plants next to or in the cornfield during this 8- to 10-day period when pollen is shed.

- What is the probability that the concentration of Bt pollen on milkweed leaves will be high enough to harm monarch larvae?

As mentioned earlier, the Cornell study does not provide data on the effects of different doses of Bt corn pollen on larval growth and survival. Studies by researchers in Iowa and Canada show that pollen density must be approximately 135 to 150 pollen grains per cm^2 for Bt corn pollen to be toxic to young monarch larvae. This level of pollen density will rarely occur if findings by researchers at the University of Maryland can be applied to the rest of the country. They collected 1,300 leaves from milkweed plants growing near cornfields and found that the average pollen density on plants within 3 meters of the cornfield edge was only approximately 20 to 30 grains per cm^2.

- What is the probability that monarch larvae will consume enough Bt corn pollen to retard development or increase mortality?

Data from the Cornell study show that monarch larvae do not readily eat milkweed leaves coated with pollen whether or not that pollen contains the Bt gene. More than likely, if a feeding larva encounters pollen, the larva will avoid it and feed elsewhere.

Safeguards to Further Decrease Risk

As we described earlier, the amount of risk depends not only on exposure to a hazard but also on safeguards that can be put into place to further decrease risk. What safeguards might decrease the risk of Bt corn pollen to monarchs?

- Because Bt corn varieties differ in the level of toxicity of the pollen, farmers in monarch-rich areas could choose varieties of lower toxicity.
- It is now possible to genetically engineer plants to limit gene expression to specific tissues. Varieties that do not express the Bt gene in the pollen would be of no risk to monarchs, because monarchs do not feed on corn plants.
- Planting a border of non-Bt corn around a Bt cornfield would further decrease the already low probability that Bt pollen would drift to milkweed plants.
- Planting milkweeds at sites distant from cornfields would increase the probability that an egg-laying female would encounter milkweeds uncontaminated with corn pollen.

Benefits of Bt corn to monarchs and others

For a fair assessment of the value of growing or not growing Bt corn, we must determine the potential benefits of Bt corn and contrast these benefits with the risks. Another way of framing the question is to consider the risks of *not* growing Bt corn, because Bt corn provides advantages to farmers, farm laborers, consumers, and the environment when compared with current corn-growing practices.

- Many feeding studies, including those submitted to U.S. regulatory agencies, have shown that the Bt protein, in contrast to chemical pesticides currently used, does not harm beneficial insects, such as ladybird beetles and honeybees, and is safe for freshwater invertebrates, earthworms, other soil organisms, birds, and mammals.
- Researchers at Iowa State University and the USDA have demonstrated that farmers using Bt corn have decreased the amount of insecticides they apply to corn. This decrease in insecticide use is beneficial to all insects, including monarch butterflies, because spray drifts of pesticides can extend to large areas owing to the small size of droplets. For chemical insecticides, the typical impact on nontarget insects in or adjacent to treated fields approaches 100%. USDA and EPA regulators who reviewed permit requests for testing and commercializing Bt corn recognized that widespread planting of Bt corn could potentially harm nontarget lepidopteran insects. However, they also felt that the benefits to lepidopterans and other insects provided by fewer insecticide applications far outweighed the risks.
- Fewer insecticide applications and a reduction in insecticide spray drift also benefit farm workers.
- Bt corn offers almost 100% protection against the ECB, providing a savings to U.S. farmers of $1 billion per year.
- Finally, studies at Iowa State University have shown that Bt corn has significantly lower amounts of mycotoxins than non-Bt corn does. These mycotoxins, which are fatal to livestock and harmful to humans, are produced by fungi that invade the corn plant at sites of insect damage.

Activities

1. Select any familiar, pervasive technology we depend on, and conduct a risk-benefit analysis.

2. Evaluate that same technology before its widespread implementation. What risks and benefits would you have predicted before that technology's development?

3. Select a future application of biotechnology or another emerging technology, and conduct a risk-benefit analysis.

4. Select a risk that is considered personal, such as skydiving, cigarette smoking, or drug use, and describe the societal ramifications of taking this individual risk.

5. Compare and contrast the media coverage of a technological development or scientific discovery with the primary literature on the subject.

6. Select a food safety issue and develop a case study, focusing on real risks and benefits.

7. Write a short essay explaining either of the following quotes:

"To be alive is to be at risk."
Daniel Koshland, Science, 1987

"Only the dose makes the poison."
Paracelsus, 16th century

Are Bioengineered Foods Safe?

(Reprinted from *FDA Consumer,* January 2000)

Since 1994, a growing number of foods developed using the tools of the science of biotechnology have come onto both the domestic and international markets. With these products has come controversy, primarily in Europe where some question whether these foods are as safe as foods that have been developed using the more conventional approach of hybridization.

Ever since the latter part of the 19th century, when Gregor Mendel discovered that characteristics in pea plants could be inherited, scientists have been improving plants by changing their genetic makeup. Typically, this was done through hybridization in which two related plants were cross-fertilized and the resulting offspring had characteristics of both parent plants. Breeders then selected and reproduced the offspring that had the desired traits.

Today, to change a plant's traits, scientists are able to use the tools of modern biotechnology to insert a single gene—or, often, two or three genes—into the crop to give it new, advantageous characteristics. (See "Methods for Genetically Engineering a Plant" [*FDA Consumer,* January 2000].) Most genetic modifications make it easier to grow the crop. About half of the American soybean crop planted in 1999, for example, carries a gene that makes it resistant to an herbicide used to control weeds. About a quarter of U.S. corn planted in 1999 contains a gene that produces a protein toxic to certain caterpillars, eliminating the need for certain conventional pesticides.

In 1992, the Food and Drug Administration published a policy explaining how existing legal requirements for food safety apply to products developed using the tools of biotechnology. It is the agency's responsibility to ensure the safety of all foods on the market that come from crops, including bioengineered plants, through a science-based decision-making process. This process often includes public comment from consumers, outside experts and industry. FDA estab-

lished, in 1994, a consultation process that helps ensure that foods developed using biotechnology methods meet the applicable safety standards. Over the last five years, companies have used the consultation process more than 40 times as they moved to introduce genetically altered plants into the U.S. market.

Although the agency has no evidence that the policy and procedure do not adequately protect the public health, there have been concerns voiced regarding FDA's policy on these foods. To understand the agency's role in ensuring the safety of these products, *FDA Consumer* sat down with Commissioner Jane E. Henney, M.D., to discuss the issues raised by bioengineered foods:

FDA Consumer: Dr. Henney, what does it mean to say that a food crop is bioengineered?

Dr. Henney: When most people talk about bioengineered foods, they are referring to crops produced by utilizing the modern techniques of biotechnology. But really, if you think about it, all crops have been genetically modified through traditional plant breeding for more than a hundred years.

Since Mendel, plant breeders have modified the genetic material of crops by selecting plants that arise through natural or, sometimes, induced changes. Gardeners and farmers and, at times, industrial plant breeders have crossbred plants with the intention of creating a prettier flower, a hardier or more productive crop. These conventional techniques are often imprecise because they shuffle thousands of genes in the offspring, causing them to have some of the characteristics of each parent plant. Gardeners or breeders then look for the plants with the most desirable new trait.

With the tools developed from biotechnology, a gene can be inserted into a plant to give it a specific new characteristic instead of mixing all of the genes from two plants and seeing what comes out. Once in the plant, the new gene does what all genes do: It directs

Societal Issues

the production of a specific protein that makes the plant uniquely different.

This technology provides much more control over, and precision to, what characteristic breeders give to a new plant. It also allows the changes to be made much faster than ever before.

No matter how a new crop is created—using traditional methods or biotechnology tools—breeders are required by our colleagues at the U.S. Department of Agriculture to conduct field testing for several seasons to make sure only desirable changes have been made. They must check to make sure the plant looks right, grows right, and produces food that tastes right. They also must perform analytical tests to see whether the levels of nutrients have changed and whether the food is still safe to eat.

As we have evaluated the results of the seeds or crops created using biotechnology techniques, we have seen no evidence that the bioengineered foods now on the market pose any human health concerns or that they are in any way less safe than crops produced through traditional breeding.

FDA Consumer: What kinds of genes do plant breeders try to put in crop plants?

Dr. Henney: Plant researchers look for genes that will benefit the farmer, the food processor, or the consumer. So far, most of the changes have helped the farmer. For example, scientists have inserted into corn a gene from the bacterium *Bacillus thuringiensis,* usually referred to as BT. The gene makes a protein lethal to certain caterpillars that destroy corn plants. This form of insect control has two advantages: It reduces the need for chemical pesticides, and the BT protein, which is present in the plant in very low concentrations, has no effect on humans.

Another common strategy is inserting a gene that makes the plant resistant to a particular herbicide. The herbicide normally poisons an enzyme essential for plant survival. Other forms of this normal plant enzyme have been identified that are unaffected by the herbicide. Putting the gene for this resistant form of the enzyme into the plant protects it from the herbicide. That allows farmers to treat a field with the herbicide to kill the weeds without harming the crop.

The new form of the enzyme poses no food safety issues because it is virtually identical to nontoxic enzymes naturally present in the plant. In addition, the resistant enzyme is present at very low levels and it is as easily digested as the normal plant enzyme.

Modifications have also been made to canola and soybean plants to produce oils with a different fatty acid composition so they can be used in new food processing systems. Researchers are working diligently to develop crops with enhanced nutritional properties.

FDA Consumer: Do the new genes, or the proteins they make, have any effect on the people eating them?

Dr. Henney: No, it doesn't appear so. All of the proteins that have been placed into foods through the tools of biotechnology that are on the market are nontoxic, rapidly digestible, and do not have the characteristics of proteins known to cause allergies.

As for the genes, the chemical that encodes genetic information is called DNA. DNA is present in all foods and its ingestion is not associated with human illness. Some have noted that sticking a new piece of DNA into the plant's chromosome can disrupt the function of other genes, crippling the plant's growth or altering the level of nutrients or toxins. These kinds of effects can happen with any type of plant breeding—traditional or biotech. That's why breeders do extensive field-testing. If the plant looks normal and grows normally, if the food tastes right and has the expected levels of nutrients and toxins, and if the new protein put into food has been shown to be safe, then there are no safety issues.

FDA Consumer: You mentioned allergies. Certain proteins can cause allergies, and the genes being put in these plants may carry the code for new proteins not normally consumed in the diet. Can these foods cause allergic reactions because of the genetic modifications?

Dr. Henney: I understand why people are concerned about food allergies. If one is allergic to a food, it needs to be rigorously avoided. Further, we don't want to create new allergy problems with food developed from either traditional or biotech means. It is important to know that bioengineering does not make a food inherently different from conventionally produced food. And the technology doesn't make the food more likely to cause allergies.

Fortunately, we know a lot about the foods that do trigger allergic reactions. About 90 percent of all food allergies in the United States are caused by cow's milk, eggs, fish and shellfish, tree nuts, wheat, and legumes, especially peanuts and soybeans.

To be cautious, FDA has specifically focused on allergy issues. Under the law and FDA's biotech food policy, companies must tell consumers on the food label when a product includes a gene from one of the common allergy-causing foods unless it can show that the protein produced by the added gene does not make the food cause allergies.

We recommend that companies analyze the proteins they introduce to see if these proteins possess properties indicating that the proteins might be allergens. So far, none of the new proteins in foods evaluated through the FDA consultation process have caused allergies. Because proteins resulting from biotechnology and now on the market are sensitive to heat, acid and enzymatic digestion, are present in very low levels in the food, and do not have structural similarities to known allergens, we have no scientific evidence to indicate that any of the new proteins introduced into food by biotechnology will cause allergies.

FDA Consumer: Let me ask you one more scientific question. I understand that it is common for scientists to use antibiotic resistance marker genes in the process of bioengineering. Are you concerned that their use in food crops will lead to an increase in antibiotic resistance in germs that infect people?

Dr. Henney: Antibiotic resistance is a serious public health issue, but that problem is currently and primarily caused by the overuse or misuse of antibiotics. We have carefully considered whether the use of antibiotic resistance marker genes in crops could pose a public health concern and have found no evidence that it does.

I'm confident of this for several reasons. First, there is little if any transfer of genes from plants to bacteria. Bacteria pick up resistance genes from other bacteria, and they do it easily and often. The potential risk of transfer from plants to bacteria is substantially less than the risk of normal transfer between bacteria. Nevertheless, to be on the safe side, FDA has advised food developers to avoid using marker genes that encode resistance to clinically important antibiotics.

FDA Consumer: You've mentioned FDA's consultative process a couple of times. Could you explain how genetically engineered foods are regulated in the United States?

Dr. Henney: Bioengineered foods actually are regulated by three federal agencies: FDA, the Environmental Protection Agency, and the U.S. Department of Agriculture. FDA is responsible for the safety and labeling of all foods and animal feeds derived from crops, including biotech plants. EPA regulates pesticides, so the BT used to keep caterpillars from eating the corn would fall under its jurisdiction. USDA's Animal and Plant Health Inspection Service oversees the agricultural environmental safety of planting and field testing genetically engineered plants.

Let me talk about FDA's role. Under the federal Food, Drug, and Cosmetic Act, companies have a legal obligation to ensure that any food they sell meets the safety standards of the law. This applies equally to conventional food and bioengineered food. If a food does not meet the safety standard, FDA has the authority to take it off the market.

In the specific case of foods developed utilizing the tools of biotechnology, FDA set up a consultation process to help companies meet the requirements. While consultation is voluntary, the legal requirements that the foods have to meet are not. To the best of our knowledge, all bioengineered foods on the market have gone through FDA's process before they have been marketed.

Here's how it works. Companies send us documents summarizing the information and data they have generated to demonstrate that a bioengineered food is as safe as the conventional food. The documents describe the genes they use: whether they are from a commonly allergenic plant, the characteristics of the proteins made by the genes, their biological function, and how much of them will be found in the food. They tell us whether the new food contains the expected levels of nutrients or toxins and any other information about the safety and use of the product.

FDA scientists review the information and generally raise questions. It takes several months to complete the consultation, which is why companies usually start a dialog with the agency scientists nearly a year or more before they submit the data. At the conclusion of the consultation, if we are satisfied with what we have learned about the food, we provide the company with a letter stating that they have completed the consultation process and we have no further questions at that time.

FDA Consumer: Since genes are being added to the plant, why doesn't FDA review biotech products under the same food additive regulations that it reviews food colors and preservatives?

Dr. Henney: The food additive provision of the law ensures that a substance with an unknown safety profile is not added to food without the manufacturer proving to the government that the additive is safe. This intense review, however, is not required under the law when a substance is generally recognized as safe

(GRAS) by qualified experts. A substance's safety can be established by long history of use in food or when the nature of the substance and the information generally available to scientists about it is such that it doesn't raise significant safety issues.

In the case of bioengineered foods, we are talking about adding some DNA to the plant that directs the production of a specific protein. DNA already is present in all foods and is presumed to be GRAS. As I described before, adding an extra bit of DNA does not raise any food safety issues.

As for the resulting proteins, they too are generally digested and metabolized and don't raise the kinds of food safety questions as are raised by novel chemicals in the diet. The proteins introduced into plants so far either have been pesticides or enzymes. The pesticide proteins, such as BT, would actually be regulated by EPA and go through its approval process before going on the market. The enzymes have been considered to be GRAS, so they have not gone through the food additive petition process. FDA's consultation process aids companies in determining whether the protein they want to add to a food is generally recognized as safe. If FDA has concerns about the safety of the food, the product would have to go through the full food additive premarket approval process.

FDA Consumer: Why doesn't FDA require companies to tell consumers on the label that a food is bioengineered?

Dr. Henney: Traditional and bioengineered foods are all subject to the same labeling requirements. All labeling for a food product must be truthful and not misleading. If a bioengineered food is significantly different from its conventional counterpart—if the nutritional value changes or it causes allergies—it must be labeled to indicate that difference. For example, genetic modifications in varieties of soybeans and canola changed the fatty acid composition in the oils of those plants. Foods using those oils must be labeled, including using a new standard name that indicates the bioengineered oil's difference from conventional soy and canola oils. If a food had a new allergy-causing protein introduced into it, the label would have to state that it contained the allergen.

We are not aware of any information that foods developed through genetic engineering differ as a class in quality, safety, or any other attribute from foods developed through conventional means. That's why there has been no requirement to add a special label saying that they are bioengineered. Companies are free to include in the labeling of a bioengineered product any statement as long as the labeling is truthful and not misleading. Obviously, a label that implies that a food is better than another because it was, or was not, bioengineered, would be misleading.

FDA Consumer: Overall, are you satisfied that FDA's current system for regulating bioengineered foods is protecting the public health?

Dr. Henney: Yes, I am convinced that the health of the American public is well protected by the current laws and procedures. I also recognize that this is a rapidly changing field, so FDA must stay on top of the science as biotechnology evolves and is used to make new kinds of modifications to foods. In addition, the agency is seeking public input about our policies and will continue to reach out to the public to help consumers understand the scientific issues and the agency's policies.

Not only must the food that Americans eat be safe, but consumers must have confidence in its safety, and confidence in the government's role in ensuring that safety. Policies that are grounded in science, that are developed through open and transparent processes, and that are implemented rigorously and communicated effectively are what have assured the consumers' confidence in an agency that has served this nation for nearly 100 years.

Debating the Risks of Biotechnology 36

Introduction

Ever since scientists produced the first recombinant DNA molecule in the early 1970s, we have heard much about the risks of biotechnology. Some claim biotechnology's benefits far outweigh its risks, because biotechnology is the panacea for all of our ills—medical, environmental, and economic. Others say it is a Pandora's box from which will spring a new set of problems, most of which we can't predict, much less control. Which side is right, or does the truth lie somewhere in between?

When "authorities" espouse disparate views on the risks of biotechnology, or any technology, deciding which view to accept poses a problem for ordinary citizens not trained as scientists. Identifying specious arguments, superfluous facts intended to mislead, and intellectual dishonesty requires understanding some of the science underlying the issue. How can nonscientists both assess the arguments and engage in productive debates as society struggles to come to terms with the risks of biotechnology or any other technically complex and potentially volatile issue?

For a debate to be satisfying and productive, it must be driven by objective facts and not irrational fears. The participants should be as clear, specific, and emotionally detached as possible. The primary goal should be not winning, but greater understanding by all parties. Remember, we usually learn the most from people whose ideas and opinions differ from our own.

Debate Methodology

What are some of the steps involved in preparing for and conducting a productive debate or discussion?

Step 1

Get clear on exactly what is being discussed. Delineate and define. Use precise language.

Ask people to define what they mean when they say the "risks of biotechnology," so that everyone will be

talking about the same thing. The issues under debate are difficult in and of themselves. Adding the additional confusion of using sloppy language will only make the debate less productive. Having a productive discussion is possible only if the people involved are talking about the same issue.

To what type of risk are people referring? Do they mean environmental and/or public health risk? Are they including other types of risks, such as social, economic, and ethical risks? Do they equate the word "risk" with the probability of death or harm, or is their view of risk less specific and complicated by variables such as control, familiarity, and freedom of choice?

What does the word "biotechnology" mean to people discussing it? In chapter 1, you learned that biotechnology is a collection of technologies such as monoclonal antibody technology, cell culture, genetic engineering, and bioprocessing. You also learned that "biotechnology" is an ambiguous term used in a variety of ways by different people. To which biotechnology is someone referring? The risks of one technology differ greatly from those of another. Usually the focus is genetic engineering. The discussion should therefore be about the risks of genetic engineering, not biotechnology. Are people concerned with only one or two applications of genetic engineering or genetic engineering in general?

Which facet of biotechnology development is the target of their concern? Biotechnology development occurs along a continuum that begins in a research laboratory and extends through product development to widespread commercial use. The risks of biotechnology laboratory research, small-scale field testing, and large-scale commercial use differ from each other both qualitatively and quantitatively.

Finally, in discussing the risks of any technology, keep in mind that technologies may be used wisely or unwisely, for good or for bad. The risks of a technology per se must be distinguished from the risks of a technology placed in irresponsible hands. The flaws of a technology and the flaws of human nature are two very different considerations.

If your goal is greater understanding, the value of using discipline in the words you choose cannot be overestimated.

Step 2

For rational risk assessments as described in the preceding chapter, get facts, not opinions. First, identify the risk; second, estimate the probability of the risk occurring; finally, assess the consequences.

In assessing the risks of biotechnology, obtain objective data that allow you to identify the risks clearly, specifically, and independently. If you want your assessment of risk to be grounded in sound science, go to primary sources such as scientific journals or scholarly articles or to review papers in publications like *Scientific American*. Talk to scientists conducting relevant research. Do not rely on news magazines, newspapers, television, or hearsay. Be particularly careful of information you gather from the Internet. Some websites are excellent sources of factual information, but others are full of inaccuracies and opinions. A general guideline in using the Internet might be to use information from primary sources, universities, and government research institutes and regulatory agencies.

Often, facts other than scientific information are important to the debate and should be clearly articulated. For example, farmers must keep their costs down if they are to make a profit. This is not a "scientific" fact, but it sure is true.

Once you have identified the risks, you must then estimate the probability of the risk occurring. Because much of biotechnology is similar to activities we have been engaged in for many years, we have abundant data on some products and applications that can be used to estimate the general probabilities of specific risks.

Then assess the consequences of a particular risk occurring. What is the significance of risk A occurring? If it occurs, will it matter? If it matters, under what circumstances? It may happen on a regular basis anyway without human interference (insects becoming resistant to chemicals produced by plants, bacteria transferring genes across species). It may be environmentally, ecologically, and economically insignificant (spraying a biopesticide in certain weather con-

ditions causes a crop plant to lose a few leaves but suffer no yield loss).

Step 3

Compare the risk with benefits, other risks, and other options for achieving the same goal.

Establishing that a risk of some type is involved is a first step in conducting a risk assessment but by itself is not particularly helpful. Every human activity carries some amount of risk with it. To weigh the value of taking a risk, we must place the risk in a meaningful context.

First, look at the benefits. An assessment of risk is meaningless unless it is compared with an equally objective analysis of the benefits. Only then can we begin to determine whether the risk is worth taking.

Second, compare risks. An assessment of the risks of one activity or product must be contrasted with the risks involved in achieving the same goal through different activities or products. In other words, if the benefits are substantial, then the goal is worthy. Is there another way of reaching the same goal that entails less risk?

This question is particularly pertinent in assessing the risks involved in biotechnology. Much of biotechnology is similar to activities we have been engaged in for centuries. Often the most appropriate question regarding the risks of biotechnology is not "Are there risks?" but "How do the risks of this new technology compare with those of existing technologies we have directed toward the same purpose?"

Activities

Many areas in biotechnology can serve as topics for developing risk assessment and debate skills. Because these topics have been discussed for a number of years, abundant reference material is available. Topics include the following:

- Labeling of genetically engineered foods
- The use of transgenic animals in the production of pharmaceuticals
- Large-scale environmental releases of genetically engineered organisms
- Patenting of genes or genetically engineered organisms
- Biotechnology and sustainability (or choose one aspect of sustainability such as genetic diversity)
- The forensic use of DNA typing

A Case Study: Herbicide-Tolerant Crops

Introduction

To conduct an objective, methodical analysis of the risks and benefits of a product of biotechnology, we suggest using herbicide-tolerant crops (HTC) for a number of reasons.

- A clear statement of the arguments against developing HTC is available in the report *Biotechnology's Bitter Harvest.*
- A clear statement of the arguments in favor of developing HTC is available in the report *Herbicide-Resistant Crops.*
- The issue is multifaceted.
- Farmers have used HTC for more than 40 years, so relevant data for conducting a risk assessment and weighing the validity of these arguments are readily available.
- The topic provides an opportunity to learn something about a technology you are totally dependent upon yet may know little about: agriculture.
- The topic provides an opportunity to interview farmers, environmentalists, public health officials, and extension agents.
- Agricultural systems are complex ecosystems, so you will be learning a lot of biology as you learn about agriculture.

Methodology

The development of HTC is one application of biotechnology that has received the negative attention we discussed in chapter 2.

A public interest group, the Biotechnology Working Group (BWG), published a report entitled *Biotechnology's Bitter Harvest* that condemns HTC as damaging to the environment and public health. The Council for Agricultural Science and Technology (CAST) counters BWG's assessment in its own report, *Herbicide-Resistant Crops.*

The primary issues the two groups disagree on are whether the use of HTC will

- Further our reliance on herbicides instead of other weed-control methods;
- Increase the amount of herbicides used;
- Cause the evolution of resistance to herbicides by weeds;
- Allow the transfer of genes for herbicide tolerance to weeds, making weeds no longer susceptible to herbicides.

These different viewpoints about the environmental effects of HTC can provide the key points for a classroom debate.

Below, you will find background information on herbicides and HTC and a summary of the main points made by the BWG in *Biotechnology's Bitter Harvest* and, when appropriate, the countering points made by the CAST in *Herbicide-Tolerant Crops.* For additional information on weed control, herbicides, and HTC, contact the U.S. Department of Agriculture, your state department of agriculture, your county extension agent, or scientists in the crop science department at a state university with specialists in agriculture.

Background on HTC

Some of the first products of agricultural biotechnology are crops genetically engineered to tolerate a variety of herbicides, the compounds farmers use to kill weeds in their fields. HTC are not new. Virtually all of the acreage of our major crop plants such as corn, soybeans, cotton, and wheat is treated with herbicides. Therefore, all of our major crops are already tolerant to some herbicides.

Where did this tolerance originate? Genetically based herbicide tolerance exists in nature. Some of our crops were naturally resistant to the herbicides we have developed. These natural HTC can serve as sources of herbicide tolerance genes for other crops.

Using existing, genetically based herbicide tolerance has been useful but, by its very nature, limited. Because of the constraints of crossbreeding, we are able to use only the herbicide tolerance genes found in crops or their close relatives. Genetic engineering now provides us with the ability to capitalize on genetic variation found anywhere in nature. Some bacteria naturally contain genes for herbicide tolerance. We have moved those genes into both major and minor crops.

Major points in the two reports

Weed Control

The BWG states that other available options for controlling weeds are just as effective as herbicides and could replace herbicides as ways to control weeds reliably.

Herbicides and the Environment

The BWG report assumes that herbicides as a group are damaging to the environment and public health.

The CAST report states that herbicides vary widely in their toxicities and implies that some are relatively harmless because they have low toxicities and do not persist in the environment beyond a few days. It also states that there are environmental benefits to using herbicides.

HTC and Herbicide Use

The BWG asserts that the development of HTC will increase the amount of herbicides used in the United States, while CAST states that the development of HTC could have any of the following effects on the amount of herbicides used: no effect, decreased amount required, or increased amount required. The CAST report states that anyone trying to determine whether or not the use of HTC will increase the amount of herbicides used by U.S. farmers must examine the issue on a case-by-case basis, taking into account both the crop and the herbicide.

HTC and the Evolution of Resistance to Herbicides

The BWG asserts that the use of HTC will cause the evolution of resistance to herbicides by weeds. The CAST report describes this phenomenon in detail. What questions do you need to ask and answer in order to assess this issue?

HTC and the Development of Weeds Resistant to Herbicides

The BWG maintains that gene transfer will occur between HTC and weeds, thus making weeds resistant to herbicides. What scientific facts do we know that support or refute this claim?

For Further Discussion

In making choices about HTC, which herbicide and crop traits would be important to consider for environmental reasons?

A Decision-Making Model for Bioethical Issues

Introduction to Bioethics

Since the middle of the 20th century, rapid improvements in technology have changed the practice of medicine profoundly. The technology often allows physicians to restore or supplant basic bodily functions. Mechanical ventilators, heart pacemakers, kidney dialysis machines, exotic drugs, organ transplantation, and artificial nutrition and hydration are some of the life-extending tools available to the modern practitioner.

Initially, the use of these techniques was seen as altogether positive, a must-do choice for the physician. It was not long, however, before difficult questions surfaced. Just because we *can* maintain a person with artificial ventilation and nutrition, *should* we? Who should answer such a question: the patient, physician, family, nurse, social worker, hospital administrator, insurance company, ethicist, politician?

Since the 1960s, questions like this one have provided us with an entirely new discipline or, more accurately, interdiscipline: bioethics. Bioethics, or biomedical ethics, has become an immensely important topic. Few major hospitals are without an ethics committee to assist patients or health care providers when they need help. Colleges offer courses of study in medical humanities or bioethics. High school students, too, must be aware of the questions bioethicists study.

Paralleling the medical developments described above were striking advances in our understanding of genetics and molecular biology. Genetic knowledge has only recently been applied to specific diseases or patients, but the ethical questions that plague us about any medical technology apply to medical biotechnology.

How do we decide what is right or wrong, what is better or worse, as medical biotechnology offers us ever more choices? Should everyone be screened to determine who the carriers of recessive alleles for genetic diseases are? Should genetically engineered products be available to everyone, or should we sometimes screen requests for them (such as requests for human growth hormone)? Who should (and who actually will) have access to a person's DNA fingerprint? Are we on the road to employment and insurance discrimination based on genetic profiles? These are but a few of the ethical questions that can be posed.

The Bioethics Decision-Making Model

In attempting to answer difficult ethical questions, it is helpful to focus on specific cases and follow a step-by-step procedure. Below, you will be presented with specific case studies to analyze and a decision-making model to be used in these analyses. Part of the decision-making process will be to gather any additional background facts you need to evaluate the situation and to find out what sort of ethical standards that apply to your case have been established. Finally, you will make a decision as to the best course of action and justify it in terms of basic ethical principles.

Identifying a dilemma

The case studies we have provided raise bioethical dilemmas. A dilemma exists when there is no "right" course of action in a certain situation but, instead, several options, none of which is wholly acceptable. Ethical dilemmas revolve around trying to find the best solution when no solution is completely good. Not every situation presents a dilemma; many times, the possible courses of action are clearly right or wrong. The following example may clarify what does and does not constitute a dilemma.

Assume a patient with a certain condition would be an appropriate candidate for a drug research study. The patient's physician places her on this drug without getting her permission. This situation is not a dilemma. It is just plain wrong. Even if the doctor believes the drug will benefit the patient, by all modern medical standards, the physician has the obligation to get the patient's informed consent to include her in the study. (The Nuremberg Code and the Declaration of Helsinki, two internationally recognized codes of ethics, specifically address the ethics of research using human beings.)

On the other hand, assume that the patient has been given all the information she needs to make a decision. She is told that the drug has the potential to help her but might also have harmful side effects. She sees benefits and costs regardless of which decision she makes. Now we have a dilemma. A dilemma exists when no choice is ideal but all options have benefits and risks that must be carefully assessed.

Once a dilemma has been identified, the next step is to pose the dilemma in the form of a question about a specific case. Often, more than one question can be formulated. You will need to choose one question at a time for analysis, although you may consider several questions, one after the other, about a single case. It is also helpful to categorize the kind of issue being discussed. In the example given previously, the issue might be human research and the question posed could be, "Should the patient agree to be part of the study or not?" Since these discussions are usually very interesting and emotional, it is easy to get off track. The decision-making model presented here reduces that risk.

Basic ethical principles

To make ethical decisions, we must agree on some basic guidelines about what constitutes moral conduct. In the field of biomedical ethics, certain guides are well established. The moral-action guides or principles can be divided into four major principles and several secondary ones. These principles are listed below. The terms in parentheses are those used in bioethical literature. You may encounter them in your library work.

Major Ethical Principles
1. Do not harm (nonmaleficence).
2. Do good (beneficence).
3. Do not violate individual freedom (autonomy).
4. Be fair (justice).

Secondary Ethical Principles
1. Tell the truth (truth telling).
2. Keep your promise (fidelity and promise keeping).
3. Respect confidences (confidentiality).
4. Use the principle of proportionality: risk-benefit ratio (how much harm can be justifiably risked to effect good).

5. Attempt to avoid undesirable exceptions, also known as the wedge principle, the slippery slope, or the camel's nose.

Although these rules are simple, they represent fundamental values associated with respect for human dignity that most people agree to. These are the principles to which you should refer when making and justifying decisions.

Basic steps of the decision-making model

Here is a summary of the steps in the decision-making model. Each step will be explained more fully later on.

1. Identify the question you want to address. Usually, for any given case, many questions could be considered. Choose the one you want to explore.
2. Identify the issue you are exploring (e.g., genetic screening, confidentiality, gene therapy, human research subjects). Naming the issue will help in the search for relevant literature.
3. State the facts in the case. Be sure to avoid inferences.
4. Think of as many possible decisions in the case as you can.
5. Gather additional information as needed.
6. Pick the decision you want to support.
7. State the ethical principle that supports your decision (your claim).
8. Identify an authority that supports your decision. Quote the authority, if possible.
9. Formulate a rebuttal. Under what circumstances would you abandon your claim?
10. How strongly do you believe your claim? What is your level of confidence, the qualifier?
11. "Box up" the case for reporting your decision.
12. Write a prose argument describing the case and your decision.

Rules for classroom discussion

Since the issues discussed in bioethics are controversial, it is important that the following rules of etiquette be observed from the very beginning.

1. Only one person at a time speaks after being recognized by the discussion leader.
2. Treat each other with respect; no name calling. Critique the argument, not the author of the argument.
3. Seek clarity by asking questions.
4. Look for gaps in the data.
5. Recognize your own biases.
6. Be true to your own position; do not jump on the bandwagon.

7. Keep emotions in check; use logic.
8. Do not follow authority blindly (the teacher doesn't know everything).
9. You must have a reason, not just an opinion. You can like pepperoni pizza better than anchovy pizza with no reason. You cannot decide bioethical issues based on opinions.
10. Be open minded and willing to be perplexed.

Example Case Study 1: Frank and Martin

This case shows an ethical dilemma but does not involve medicine or biotechnology. It serves as a useful example of how the decision-making model can be applied to many problems outside the bioethics arena.

Frank is an 11th-grade student at a small public high school that prides itself on its family atmosphere and strong academic reputation. The school has an honor code. Each student agrees to abide by a published list of rules. One part of the code obligates students to report observations of fellow students who may be cheating. A judiciary council composed primarily of students decides what should happen in each case. Frank works hard at school. He has a B+ average and hopes to go to a good college. He also works part-time, runs track, and helps at home with two younger brothers. Both of his parents work. Frank is a close friend of Martin. Martin is very bright and does well in school, although he does not have to put in as much time studying as Frank does. Martin does not have many extracurricular activities or responsibilities. During a history test, Martin notices that Frank appears to be cheating off a note card.

I. Identify the question

Think of as many questions as you can. It may help you to think of questions that start with "Should. . . ." For example, one question you might ask could be, "Should Martin ask other students if they saw anything?" List as many questions as you can. After you have done so, choose one question for analysis. Remember, you can consider other questions later.

II. Identify the issue

What general problem does the case demonstrate? In this case, it might be cheating or school rules. (Although the principal figures in the case are students, students per se are not the issue.) This step is an attempt to categorize the case in a way that would help you find additional information when you search the bioethical literature.

III. State the facts of the case

What are the facts in the case? Most of us have a tendency to draw conclusions based on some information and then believe that our conclusions are facts, too. Be sure to distinguish between the facts and your inferences. Once the facts have been established, list them in an accurate but concise manner.

IV. List as many possible decisions as you can

What are the possible answers to the question you are addressing? Now is the time to be creative. Think of as many possible answers as you can. At this point, don't worry about which answer is best; that comes later.

V. Gather additional information as needed

Obviously, you don't have access to any additional information about Frank and Martin, but often, background information on the important issues involved in a case study is available. Frank and Martin's case will probably not require gathering additional background information. However, other cases may. For example, in a case involving the use of an experimental drug to treat a patient, you may want to know just how severe the patient's disease is. It might also be appropriate to find out what sort of safety testing a drug must undergo before it is approved for experimental treatment of humans.

The bottom line is that you should be as well informed as possible before you make an ethical decision.

VI. Pick the decision you wish to support

Consider all the possible decisions listed in step IV. Since ethical problems are usually complex, it is important to take time for thoughtful, honest reflection. You must have a reason for the option you choose, and that reason should be related to one of the principles listed at the beginning of this chapter.

VII. Identify the ethical principle that supports your decision

An ethically justifiable decision can be based on alternative principles. In a dilemma, adherence to one principle often results in the breach of another; dilemmas exist because principles often conflict. Application of different basic values can lead to different responses to a situation. Part of the decision-making process is realizing to which principle you are giving preference.

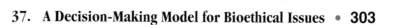

For example, someone who believes it is more important in this case to do no harm (principle 1) might make a different decision from someone who believes it is more important to uphold justice (principle 4). Neither person would be wrong. Both are adhering to high moral principles, but their principles are different.

Contrast these decisions to those of a person who decides to do nothing because someone might get angry or because he doesn't want to get involved. Keeping people from getting mad at you or keeping from getting involved are not high moral principles.

VIII. Identify a supporting authority

What experts or authorities would back up our position on this case? Normally, it is appropriate to look to professional codes of ethics. In this case, the honor code itself would be the authority. In bioethical decisions, several codes of ethics are used; these include the Hippocratic Oath, the American Medical Association Code of Ethics, the American Hospital Association's Bill of Rights for Patients, and the American Nursing Association Code of Ethics as well as the Nuremberg and Helsinki statements.

IX. Formulate a rebuttal

Under what circumstances would you change your decision about what to do? Try to imagine a circumstance or new information that could make you change your mind.

X. State your level of confidence in your decision

Use a one- or two-word statement to describe how strongly you believe in the argument you have made for your decision (your claim). One way to assess the strength of the argument is to gauge the likelihood of rebuttal. If a rebuttal is highly unlikely and the rest of the argument makes sense, then the argument is a strong one. Also, if the principle ties the claim and the facts tightly together, then the argument is strong.

Indicate the degree of confidence you have by using one of the following qualifiers or a similar one: "moderately confident," "absolutely confident," or "questionably confident."

XI. Box up the case for reporting

Your teacher will show you how to summarize your case in the box outline provided at the back of the book.

XII. Prepare a prose argument

Write up your argument, using the box as an outline. You should produce a paper that can be understood by someone who is completely unfamiliar with the original case. You should also explain more in the paper than the box can show. For instance, you should elaborate on why you selected the decision you chose and why the ethical principle justifying that decision was most important in this incident.

Example Case Study 2: Mr. Johnson

Mr. Johnson, age 76, was admitted to a medical unit with pneumonia. He had a history of severe emphysema, for which he had been hospitalized twice in the past year because of secondary pneumonia. During the evening of the third day of intravenous antibiotic treatment, Mr. Johnson's complexion took on an increasingly bluish tint, and he was short of breath. Because the attending physician was out of town, the physician taking calls was notified by the evening nursing supervisor. The physician found the patient to be severely oxygen deficient. He tried unsuccessfully to reach Mrs. Johnson to notify her of his plan to transfer Mr. Johnson to the intensive care unit for ventilator support. During the process of arranging for the transfer, a nursing assistant said that she thought that Mr. Johnson and his wife had said that they would refuse life support measures if Mr. Johnson ever needed them. No documentation of these comments could be found.

When Mr. Johnson arrived in intensive care, he was anxious, and his breathing was so labored that he could not talk. He shook his head "no" when the physician explained the procedure for inserting a tube to connect him to the ventilator, and he attempted to push the physician away. Mr. Johnson's competence to make decisions about his treatment was questioned because his arterial blood oxygen level was so low. Another unsuccessful attempt was made to reach Mrs. Johnson. Mr. Johnson's condition deteriorated as the physician and staff deliberated about proceeding with the intubation. Finally, the patient was sedated and placed on the ventilator.

Mrs. Johnson arrived the next morning and was shocked that her husband's condition had deteriorated so rapidly. She was upset that her husband was on the ventilator, explaining that they had agreed that life support measures would not be used for him. When the on-call physician asked, "What do you want me to do?," Mrs. Johnson replied, "I don't know." The physician asked for assistance from the ethics committee.

I. Identify the question

What question will you address about this case? List possible questions ("Should . . . ?") and then select one.

II. Identify the issue

The general category of problem could be called withdrawal of life support or end-of-life decisions or extraordinary means. This step helps in grouping cases and may make it easier to find references in the literature.

III. State the facts of the case

Only the facts as they are known may be used. List the facts in a concise manner, avoiding inferences or conclusions you may have drawn.

IV. List the possible decisions

What are the possible courses of action in this case? List as many as you can, without worrying for now about which is best.

V. Gather needed background facts

Obviously, no further information is available about the Johnsons, but you may have other questions. What is emphysema? What is secondary pneumonia? What is arterial blood oxygen, and why would low oxygen levels make a doctor think Mr. Johnson wasn't competent to make decisions? What is a ventilator? Be sure you have a good understanding of the case before you attempt to make a decision.

VI. Make a decision

VII. Identify the guiding principle

VIII. Cite a supporting authority

As time permits, read about similar cases. What do recognized codes of ethics say about these situations? A list of recognized codes of ethics is given in the discussion of the Frank and Martin example.

IX. Formulate a rebuttal

Under what circumstances would you abandon your decision?

X. State your level of confidence

XI. Box up the case for reporting

Summarize the case and your claim in the box form provided in Appendix A.

XII. Prepare a prose argument

Write up your argument in paragraph form so that someone who has never seen the case before can understand it and the reasons for your decision.

Societal Issues

Bioethics Case Study: Gene Therapy

Background Information

Advances in genetic engineering technology offer hope to people afflicted with a number of genetic diseases. In the first approved human gene therapy experiment, W. French Anderson attempted using genes to treat the genetic disease called severe combined immunodeficiency disease. You may be familiar with the movie (starring John Travolta) about the "bubble boy," a young man with this disorder. The disease is caused by the lack of the enzyme adenosine deaminase (ADA).

In 1990, Anderson inserted the correct sequence for ADA into the white blood cells of a young child who was not producing it. Before this experimental treatment could begin, Anderson's proposal went through extensive review by the Recombinant DNA Advisory Committee at the National Institutes of Health (NIH), the government agency that funds most of the country's biomedical research. In fact Anderson's proposal was reviewed 15 times. The need for a standard review procedure for proposed human gene therapy experiments was made clear by an incident in 1979–1980, when a UCLA researcher tried a human gene therapy experiment on two patients without getting approval from the appropriate review committee at his institution. He was eventually demoted, and his case demonstrated the need for a national policy on human gene therapy experiments.

A standard procedure is now in place, and researchers are using gene therapy to treat not only genetic diseases but other diseases such as cancer (see chapter 1 for more details).

What is gene therapy, and what ethical issues, if any, does it raise? At first glance, gene therapy, like any medical treatment that proposes to benefit the patient, seems free of any ethical implications. Closer examination raises a number of important questions regarding the acceptability of different kinds of gene therapy. Leroy Walters, of Georgetown University's Kennedy Institute of Bioethics, divided gene therapy into four possible categories.

1. Somatic-cell gene therapy for cure or prevention of disease. *Example:* Insertion of a DNA sequence into a person's cells to allow production of an enzyme like ADA.
2. Germ line gene therapy for cure or prevention of disease. *Example:* Insertion of an ADA sequence into early embryo or reproductive cells; will affect not only the individual but all of his or her offspring.
3. Somatic-cell enhancement. *Example:* Insertion of a DNA sequence to improve memory, increase height, or increase intelligence; would affect only that individual.
4. Germ line enhancement. *Example:* Insertion of a DNA sequence for enhancement into blastocyst, sperm, or egg; would affect future generations.

One of the ongoing issues concerning gene therapy is whether any or all of the four types of manipulation are ethically acceptable. At present, only somatic-cell gene therapy for cure or prevention of serious disease is considered ethically appropriate even by researchers like Anderson. Germ line therapy (altering disease genes so that the individual will not only be healthy but will pass on the healthy genes to his or her offspring) is considered desirable by some, but the techniques used to alter animal embryos have far too high a failure rate to consider their application to humans at this time. Human germ line therapy will probably be feasible in the future.

Enhancement therapy (whether somatic or germ line) is generally viewed as less acceptable. Anderson has stated, "I will argue that a line can and should be drawn to use gene transfer only for the treatment of serious disease, and not for any other purpose. Genetic transfer should never be undertaken in an attempt to enhance or 'improve' human beings." Anderson, a pioneer in this field, has maintained his position since 1980.

Guidelines for Gene Therapy

In considering any application of gene therapy, basic respect for human dignity is, as always, the underlying

moral principle. Like any other experimental medical treatment, gene therapy should be used to benefit the patient. Harm should be avoided. Certain factors must be considered. An NIH committee suggested the following items in 1985.

1. What is the disease to be treated?
2. What alternative treatments for the disease exist?
3. What are the potential benefits of gene therapy for human patients?
4. How will patients be selected in a fair and equitable manner?
5. How will a patient's voluntary and informed consent be solicited?
6. How will the privacy of patients and the confidentiality of their medical information be protected?

These six questions address some ethical concerns that have been considered in evaluations of experimental proposals. You may think of additional ones.

Case Study 1: Anne B.

Gaucher's disease, first identified in the late 19th century, is an autosomal recessive disorder. The disease results from the production of a structurally altered form(s) of the enzyme β-glucocerebrosidase, which breaks down a lipidlike substance, glucocerebroside. Failure to properly metabolize this substrate results in its accumulation in the spleen, liver, and bone. The disease varies widely in severity: some afflicted individuals die in early childhood, while others are diagnosed on autopsy following death from other causes at an old age. The severest form (the infantile form) follows a course somewhat similar to that seen in a related lipid disorder, Tay-Sachs disease.

Because of the variation in severity, Gaucher's disease is classified into forms I, II, and III: adult noncerebral, acute neuropathic, and subacute neuropathic, respectively. Studies suggest that these forms of the disease are allelic; that is, they are caused by different mutations within the same gene. The resulting structural defect in the protein is related to the severity of the disease.

The most common of the three types is form I (the least severe adult form). Form I Gaucher's disease is known to occur frequently in Ashkenazic Jews; approximately 1 in 13 members of this population is a carrier (heterozygote). Patients with type I disorder have about 15% of the enzyme activity of normal individuals. There are estimated to be 20,000 cases in the United States. Although form I Gaucher's disease is the least severe of the three forms of the disease, it,

too, varies in severity: some afflicted individuals die in early adulthood, while others live essentially symptom free to old age. Form I Gaucher's disease is therefore subdivided into acute, moderate, and mild courses.

Form I Gaucher's disease may first produce symptoms in a patient's late teens, with an enlarged spleen often being an early sign. Blood studies show reduced platelet, white blood cell, and red blood cell counts. Anemia and enlargement of the spleen may require removal of the spleen. Although this surgery decreases the anemia and relieves abdominal stress, the unmetabolized material is deposited in the bone more rapidly after splenectomy. Bone pain and pathologic fracture may occur.

In 1989, Joan B., age 32, died from the secondary effects of Gaucher's disease. Joan's 23-year-old sister, Anne B., also has the condition. However, she has been receiving enzyme replacement therapy each week since April 1996.

The biotechnology firm Genzyme marketed the β-glucocerebrosidase enzyme in June 1991. Anne receives the enzyme every week by intravenous infusion at the office of her physician. Her physician is considered one of the most knowledgeable practitioners with regard to this particular disorder. The infusion process takes about 4 h each week, including 2 h of travel time. The efficiency of the enzyme replacement therapy is not completely known, but early studies have been very encouraging. There have been no harmful side effects.

Anne recently read about a new experimental treatment that involves removing some of her own cells, treating them with a retrovirus that carries the correct sequence for glucocerebrosidase, and replacing the cells. The cells should reproduce, creating a colony of cells that make the proper enzyme. The hope is that this treatment could be done several times, resulting in enough cells to produce permanently adequate amounts of normal enzyme. The technology is experimental and there may be risks, some known and some unknown. For instance, since the sequence is added by using a retrovirus, some of Anne's normal DNA might be altered. This treatment is also more expensive than Anne's current treatment. Anne requests the experimental treatment from her physician.

Implementing the decision-making model

First of all, does an ethical dilemma exist? Anne has requested an experimental treatment that may harm

Societal Issues

her or may help her and improve her quality of life. Should her physician help her get the treatment she wants? Is it medically indicated? Is her disease as serious as French Anderson says it should be for her to qualify for gene therapy? There are certainly enough questions to indicate that this case does present bioethical issues.

I. Question
Several possible questions are listed above. List as many others as you can think of, and then choose one for consideration.

II. Issue
What is the issue you chose? Experimental treatment? Medical necessity? Something else?

III. Facts
There is a lot of background material in this case. For the box summary, try to select only the information that is necessary for presenting the dilemma. You could reasonably omit background information about Gaucher's disease from the box, for example.

IV. Possible Decisions
List as many alternatives as you can without trying to decide at this time which is best.

V. Additional Information
Do you need additional information to make a good decision? For example, if you are trying to decide whether the doctor should give Anne the treatment, do you need to know more about retroviruses, how they work, and how they might alter Anne's DNA? Do you need more information about Gaucher's disease to decide whether it is serious enough to warrant gene therapy?

VI. Decision
Choose what you believe to be the best solution to your question.

VII. Principle
Which of the major ethical principles justifies your decision?

VIII. Authority
What experts would back up your option? Find relevant passages in documents such as the Hippocratic Oath, the Nuremberg Statements, and the Patient Bill of Rights.

IX. Rebuttal
Under what circumstances would you abandon your choice?

X. Level of Confidence
How strongly do you believe your position?

XI. Box
Use the outline in Appendix A.

XII. Prose Format
Explain on paper why you chose one option rather than the other and why the principle supports the option. Your written argument should reflect the thinking necessary to construct the box and complete enough that someone who has never read the case study can understand the situation and your analysis.

Additional case studies follow.

Case Study 2: Bobby K.

Bobby K. is a healthy 10-year-old boy. He is very agile and quick and loves sports. The coach of his city league basketball team has told Bobby's parents that their son's skills on the court are astounding for a child of his age.

Bobby's father, Mr. K., is a healthy man who is 5'3" tall. Bobby's mother is also short, only 5'0". Bobby's pediatrician has predicted that Bobby will attain an adult height of about 5'3" but has emphasized to his parents that he is a normal, healthy boy. Mr. K. remembers being teased constantly about his size and recalls that his lack of height kept him off all the varsity sports teams at his high school. He has often wondered if his shortness is a disadvantage to him in business dealings, too. Mr. and Mrs. K. both anticipate that their son will be the recipient of more and more pointed teasing as he reaches his teenage years. They also fear that he will not be selected for the basketball team when he reaches junior high or high school.

Bobby's coach has heard of a program at the local university in which gene therapy is being conducted on children who have a disease that results in the inadequate production of growth hormone. The children are being given working copies of the gene for human growth hormone, and the levels of growth hormone in their bodies have increased. These children's growth rates have also increased.

Bobby's coach tells the K. family about what he has heard. Bobby is thrilled, because he might be able to keep playing basketball, maybe even on a professional level. Bobby's parents are more cautious but would like their son to be spared the pain of being much smaller than his classmates.

The K.'s go to Bobby's pediatrician and request that Bobby be given the growth hormone gene therapy.

Using the decision-making model

This case obviously presents many possible questions for consideration. For example,

- Should Bobby's doctor allow him to have the gene therapy?
- Should Bobby's parents let him have the therapy?
- Should Bobby's coach have told the K.'s about the program?

There are probably many background questions you will need to investigate. For example,

- Have there been any studies to determine whether it is safe to give additional growth hormone to normal children?
- What height range is considered normal? How can you tell whether a short person is normal?

Choose a dilemma question for analysis, and use the model to generate a solution.

Case Study from Real Life

This case study also revolves around human growth hormone but concerns administration of the hormone as a drug, not gene therapy. It is a summary of the article "NIH stops enrolling children in growth hormone studies" in *Biotechnology Newswatch,* August 17, 1992.

In 1992, because of questions about the safety and appropriateness of the study, the NIH temporarily stopped doctors from enrolling more children in a study of the effects of growth hormone on healthy short children. Children already in this trial will continue. Without treatment, the boys in this study were expected to reach 5'6" or less and the girls were expected to reach 5'0" or less as adults. Half of the children are being given genetically engineered human growth hormone (the protein, not the gene), and the other half are receiving a placebo. These children do not know that their treatment is a placebo. They will serve as a control group for the children who are receiving the hormone. The physicians hope to learn how much additional growth the hormone produces in healthy short children.

Doctors normally prescribe growth hormone to children who fail to produce enough of it naturally and are expected to reach an adult height of 4'0" or less. The hormone has been on the market for over 10 years, and Genentech (one of the companies that make the hormone) is convinced that no safety problems are associated with its long-term use.

Scientists who defend the study present several points. One is that many parents obtain human growth hormone for their children on the black market. Another is that studies to determine the safety and efficacy of this treatment for short stature are needed. Another argument is that our society is preoccupied with height and that short stature can cause anxiety and poor self-image to the point that mental health is compromised. One of the scientists who defend the study is Dr. Arthur Levine, scientific director of the National Institute of Child Health and Human Development. At 5'4", Levine says he probably would have enrolled himself in the study if it had been available to him as a child.

Critics of the study argue that shortness is not a disease and should not be defined as requiring treatment. They question the use of a relatively experimental treatment in healthy children for apparently cosmetic reasons. They are concerned that virtually any cosmetic defect could eventually be redefined as illness.

Using the model

Apply the decision-making model as outlined previously. Some questions that could be asked include the following.

- Should this study be conducted?
- Should a parent enroll a short child in this study?
- Should doctors prescribe growth hormone to short but otherwise normal children?
- Should people be able to make their own decisions about taking growth hormone?
- Should any treatment be available for shortness?
- Should NIH prevent children who want to enroll from participating?

You may be able to think of more questions. Choose one for analysis, using the decision-making model.

Bioethics Case Study: Genetic Screening

Background Information

Genetic screening is examining a person's DNA to determine whether he or she carries a gene of interest. It is usually thought of in the context of determining whether someone has a genetic disease or is a heterozygous carrier of a recessive disease gene.

Although the ability to look directly at a person's DNA is relatively new, it has been possible for several years to screen for some genetic disorders by using other approaches. For example, Down's syndrome is caused by the presence of an extra chromosome 21, which can be detected by microscopic examination of an individual's chromosomes. Heterozygous carriers of diseases such as Tay-Sachs and sickle-cell anemia have also been identifiable for some time. Even though these individuals do not have the disease in question, their cells display characteristic differences from fully normal cells.

However, as we enter the new age of molecular genetics, it will be possible to screen for more and more genetic conditions. This change results from our new ability to analyze DNA directly and from research that is identifying more and more genes involved in human disease. With our current DNA technology, once a disease gene has been identified, it is usually possible to devise a test that will reveal whether a person carries a disease-causing mutation. In the past several years, genes responsible for cystic fibrosis (CF), muscular dystrophy, and many other diseases have been identified. The Human Genome Project, which proposes to map all 50,000 to 100,000 genes in the human genome, should give us the power to screen for many more of the 3,000 to 4,000 known genetic defects.

Many other diseases, such as colon cancer, Alzheimer's disease, multiple sclerosis, emphysema, and diabetes, are not simple genetic diseases but clearly result from an interaction between many genes, a person's genotype, and the environment. Research is currently focused on identifying genes that predispose individuals to these conditions. It is reasonable to assume that we will eventually be able to determine (through genetic screening) a person's genetic risk of developing these diseases as well.

What ethical questions does genetic screening present? We have more experience with screening issues than with gene therapy. Some of that experience has made it clear that we must proceed with caution and with a clear policy. In the 1970s, screening for sickle-cell anemia resulted in confusion and fear. It is evident that many people did not understand the difference between being a carrier and having the disease. People who merely carried the sickle-cell gene were threatened with loss of jobs and insurance. The initial purpose of helping people understand their possible risk of bearing children with sickle-cell disease was lost in a mire of misunderstanding.

Another problem with genetic screening is that we can identify many more diseases than we can treat. Huntington's disease (HD) is a good example. People who carry this dominant genetic disease gene live their early lives normally and do not develop symptoms until middle age. At this point, however, the individual begins to lose the ability to think and develops uncontrollable body movements such as twitching and shaking. These conditions become worse and worse over a period of years until the victim dies. There is no treatment or cure. Unfortunately, since the disease is dominant, all of the children of a victim have a 50% chance of inheriting the disease gene. And since the disease does not reveal itself until middle age, most victims have already had children before they know whether or not they themselves have the gene.

Therefore, should a 20-year-old be able to find out that he carries a dominant gene for a disease that will first debilitate and finally kill him in middle age? If people should be given this information, at what age is it best revealed? Does respect for human dignity in this situation mean that we try to prevent harm by not telling or try to promote autonomy by telling the genetic status? Because of the experiences with sickle-cell screening 25 years ago, screening for HD is being done under very strict guidelines at 14 centers in the United States. The centers require clients to undergo extensive psychological testing before screening and counseling after screening. If a cure or even a treatment were available, then these questions would not be nearly as difficult, but as mentioned earlier, neither treatment nor cure is available. In the case of sickle-cell anemia, the screening test has been available for almost 40 years, but there is still no cure.

The ethical questions surrounding screening are apparent. Three percent of the $3 billion budget for the Human Genome Project has been set aside to study the ethical considerations. The Working Group of Ethical Legal and Social Implications of the Human Genome Project has proposed nine areas that merit attention.

1. Fairness in the use of genetic information
2. Impact of knowledge of genetic variation on the individual
3. Privacy and confidentiality
4. Impact on genetic counseling
5. Impact on reproductive decisions
6. Issues raised by the introduction of genetics into mainstream medical issues
7. Uses and misuses of genetics in the past
8. Questions raised by commercialization
9. Conceptual and philosophical implications

It is easy to see that different genetic conditions present different problems. With HD, there is currently no medical benefit to the individual in question in knowing that he or she has the disease gene (though such knowledge might affect the choices that person makes about how to live his or her early life). On the other hand, for conditions that have both environmental and genetic causes, such as high blood pressure and emphysema, early detection of susceptibility may save many lives and significant dollars. If people do make appropriate lifestyle changes, good has been achieved.

But who will be tested for these kinds of traits? Will the tests be voluntary? Who will pay for the test? (Remember that when anyone says, "The government pays," they really mean you pay, since the government gets its money from taxing people who work for a living. Middle class people pay about one-third of their incomes in various taxes.) Will we deny insurance or employment to people who refuse testing or to people who refuse to make lifestyle changes after tests indicate they have increased risks for certain conditions?

These are not easy questions. The medical profession and insurance industry are considering them now, and you will probably be affected by them during your lifetime.

Case Study 1: James and Carol H.

Refer to the step-by-step methodology for the decision-making model in chapter 37.

Alice is the 6-year-old daughter of James and Carol H. Alice was diagnosed with CF at age 2. Alice has a moderately severe case of CF and receives postural drainage, a carefully specified, time-consuming program of thumping to loosen mucus in the lungs, from her mother three times a day. In addition, she is on a regimen of vitamins and enzymes that requires her to take about 25 pills each day. Carol spends much of her day caring for Alice. At least twice each year, Alice requires hospitalization to fight respiratory tract infections.

James and Carol have been talking about having another child. Alice is cared for at a public clinic associated with a medical school that has a research project on CF. The clinic physicians suggest that James and Carol undergo screening for the CF gene as part of the project.

When the results come back to the clinic, Alice's physician is quite surprised to learn that Carol, but not James, carries the CF gene. Since James is not a carrier, it is virtually impossible for him to be Alice's biological father.

I. Identify the question

First of all, is there a dilemma, and if there is, what is it? List some possible questions (Should . . . ?)

II. Issue

What is the general issue in this case?

III. Facts

What are the facts in the case? It is easy to jump to some conclusions in this case, so try hard to stick to what you actually know (as opposed to what you have concluded or what you suspect). Depending on the question you are considering, different facts may be more relevant.

IV. Possible decisions

What are some possible solutions to the problem? Brainstorm as many as possible.

Societal Issues

V. Additional information

Do you need additional background information to inform your decision?

VI. Decision

Formulate your decision (or your group's decision) in the form of an "I" or a "we" statement. For instance, "I believe the physician would be morally justified to. . . ."

VII. Principle

Which of the major ethical principles justifies your decision?

VIII. Authority

What authorities would back you up on this point? One of the first places to look is in professional codes of ethics. For instance, the Hippocratic Oath states that the physician should do no harm. Similar statements are found in the American Medical Association Code of Ethics.

IX. Rebuttal

Under what circumstances would you change your minds about what to do? Think freely.

X. Level of confidence

Formulate a one- or two-word statement to describe how strongly you believe your argument.

XI. Box

Draw up a box as described previously (chapter 37).

XII. Prose argument

Write out an argument as described previously (chapter 37).

Case Study 2: Angela

Angela is a healthy 32-year-old certified public accountant who is married with no children. She has been working for 8 years in a small accounting firm. She has decided that she would like to move to a nearby large city and work for a major accounting firm.

Five years ago, Angela's mother was diagnosed with HD, an autosomal dominant disease. Angela's mother is in a nursing home.

Angela's job interviews go well, and she is offered an excellent position that she readily accepts. However, as Angela is filling out the required paperwork in the personnel office, she notices that she must sign a consent form for a physical that includes a genetic screening test of her DNA. When Angela questions whether she must have such a test done, she is told it is required by the company.

A number of questions could be raised about this case. Here are a few examples.

1. Should the accounting firm require the testing?
2. Who should have access to Angela's results if she agrees to be tested?
3. Should Angela's employment be contingent on the results of her DNA test?
4. If Angela has a sister who is also at risk, should she be told Angela's results?

Think of more questions, and choose one for analysis. Apply the model as you have done before.

Careers in Biotechnology

Introduction

Because of the relatively recent development of biotechnology, many people are not aware of the wealth of employment opportunities that are available in new biotechnology companies and in existing industries that are incorporating biotechnologies into their operations. Becoming acquainted with the biotechnology industry, job types, educational requirements, and desired skills will put you in an excellent position for finding employment in this exciting and expanding field.

The Biotechnology Industry

Because biotechnology is a collection of technologies that use cells and molecules isolated from cells, the phrase "biotechnology industry" is misleading. To be more accurate when describing possible careers in biotechnology, we should talk about "biotechnology companies (which are usually small, relatively young companies founded by scientists who have made a biological discovery with commercial potential) and the existing, long-established industries, such as the agricultural, chemical, and pharmaceutical industries, that are incorporating various biotechnologies into their research, development, and manufacturing processes." But rather than being perfectly accurate, we are opting for the term "biotechnology industry" when describing career possibilities, and you will know what we really mean.

Since 1986, the number of biotechnology companies (as defined above), the average number of employees per company, and the sales revenues generated by these companies have grown significantly (Table 40.1).

Table 40.1   Growth of the biotechnology industry[a]

Year	Number of companies	Number of employees	Worldwide sales (billions of dollars)
1986	850	40,000	1.1
1991	1,107	66,000	2.9
1996	1,308	108,000	9.3
1999	1,283	153,000	13.4

[a]Data from Ernst and Young, LLP, Annual Biotechnology Industry Reports.

Companies involved in biotechnology range in size from small entrepreneurial start-ups to large multinational corporations employing thousands of people. Over time, new companies will emerge and existing companies will grow. Because the applications of biotechnology are so varied, the industry is growing rapidly, and the work environments are diverse, a career in biotechnology can be exciting and rewarding for those with an interest in science or engineering. If you acquire technical skills that are useful in one sector of the biotechnology industry, you will find those skills are readily transferable to other sectors.

Types of Jobs in Biotechnology Companies

A well-established technology-based company usually has several divisions that carry out different functions. In each of these divisions, jobs are available for people with varying skills and educational levels.

Research and development division

Scientific research is the basic foundation of any high-technology industry. New ideas represent the company's future, because they lead to a continuing line of new products. In some companies, research focuses on specific applications or products: how to apply scientific knowledge in new ways or how to improve an existing product. In other companies, some research teams carry out basic scientific research with no immediate application. These companies believe that simply acquiring new knowledge and understanding of how living systems work will pay off in the long run with new product ideas. Even though biotechnology companies have their own research teams and often contract with other companies for specialized work, much of the research that drives progress in biotechnology is also carried out in universities by academic scientists.

Once a promising idea is generated, it is tested, refined, and made practical in a process known as product development. During the product development phase, scientists and engineers address issues such as safety, efficacy, and the most efficient and profitable way to manufacture the product.

The entire process of research and development is a substantial effort. The typical new pharmaceutical product can take as long as 10 years to develop and can cost up to $500 million.

In the research and development division of a biotechnology company, the research is usually directed by scientists with doctoral degrees. Research associates are typically college graduates with bachelor's degrees in science, although some graduates of 2-year technical schools (Associate of Applied Science [A.A.S.] degree) with specialized training are hired for these positions. In research and development divisions, individuals with high school diplomas might find employment as media preparation technicians, animal caretakers, or greenhouse assistants.

Production and quality control division

Workers in production actually make the products or deliver the services the company sells. Large-scale production and manufacturing often require not only people with scientific expertise but also people with a knowledge of engineering and industrial manufacturing technology and people with good mechanical skills.

Workers in quality control divisions make sure that the product meets specifications. Quality assurance personnel monitor the entire production process, and sometimes the research and development operations as well, to ensure that good manufacturing practices and standard operating procedures are followed at all times. Quality assurance is particularly important in the pharmaceutical industry, for which the Food and Drug Administration has established especially stringent guidelines for the testing and manufacture of medicines that will be used by humans.

There is less demand for advanced degrees in the manufacturing and production division of companies than in the research and development division. Conversely, there is more demand for people with A.A.S. degrees or high school diplomas in this aspect of company operations than in research and development. Examples of entry-level jobs include chemical operator, manufacturing technician, instrument technician, and packaging operator. If you have a college degree in science or engineering, typical jobs would be quality assurance manager, process development scientist, or manufacturing engineer. Even

though high school graduates and holders of A.A.S. degrees may be hired in production and quality control, room for advancement is limited for those without a college degree.

Sales and marketing division

On the basis of its scientific research, a company may think it has a terrific product idea. But will it sell? Market researchers try to answer this question by assessing the need for the product, the number of people likely to buy it, and the price they might be willing to pay for it. Marketing personnel also try to find new markets for a product already being sold by the company and seek new ways to advertise and promote the product.

Salespeople are in the front lines, dealing directly with customers and selling the product. Sales personnel not only make sales but are also highly visible representatives of their company. They are often asked for technical advice about their products, and they relay feedback from customers to the company.

In the biotechnology industry, sales and marketing employees who have a scientific education have a competitive edge in securing jobs. To be good at sales and marketing, these employees must understand the nature of the highly technical products they sell and must know how to communicate with their customers, who often are scientists or medical professionals.

Regulatory affairs division

Various activities of companies of all types are regulated by federal and state agencies. This is particularly true of agricultural and pharmaceutical biotechnology companies, which must comply with intricate regulations imposed by the Food and Drug Administration, Environmental Protection Agency, and U.S. Department of Agriculture concerning the nature and manufacture of products. Many companies have teams of specialists, often with scientific backgrounds, who keep track of all federal and state regulations that apply to the company and make sure the company complies with them.

Legal affairs division

One of the most important tasks for a biotechnology company is to secure patent protection for its inventions. Without a patent, a new product idea may not be profitable for a company because competitors can make the same product. Having exclusive rights to market a new invention during the term of a patent is often the only way a company can be assured of sales sufficient to pay for the high costs of research, development, and production of the product.

Consequently, companies may hire specialists to prepare and track patent applications, and larger corporations may have their own patent attorneys on staff. Many law firms now have attorneys who specialize in biotechnology patent law.

Public relations, communications, and training

Every company needs people who are effective communicators. This is particularly true in the biotechnology industry, where companies must be able to offer technical product information to the lay public in an easily understandable way. Technical writers may be employed to write internal or external scientific reports.

By their very nature, high-technology industries are involved in sciences that are breaking new ground rapidly. This rapid expansion of knowledge requires that employees be able to learn new technologies quickly. Large corporations may employ full-time staff development personnel who organize technical training within the company to keep their employees abreast of new developments.

Management and support functions

Managers at different levels in all divisions organize and supervise the activities described previously. In most of the technical divisions of a company—research and development, production, and quality control, for example—people who become managers most often start out as scientists or engineers and work their way up. In entrepreneurial biotechnology companies, the chief executive officer or other high-level managers are often Ph.D.-, M.D.-, or D.V.M.-holding research scientists whose ideas for new products provided the initial impetus to start the company. In other companies, these scientists remain in charge of research divisions, while managers with business training and experience assume other executive positions.

Like all companies, biotechnology companies need a variety of support personnel such as administrative assistants, accountants, information management specialists, and computer technicians.

Preparing for Careers in the Biotechnology Industry

People with all levels of education may be hired for various jobs in the research, development, or produc-

tion divisions in biotechnology companies. Types of jobs and job expectations related to different levels of education are discussed below.

The starting salaries for technical jobs vary widely with the type of job, the type of degree the applicant has, his or her previous job experience, and the geographical location of the job. Currently, typical salary ranges for technical positions in a biotechnology company are as follows:

Technical staff	$20,000–60,000
Technical middle management	$60,000–80,000
Vice president, research and development	$70,000–100,000+

High school diploma

Many people working in biotechnology-related manufacturing facilities have only a high school diploma. They are trained on the job to perform routine technical operations as manufacturing process operators or quality control technicians, or they work in more conventional jobs typical of those in any manufacturing facility (assembling, packaging, transporting, warehousing).

Less frequently, people with high school diplomas may be hired in research laboratories as laboratory assistants. They may perform routine experimental operations, prepare solutions and media, care for laboratory animals or greenhouse plants, or wash glassware.

Typical salary: $15,000–25,000

Typical job titles: media prep technician, greenhouse assistant, laboratory assistant, process operator

Opportunity for advancement: Limited. In industry or academia, advancement beyond a fairly low level is difficult or impossible without a college degree. However, a responsible, adaptable individual who can learn new skills quickly can earn job stability and pay raises.

Getting a job: Entry-level positions are advertised in local newspapers or through the local Employment Security Commission. The toughest competition is from the many other high school graduates seeking jobs. But since employers routinely find that only about 10 to 20% of applicants for entry-level positions have the basic skills to be considered after the first round of the selection process, job applicants are already way ahead of the crowd if they do nothing more than study hard enough to maintain a B

Societal Issues

average in high school. Good basic skills—reading, writing, speaking, and mathematics—and interpersonal skills are absolutely essential. A willingness to work the night shift and a good mechanical aptitude are also important for those seeking employment in biomanufacturing.

A.A.S. degree

People with a 2-year-college degree in an appropriate field are very employable. Those graduating from biotechnology technician training programs at technical colleges readily find jobs in the industry, because such training programs focus on practical, hands-on skills that are in demand. The pragmatic training offered by technical colleges gives A.A.S.-degree holders more laboratory experience and "how-to" knowledge than many graduates with bachelor of science (B.S.) degrees acquire.

Two-year-college graduates are frequently employed in biotechnology manufacturing facilities. Some also find jobs as technicians in research laboratories in industry, universities, and medical centers. Since technical college graduates with specific training in biotechnology are few, industries do not usually specify a 2-year degree as a job requirement. Nonetheless, technical college graduates with training in biotechnology compete successfully for many jobs in the biotechnology industry that are usually held by people with a bachelor's degree.

Typical starting salary: $17,000–20,000, advancing to $22,000–34,000

Typical job titles: Same as for high school diploma (given above); may also include quality control inspector, instrumentation technician, laboratory technician, research assistant.

Opportunities for advancement: Somewhat limited with only a 2-year degree but better than opportunities for those with no postsecondary training. However, people with an associate degree who work for awhile and then go back to school and get a bachelor's degree have a competitive advantage over new college graduates because of their hands-on training and industrial experience.

Getting a job: Same as for high school diploma (described above). Most jobs at this level are advertised in local newspapers. For A.A.S. graduates, the chances for employment as an academic research technician are better than the chances for high school graduates. Job seekers should visit the employment offices of universities and research centers and send résumés

and letters to companies, particularly those with production facilities.

B.S. degree

For a career in the biotechnology industry, receiving a 4-year degree in the physical or biological sciences is a logical beginning. The biological sciences are organized differently in different colleges and universities, so depending on the institution, you might major in general biology, genetics, biochemistry, microbiology, botany, or zoology. If you major in general biology, botany, or zoology, choose as many courses as possible in molecular and cellular biology. A chemistry degree is also an excellent foundation, since so much of modern molecular biology is based on chemistry. Chemical engineering is a very marketable specialty. Other specialties useful in the industry are environmental science, toxicology, pharmacology, computer science, and agricultural science.

Employers value the B.S. degree because it gives the student a well-rounded background in general principles of science, provides a good knowledge base in a particular area (for example, biology or chemistry), and enhances basic communication and problem-solving skills. A B.S. in science or engineering is not only essential preparation for a scientific career but increasingly in demand by biotechnology companies as a basic credential for those who want to work in sales, marketing, or other business areas not traditionally considered scientific.

Typical starting salary: $23,000–35,000, advancing to $30,000–45,000

Typical job titles: research associate, technician, technologist, manufacturing or production associate, product development engineer, quality control analyst or engineer, microbiologist, chemist

Opportunities for advancement: Good. In industry, if you are an enterprising, talented employee with a B.S. in science or engineering, you can advance to supervisory positions with higher pay. You may also be able to move from one division to another within a company, thus acquiring a new set of skills and improving your prospects for advancement.

Although the emphasis in this discussion is on research and production jobs, a B.S. in science or engineering is good preparation for jobs in sales, marketing, regulatory affairs, and other aspects of a company's business operations.

Getting a job: Jobs at this level are often advertised in local newspapers as well as in scientific journals. One

of the main sources of scientific want ads is the weekly journal *Science,* published by the American Association for the Advancement of Science. *Science* runs ads for all levels of positions, from technicians to research directors, in universities as well as industry. Other sources for national ads are *The Scientist, Chemical & Engineering News,* and other trade journals. Entry-level jobs for which a B.S. is sufficient qualification are frequently filled by local applicants. However, jobs for people with specialized industrial experience or postgraduate training are often advertised nationally.

For positions at the B.S. level, job seekers should answer want ads, identify companies that might need their skills, and send out many résumés (100 is not too many). If there are only a few companies in your area, you may have to relocate.

Industrial employers place a premium on previous industrial experience and often prefer an applicant with industrial work experience over a college graduate fresh out of school. Therefore, seek out cooperative education opportunities, internships, or other summer jobs in industry during college. Or identify a mentor in an academic department and carry out an undergraduate research project or get a part-time job in a university research laboratory. These experiences will add great value to your résumé.

Grades matter. Some companies award a higher starting salary to an applicant with better grades, and some will not even consider applicants who have less than a B average. Good grades are important even for obtaining summer internships.

M.S. degree

In general, people with a master of science (M.S.) degree hold jobs similar to those held by people with bachelor's degrees. Pay is about $5,000 to $7,000 more, and advancement to supervisory positions may be more likely. The increased value of the M.S. degree is that it gives students more hands-on laboratory experience and, through work on a thesis project, teaches them how scientific research is actually conducted. Also, someone with a B.S. degree who did not have much specific biotechnology training in the undergraduate program can acquire advanced skills in an M.S. degree program. Obtaining both a B.S. and an M.S. is also a way to combine two specialties, such as biology and business.

Doctoral degree

A doctoral degree is the primary qualification for a scientific research career. Ph.D.'s, M.D.'s, and D.V.M.'s

design the research process, directing the activities of B.S. and M.S. technicians.

Typical starting salary: $40,000–55,000, advancing to $60,000–100,000+

Typical job titles: research scientist, senior scientist, principal scientist, research or scientific director, vice president for research (listed in order of increasing seniority)

Opportunities for advancement: Excellent. Companies have different promotion pathways for those with doctoral degrees. In many, promotion to higher levels traditionally involves more management and less science. But since many scientists want to continue practicing science and not become involved in company management, some companies are finding ways to reward their top-notch researchers other than promotion to management positions.

Getting a job: If you aspire to become a research scientist, take all the science and mathematics you can in high school, and seek out the best undergraduate education possible. If you excel, faculty at your college will counsel you in selecting a graduate school and research area. Taking summer jobs in industry or working on high school or undergraduate research projects are also important preparation for a career as a research scientist.

Academic Careers

Much of the basic research in biotechnology is carried out in universities and government research laboratories, not by companies. Another career choice open to students interested in science is that of academic research scientist. Traditionally, pay across all levels is substantially lower in universities than in industry, and for technicians with B.S. or M.S. degrees, the opportunities for promotion are limited.

In the typical university or government research laboratory, there are three classes of scientists: technicians, scientists-in-training, and professors. Technicians may be hired with B.S. or M.S. degrees. Frequently, undergraduates working on their B.S. degree can get part-time positions. Getting your first job as a technician is sometimes easier at a university than in industry, particularly if you have worked in that laboratory on a part-time basis while still in school. Even if your ultimate goal is to work in industry, acquiring technical skills and work experience in a university laboratory can provide a solid foundation and a competitive edge for seeking a job in industry.

Scientists-in-training may be graduate students working on Ph.D. thesis research or postdoctoral fellows. The postdoctoral fellowship is a temporary research position in which new Ph.D.'s get more extensive research experience. A "postdoc" is now considered a standard part of the career path for Ph.D. scientists. Some companies offer postdoctoral fellowships.

Professors at research universities and most 4-year colleges are Ph.D.'s, D.V.M.'s, and/or M.D.'s. Smaller 4-year colleges and community colleges may hire scientists with M.S. degrees as faculty members or, less often, those with B.S. degrees as laboratory instructors.

Essential Skills of a Biotechnician

Although some students might be interested in pursuing a Ph.D., the great majority who are interested in a career in biotechnology will have a B.S. or A.A.S. degree. What are the specific skills you should acquire to increase your competitiveness in the job market?

A number of nationwide studies have addressed this question for an entry-level job as a biotechnician. Although each report describes some skills that are industry specific, other sets of skills appear to be common to all industries utilizing some aspect of biotechnology. The skills that seem to be in universal demand by very diverse companies are listed below.

Technical skills

Basic Laboratory Skills
Be familiar with standard laboratory equipment (balances, pH meters, glassware, centrifuges) and procedures (filtration, distillation, weighing, measuring). Maintain and calibrate laboratory equipment and instruments. Prepare buffers, reagents, and solutions. Prepare dilutions of solutions correctly. Perform basic separation techniques. Use aseptic techniques when appropriate. Follow protocols and standard operating procedures.

Biological Laboratory Techniques
Basic microbiological techniques such as identifying, screening, and quantifying microorganisms; isolating and maintaining pure cultures; analyzing products; and harvesting microorganisms. Basic cell biology techniques such as using microscopes, preparing cells for microscopic analysis, and propagating cells.

Safety Skills
Monitor, use, store, and dispose of hazardous materials properly. Use protective equipment and hoods. Main-

tain, understand, and follow Material Safety Data Sheets. Know and comply with current federal, state, and local regulations. Maintain safe work area.

Quality Control Skills
Perform validation testing. Document product specifications. Use analytical equipment. Compare results with government and/or company standards. Collate large volumes of data. Perform statistical tests and data analysis.

Instrumental Analysis Skills
Obtain representative samples. Prepare samples for analysis. Use and calibrate standard analytical instruments such as gas chromatographs, high-pressure liquid chromatographs, and spectrophotometers.

General employability skills

Communication Skills
Create, follow, and record protocols and standard operating procedures. Maintain accurate and clear records in laboratory notebook. Summarize experiment results for reports. Write business letters, memos, and technical reports. Proofread and edit written materials for correct spelling and grammar. Organize and present oral summaries. Locate and review scientific reference material. Comprehend and follow verbal instructions. Conduct literature searches.

Mathematical and Statistical Skills
Read, construct, and comprehend data in graphical form. Calculate and interpret ratios. Perform basic statistical tests (mean, standard deviation). Recognize anomalies in data collection. Perform calculations using exponents, roots, and logarithms. Solve simple algebraic equations. Convert units between metric and English systems. Express a number in scientific and standard notation. Convert word problems to mathematical expressions. Use linear regression to forecast data.

Computer Skills
Use basic word processing systems. Enter, store, and retrieve data. Create and use a spreadsheet. Manipulate data electronically with graphical software. Use electronic communication techniques and systems. Develop and maintain a database. Transfer data to and from remote databases and instruments.

Interpersonal Skills
Develop and use listening skills. Participate as a team member. Demonstrate an understanding of team planning, problem solving, and the value, roles, and responsibilities of individuals. Develop conflict resolution

and consensus-building techniques. Recognize and respect organizational structure and goals. Be open and adaptable to new technology and applications. Develop initiative taking and observation skills. Work independently.

In 1994, the U.S. Department of Labor and the U.S. Department of Education initiated a number of national skills standards projects, three of which relate to biotechnology: agricultural biotechnology, chemical processing, and biomedical sciences. If you are interested in obtaining detailed information on the skills employers expect entry-level biotechnicians to have, you may want to contact the organizations that conducted the studies and request a copy of their reports. They are as follows.

Agricultural biotechnology: National FFA Foundation, 5632 Mt. Vernon Highway, Box 15160, Alexandria, VA 22309; (703) 360-3600

Chemical processing: American Chemical Society, Education Division, 1155 16th Street, NW, Washington, DC 20036; (202) 872-8734

Biomedical sciences: Education Development Center, 55 Chapel Street, Newton, MA 02160; (617) 969-7100

For additional information on careers in biotechnology, contact the Biotechnology Industry Organization (BIO) at their mailing address (1625 K Street, NW, Washington, DC 20006-1604) or through their website (http://www.biocareers.org). Also see their home page (http://www.bio.org).

Appendix A: Templates

This appendix contains 53 templates that can be cut out from this section or photocopied as needed for use with the *Student Activity* pages.

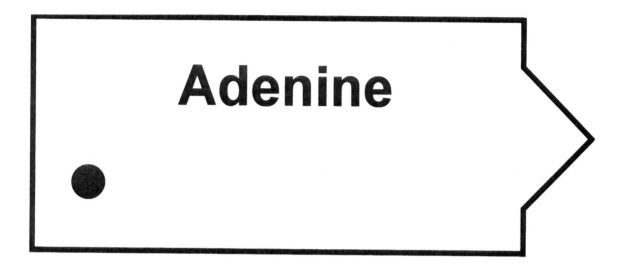

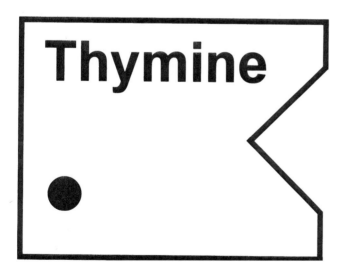

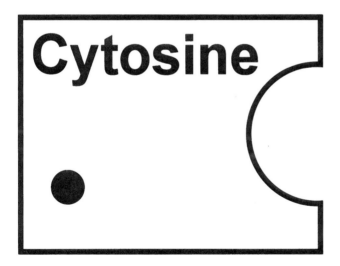

Guanine

Cytosine

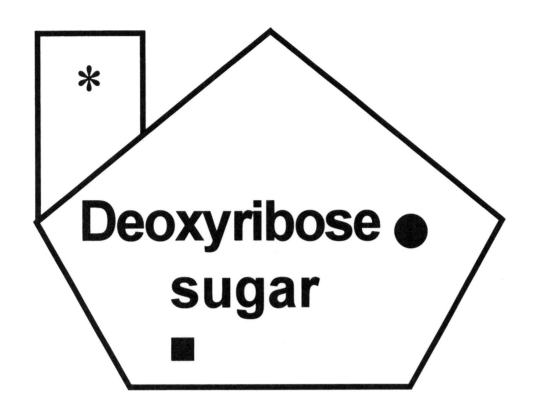

327

DNA, mRNA, tRNA, and amino acid cards for *From Genes to Proteins*

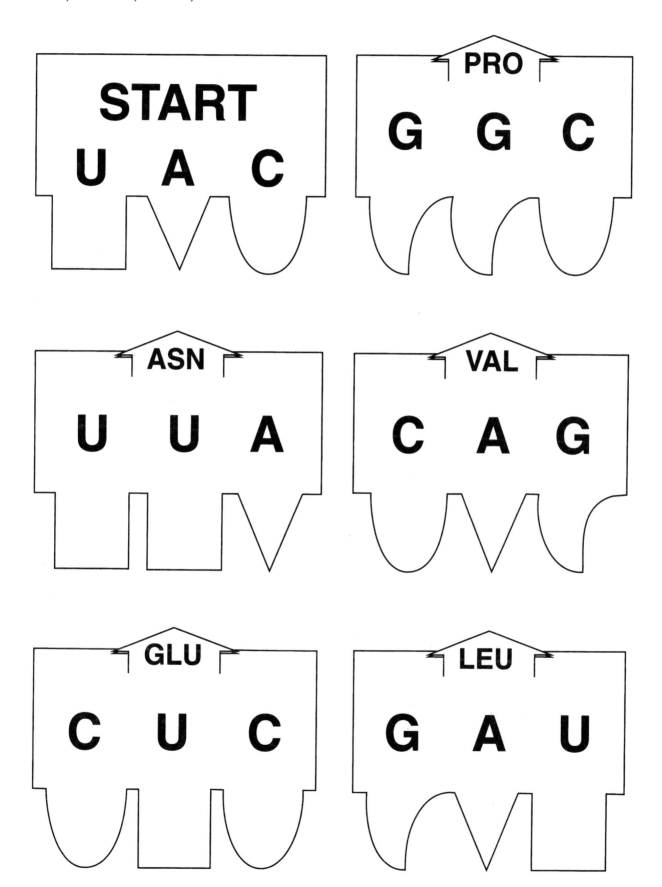

tRNA

tRNA

tRNA

tRNA

tRNA

tRNA

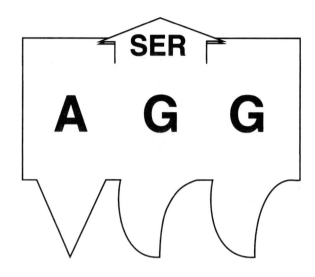

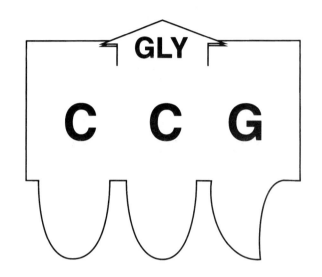

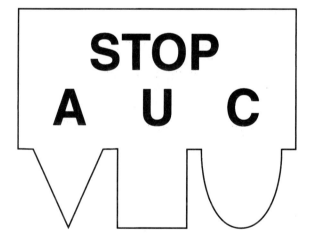

tRNA

tRNA

tRNA

332

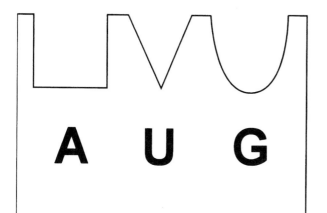

A U G

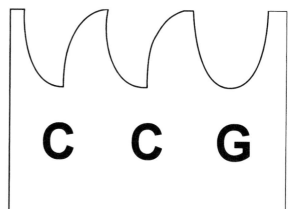

C C G

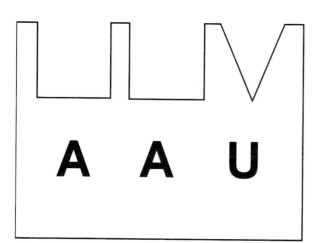

A A U

G U C

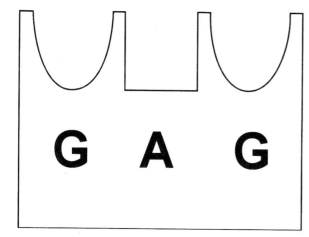

G A G

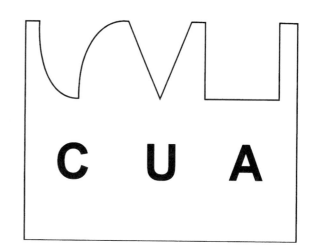

C U A

mRNA

mRNA

mRNA

mRNA

mRNA

mRNA

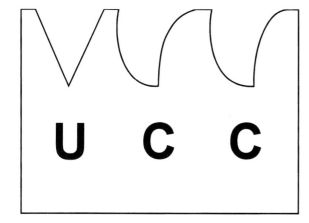

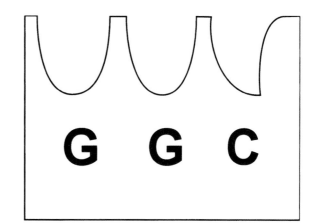

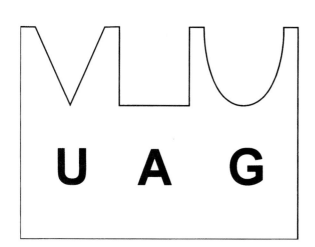

mRNA   mRNA

                       mRNA

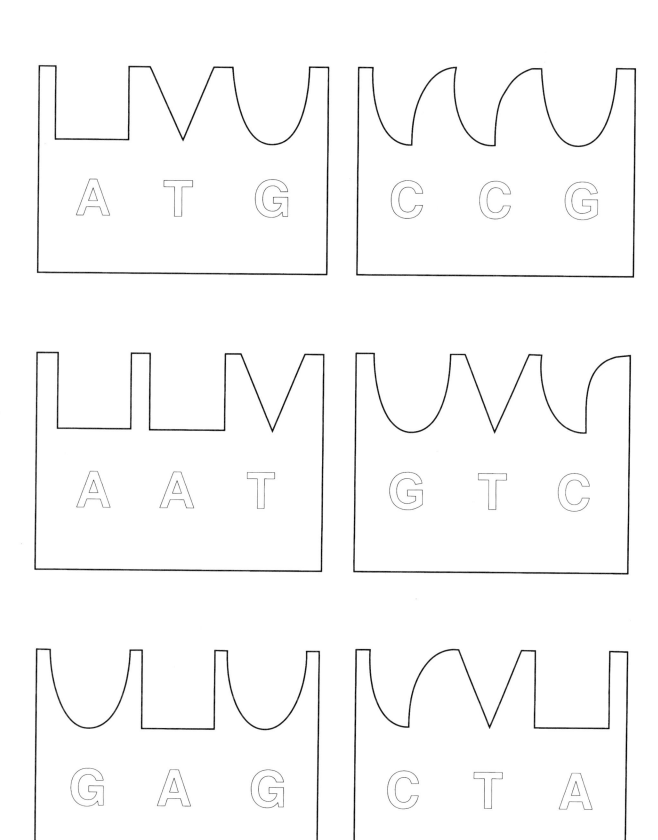

DNA

DNA

DNA

DNA

DNA

DNA

338

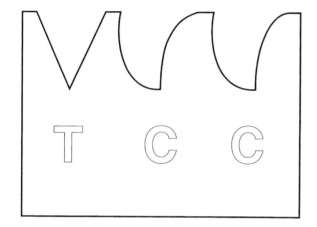

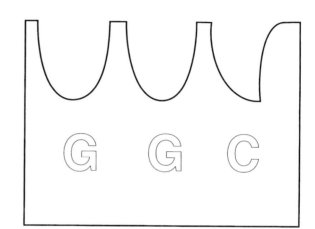

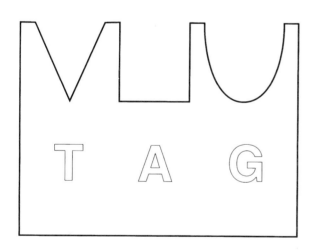

DNA   DNA

                   DNA

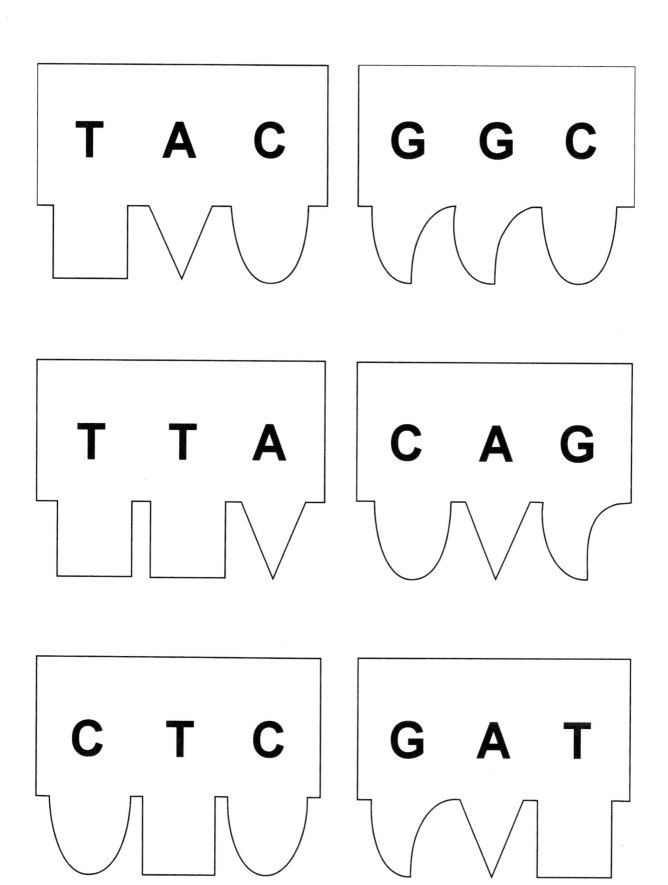

DNA

DNA

DNA

DNA

DNA

DNA

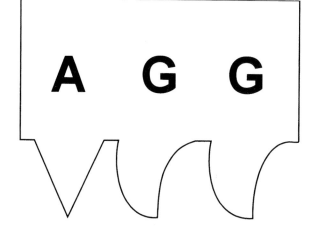

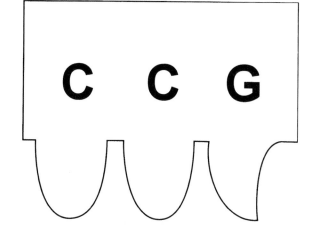

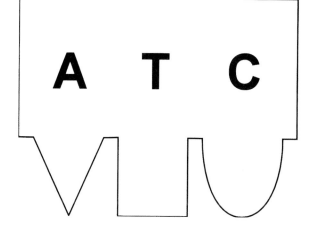

343

DNA

DNA

DNA

DNA

344

PROLINE

ASPARAGINE

VALINE

GLUTAMIC ACID

LEUCINE

SERINE

GLYCINE

Amino Acid   **Amino Acid**

Amino Acid   **Amino Acid**

Amino Acid   **Amino Acid**

                                  Amino Acid

Start
Splice
(RNA)

Start
Splice
(DNA)

End
Splice
(DNA)

End
Splice
(RNA)

RRE (DNA)

RRE (RNA)

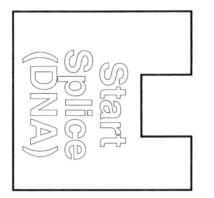

Start Splice (DNA)

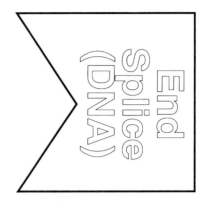

End Splice (DNA)

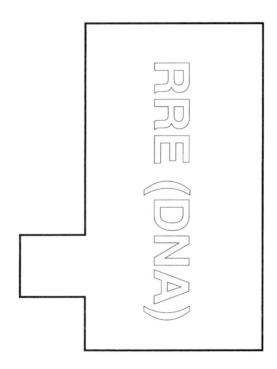

RRE (DNA)

Promoter
(no RNA transcript)

Terminator
(no RNA transcript)

Promoter
(no RNA transcript)

Terminator
(no RNA transcript)

DNA sequence strips for *DNA Scissors*

```
┌─1─────────────────────────────────────1─┐
│                                          │
│   5'-TAGACTGAATTCAAGTCA-3'               │
│      | | | | | | | | | | | | | | | | |  │
│   3'-ATCTGACTTAAGTTCAGT-5'               │
│                                          │
└──────────────────────────────────────────┘
```

```
┌─2─────────────────────────────────────2─┐
│                                          │
│   5'-ATACGCCCGGGTTCTAAA-3'               │
│      | | | | | | | | | | | | | | | | |  │
│   3'-TATGCGGGCCCAAGATTT-5'               │
│                                          │
└──────────────────────────────────────────┘
```

```
┌─3─────────────────────────────────────3─┐
│                                          │
│   5'-CAGGATCGAAGCTTATGC-3'               │
│      | | | | | | | | | | | | | | | | |  │
│   3'-GTCCTAGCTTCGAATACG-5'               │
│                                          │
└──────────────────────────────────────────┘
```

```
┌─4─────────────────────────────────────4─┐
│                                          │
│   5'-AATAGAATTCCGATCCGA-3'               │
│      | | | | | | | | | | | | | | | | |  │
│   3'-TTATCTTAAGGCTAGGCT-5'               │
│                                          │
└──────────────────────────────────────────┘
```

Restriction maps for *DNA Goes to the Races*

Below are three representations of a 15,000-base-pair DNA molecule. Each representation shows the locations of different types of restriction site, with vertical lines representing the cut site. The numbers between the cut sites show the sizes (in base pairs) of the fragments that would be generated by digesting the DNA with that enzyme.

EcoRI sites

4,000	3,500	2,500	5,000

BamHI sites

6,000	4,000	3,000	2,000

HindIII sites

8,000	4,500	2,500

Gel outline for *DNA Goes to the Races*

EcoRI   HindIII   BamHI

Sample wells

Size scale in base pairs

8,000 —

6,000 —

4,000 —

3,000 —

2,000 —

Paper pAMP plasmid model for *Recombinant Paper Plasmids*

paste 2

paste 3

paste 1

Ampicillin resistance gene

1
5' AATTCGATGAATTCXXXXXXXXXXXXXXXXXXXXXXXXXXXXXGAATTCTGAAGGTTCGAAGCGCTAT
3' TTAAGCTACTTAAGXXXXXXXXXXXXXXXXXXXXXXXXXXXXXCTTAAGACTTCCAAGCTTCGCGATA

2
5' GTCGGATCCAGATCCGAAGTCTCTCTAGGACCTTGCGAAGCCACGTAGTTCAGATTAATGCCTGAT
3' CAGCCTAGGTCTAGGCTTCAGAGAGATCCTGGAACGCTTCGGTGCATCAAGTCTAATTACGGACTA

Origin of replication

3
5' CGCTACAAGCTTATAGGGCCXXXXXXXXXXXXXXXXXXXXXAATATTGCGCAGTCTTAGCACTCC
3' GCGATGTTCGAATATCGCCGGXXXXXXXXXXXXXXXXXXXXXTTATAACGCGGTCAGAATCGTGAGG

Paper pKAN plasmid model for *Recombinant Paper Plasmids*

1

Origin of replication

5' TACTCGATGAAATCXXXXXXXXXXXXXXXXXXXXXXXXXXXXXXAGCTATGTTCTGAAGGATCCATATAGCGC

3' ATGAGCTACTTTAGXXXXXXXXXXXXXXXXXXXXXXXXXXXTCGATACAAGACTTCCTAGGTATATCGCG

paste 2

2

Kanamycin resistance gene

5' ATGACCGTCAGATCCGATGCTTCXXXXXXXXXXXXXXXXXXXXXXXXXXXTCGAACGTACGGTCCGA

3' TACTGGCAGTCTAGGCTACGAAGXXXXXXXXXXXXXXXXXXXXXXXXXAGCTTGCATGCCAGGCT

paste 3

3

5' GATCACATGCTTATAAATATTGCGAAGCTTCAGTCAGCGGTAGCACTCCTTAACGGCGATGCATTAA

3' CTAGTGTACGAATATTTATAACGCTTCGAAGTCAGTCGCCATCGTGAGGAATTGCGCTACGTAATT

paste 1

Worksheet 16.I: single-stranded DNA sample sequence and probe for *Detection of Specific DNA Sequences: Hybridization Analysis*

Probe:

3′ GGATGCTACCATAGC 5′

3′ GGATGCTACCATAGC 5′

Sample DNA sequence, written 5′ to 3′

```
1
  GATCAGACTTCTAGCAGGCTCTTGACCAATGATCACAGCTTCCGATCTAGAGCTCGATCTCGTGTCGGAATCTAG

91
  CCGGGCGTGAATTCTAGCCGGGTCAGCTATGCTAAGATAGACCGGAATCGAGAATTCCGGATATCGATTGTGCGACCGGCATTAT

181
  CGATCGTTTGCCCGGGGATCTAGCTTTCCGATCTAGCTGTGGCGATCTCGGGGATCGATTCCCGGGGATCTAGGCCTACGATGGTATCGTTAG

271
  TAGCTCTCTAGCTTAGCTCTTCAAGTGATCTACCCGGGTAGATCTAGTATATTGTATCGATATTTGGGCCCCCCTAGCTCGGAGCTAGCT

361
  TCTCTAGCTAATAGATAG
```

365

Worksheet 16.II: outlines for gel and results of hybridization analysis for *Detection of Specific DNA Sequences: Hybridization Analysis*

Stained electrophoresis gel

Fragment size, bp

Sample well

100 —
90 —
80 —
70 —
60 —
50 —
40 —
30 —
20 —

Results of hybridization analysis

100 —
90 —
80 —
70 —
60 —
50 —
40 —
30 —
20 —

Worksheet 16.IIIA: restriction maps and hybridization analysis of virus X for *Detection of Specific DNA Sequences: Hybridization Analysis*

The map at the top shows the *Eco*RI restriction sites. The bottom line shows the *Bam*HI restriction sites. Fragment sizes are in base pairs. The stained elec-trophoresis gel shows *Eco*RI and *Bam*HI fragments of the virus. The hybridization analysis shows which of the bands in the stained gel hybridized to the probe.

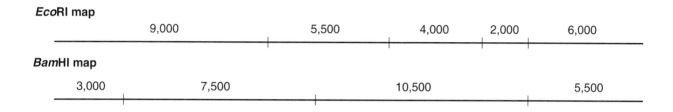

EcoRI map

| 9,000 | 5,500 | 4,000 | 2,000 | 6,000 |

BamHI map

| 3,000 | 7,500 | 10,500 | 5,500 |

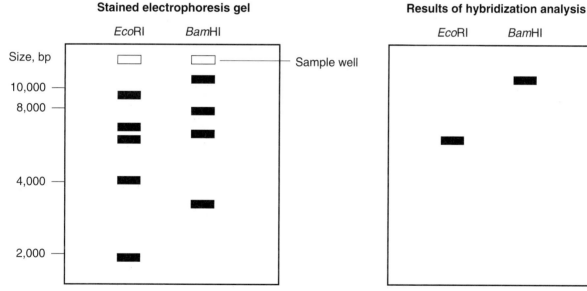

Stained electrophoresis gel

*Eco*RI   *Bam*HI

Size, bp — Sample well

10,000
8,000

4,000

2,000

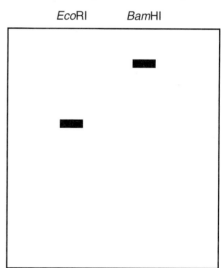

Results of hybridization analysis

*Eco*RI   *Bam*HI

Worksheet 16.IIIB for *Detection of Specific DNA Sequences: Hybridization Analysis*

Shown below are restriction maps for bacteriophage lambda, showing the *Bam*HI, *Hin*dIII, and *Eco*RI sites. Fragment sizes are in base pairs. The gel and hybridization analysis outlines are provided for your use.

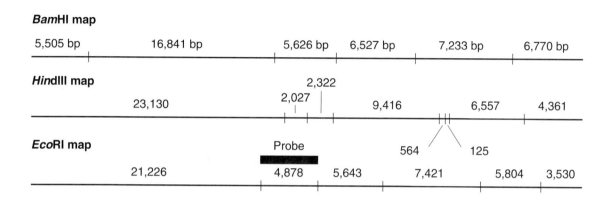

BamHI map

5,505 bp   16,841 bp   5,626 bp   6,527 bp   7,233 bp   6,770 bp

HindIII map

2,322
23,130   2,027   9,416   6,557   4,361

EcoRI map

Probe   564   125
21,226   4,878   5,643   7,421   5,804   3,530

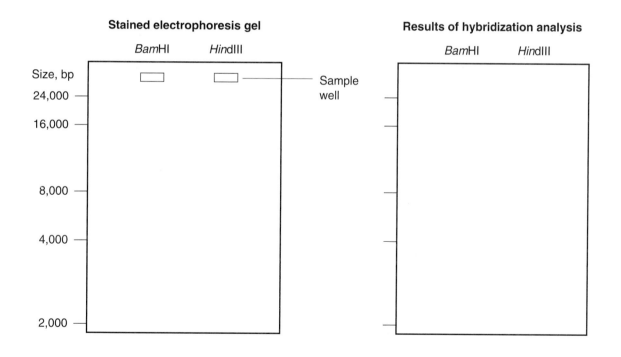

Stained electrophoresis gel

BamHI   HindIII

Size, bp — Sample well
24,000
16,000

8,000

4,000

2,000

Results of hybridization analysis

BamHI   HindIII

371

Template-primers for *DNA Sequencing*

5' CGAACATC
3' GCTTGTAGAAGCTGCATTGACGCT 5'
1 2 3 4 5 6 7 8 9 10 11 12 13 14 15 16

5' CGAACATC
3' GCTTGTAGAAGCTGCATTGACGCT 5'
1 2 3 4 5 6 7 8 9 10 11 12 13 14 15 16

5' CGAACATC
3' GCTTGTAGAAGCTGCATTGACGCT 5'
1 2 3 4 5 6 7 8 9 10 11 12 13 14 15 16

5' CGAACATC
3' GCTTGTAGAAGCTGCATTGACGCT 5'
1 2 3 4 5 6 7 8 9 10 11 12 13 14 15 16

Parental DNA molecule and primers for
The Polymerase Chain Reaction

5' **TACGACCCGGTGTCAAAGTTAGCTTAGTCA** 3'

5' **TACGACCCGGTGTCAAAGTTAGCTTAGTCA** 3'

5' **TACGACCCGGTGTCAAAGTTAGCTTAGTCA** 3'

5' **TACGACCCGGTGTCAAAGTTAGCTTAGTCA** 3'

5' **CCCGG** 3'   5' **CCCGG** 3'

5' **CCCGG** 3'   5' **CCCGG** 3'

3' ATGCTGGGCCACAGTTTCAATCGAATCAGT 5'

3' ATGCTGGGCCACAGTTTCAATCGAATCAGT 5'

3' ATGCTGGGCCACAGTTTCAATCGAATCAGT 5'

3' ATGCTGGGCCACAGTTTCAATCGAATCAGT 5'

3' TCGAA 5'   3' TCGAA 5'

3' TCGAA 5'   3' TCGAA 5'

Simulation of PCR diagnostic test for
The Polymerase Chain Reaction

Primers: | 5´ TTCCAGCC 3´ ⟩ | ⟨ 3´ CGGAATAC 5´ |

Sample 1

5´ GCGTAATCGGATGCCGTAATAGGTATGCGCGAATTTGTG
3´ CGCATTAGCCTACGGCATTATCCATACGCGCTTAAACAC

5´ ATGCATCCGATAGCGCGGGCCTATATTGTAACTGGCATC
3´ TACGTAGGCTATCGCGCCCGGATATAACATTGACCGTAG

5´ ATGCGTTAACGCGTAATGGCCTAATCTATGTAGCGCGAA
3´ TACGCAATTGCGCATTACCGGATTAGATACATCGCGCTT

5´ AATGCGCTTCCAGCCAGAGTCTCGGAACTAGCCTTATG
3´ TTACGCGAAGGTCGGTCTCAGAGCCTTGATCGGAATAC

Sample 2

5´ GAATTCCTCATGATCCAGGTCACTAATGCACGGTTACAC
3´ CTTAAGGAGTACTAGGTCCAGTGATTACGTGCCAATGTG

5´ GGGCCCTATAGCTACTCTAGAATCTAGCGAATATTGCGC
3´ CCCGGGATATCGATGAGATCTTAGATCGCTTATAACGCG

5´ GAATTACGCTAGCGATCGGCTATTCGAATTCGGTTATCG
3´ CTTAATGCGATCGCTAGCCGATAAGCTTAAGCCAATAGC

5´ AGGCCTCGCATGAATCTCGATTTAAATGCGCATCGATAT
3´ TCCGGAGCGTACTTAGAGCTAAATTTACGCGTAGCTATA

Worksheet 27.1 for *Testing for Amylase Activity*

Use additional sheets if needed.

Material	Prediction: Will there be amylase activity and why do you think so?	Result: Was there activity?	Possible explanations for results different from expectation
Germinating bean			
Leaf extract			
Root extract			
Dog saliva			

381

Worksheet 29.1: amylase sequences for *Constructing an Amylase Evolutionary Tree*

A NNNGV–IKEVTINPDTTCGND . . . GRGNRGFIVFNNDDWSFSLTLQTGLPAGTYCDVISGDKINGNCTGI

B NNNGV–IKEVTINADTTCGND . . . GTGNRGFIVFNNDDWQLSSTLQTGLPGGTYCDVISGDKVGNSCTGI

C NSDGS–TKSVTINADTTCGND . . . GRGDRGFIVFNNDDWYMNVDLQTGLPAGTYCDVISGQKEGSACTGK

D HDGSFNIISPSFNADGSCGNG . . . CRGNKGFLAINNDGWDLKETLQTCLPAGTYCDVISGSKNGGSCTGK

E TTDGHNIASPIFNSDNSCSGG . . . SRGSRGFVAFNNDNYDLNSSLQTGLPAGTYCDVISGSKSGSSCTGK

F NNNGK–TKEVSINPDSTCGND . . . GRGNKGLIVFNNDDWALSETLQTGLPAGTYCDVISGDKVDGNCTGI

G TTDGQNIASPVFNSDSSCSGG . . . SRGSRGFVAFNNDNYDLNSSLQTGLPAGTYCDVISGSKSGSSCTGK

Worksheet 31.1 for *An Adventure in Dog Hair, Part I*

Label the coat color alleles that Cocoa and Midnight could contribute to their gametes on the diagram. Draw the puppies' chromosomes with coat color genes, and label the genes. Color Cocoa, Midnight, and the puppies.

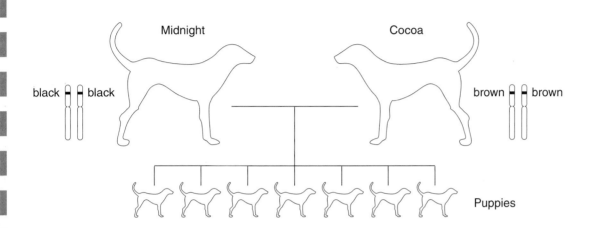

Table 1

Individual	Coat color genes	Coat color
Midnight		
Cocoa		
Puppies		

Table 2. Coat color and TRP-1

Individual	Coat color genes	Do melanocytes make TRP-1?	Coat color
Midnight			
Cocoa			
Puppies			

Worksheet 31.2 for *An Adventure in Dog Hair, Part I*

Genetics of roundbuds

Variety	Genotype	Enzyme Q produced?	Color
Red			
White			

Why might the offspring of a red × white cross be pink instead of red?

Worksheet 32.1 for *An Adventure in Dog Hair, Part II: Yellow Labs*

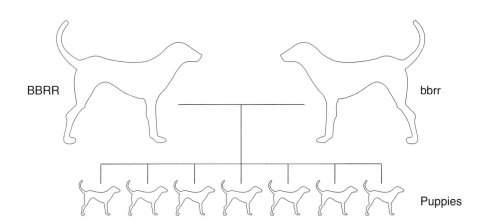

BBRR   bbrr

Puppies

Fill in the blanks, and then color the parents and puppies the correct colors for their genotypes.

Male gametes' genotype(s)   _____

Female gametes' genotype(s)   _____

Puppies' genotype(s)   _____

Puppies' phenotype(s)   _____

Worksheet 32.2: Punnett square for *An Adventure in Dog Hair, Part II: Yellow Labs*

Parents: BbRr × BbRr

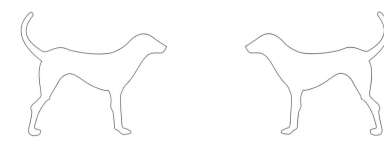

Genotype	No. of occurrences in Punnett square	Does this genotype synthesize		Phenotype	
		MSH-R?	TRP-1?	Hair	Lips

Summary

Phenotype   No. of occurrences in Punnett square

Black Lab   _____

Chocolate Lab   _____

Yellow Lab, black lips   _____

Yellow Lab, brown lips   _____

Worksheet 33.1 for *Human Molecular Genetics*

Fill in the individuals' genotypes in the blanks. If the individual could have more than one genotype, write each genotype that applies.

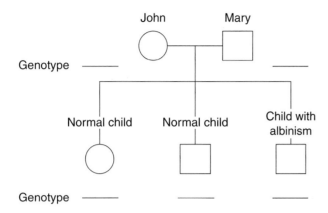

1. What protein does the "albinism gene" encode?

2. Explain how the "albinism gene" causes albinism.

3. Is the "albinism gene" dominant or recessive? What is the molecular explanation of your answer?

Worksheet 33.2 for *Human Molecular Genetics*

Write the genotype of the individuals in the blanks provided. If an individual could have more than one genotype, write all that apply.

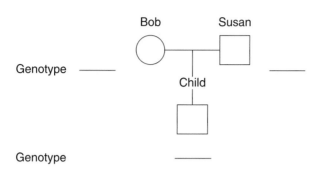

1. What protein does the "PKU gene" encode?

2. How can two normal parents have a child affected with PKU?

3. If Bob and Susan have another child, will it also have PKU?

4. Explain how the "PKU gene" causes PKU and why Bob and Susan don't have the disease but their child does. Use your knowledge of cellular biochemistry.

5. Is PKU a dominant or recessive genetic disease? What is the molecular explanation of your answer?

Worksheet 33.3: genotypes in sickle-cell anemia for *Human Molecular Genetics*

1. What is the relationship between sickle-cell anemia and the sickle-cell trait?

2. Why do the parents of children with sickle-cell disease have the sickle-cell trait? Why do many of the siblings of these children have the trait?

3. Why did individuals with sickle-cell disease have one form of hemoglobin in the electrophoresis test, normal individuals have a second form, and people with the sickle-cell trait have both?

4. Based on the story in the text, what do you think Dr. Neel's conclusion about sickle-cell disease was, and why did the protein electrophoresis results confirm it?

"Box-up" forms for *A Decision-Making Model for Bioethical Issues*

Issue: *Question:*

Facts

Authority

Qualifier

Principle

Decision

Rebuttal

Issue: *Question:*

Facts

Authority

Qualifier

Principle

Decision

Rebuttal

401

Issue: *Question:*

Facts

Authority

Qualifier

Decision

Principle

Rebuttal

403

Issue: *Question:*

Facts

Authority   *Qualifier*

Principle   *Decision*

Rebuttal

405

Issue: *Question:*

Facts

Authority   *Qualifier*

Principle   *Decision*

Rebuttal

407

Issue: *Question:*

Facts

Authority

Qualifier

Decision

Principle

Rebuttal

409

Glossary

Abzyme   An antibody designed to catalyze a specific chemical reaction. Also called a catalytic antibody.

Adenine   A nitrogen-containing base found in DNA and RNA.

Agrobacterium tumefaciens   A common soil bacterium that causes crown gall disease by transferring some of its DNA to the plant host. Scientists alter *A. tumefaciens* so that it no longer causes the disease but is still able to transfer DNA. They then use this altered organism to ferry desirable genes into plants.

Allele   One of several alternative forms of a specific gene that occupies a certain locus on a chromosome.

Alpha helix   One of a small number of stable arrangements of the peptide backbone within proteins.

Amino acids   The fundamental building blocks of a protein molecule. A protein is a chain of hundreds or thousands of amino acids. Our bodies can synthesize most of the amino acids from their component parts (carbon, nitrogen, oxygen, hydrogen, and sometimes sulfur). However, eight amino acids (called essential amino acids) must be obtained from food.

Antiangiogenesis   Angiogenesis is the growth of new blood vessels.

Antibiotic resistance marker   A gene encoding a protein that renders a cell resistant to an antibiotic. The organisms, usually bacteria, that have this gene are not susceptible to that antibiotic. Often used to identify organisms that have been successfully transformed.

Antibody   A protein produced in response to the presence of a specific antigen.

Anticodon   A triplet of nucleotides in a transfer RNA molecule that is complementary to a codon in a messenger RNA molecule.

Antigen   A foreign substance that elicits the production of antibodies.

Antisense molecule   A single-stranded nucleic acid that is complementary to a gene and, therefore, to the messenger RNA transcribed from that gene. It blocks the expression of that gene by interfering with protein production.

Assay   A method for determining the presence or quantity of a component.

Autosome   A chromosome that is the same in males and females of the species (as opposed to the sex chromosomes). Humans have 22 pairs of autosomes and 1 pair of sex chromosomes. Mutations in or traits encoded by genes on these chromosomes can be described as autosomal. For example, cystic fibrosis is an autosomal recessive trait.

Bacillus thuringiensis   A naturally occurring bacterium with pesticidal properties. *B. thuringiensis* produces a protein (Bt toxin) that is toxic only to certain insect larvae that consume it.

Bacteriophage   A virus that infects bacteria. Also called a phage.

Baculovirus   A class of insect viruses used as cloning vectors for eukaryotic cells.

Beta sheet   One of a small number of stable arrangements of the peptide backbone within proteins.

Bioassay   A method of determining the effect of a compound by quantifying its effect on living organisms or their component parts.

Biolistics   A method of getting DNA into cells by using small metal particles coated with DNA. These particles are fired into a cell at very high speed.

Biological control   The use of one organism to control the population size of another organism.

Biological molecules   Large, complex molecules, such as proteins, nucleic acids, lipids, and carbohydrates, that are produced only by living organisms. Biological molecules are often referred to as macromolecules or biopolymers.

Bioprocess   A process in which microorganisms, living cells, or their components are used to produce a desired end product.

Bioreactor   A container used for bioprocessing.

Bioremediation   The use of organisms, usually microorganisms, to clean up contamination.

Biosensor   An electronic system that uses cells or biological molecules to detect specific substances.

Consists of a biological sensing agent coupled to a microelectronic circuit.

Biosynthesis   Production of a chemical by a living organism.

Biotechnology   (Broad definition) The use of living organisms to solve problems and make useful products. (Modern definition) A collection of technologies that use living cells and/or biological molecules to solve problems and make useful products.

B lymphocyte   A type of immune system cell that is responsible for the production of antibodies.

Bovine somatotropin (BST)   The proteinaceous growth hormone found naturally in cattle; very similar in structure and function to human growth hormone. BST is chemically similar to human growth hormone and is being used commercially to increase growth rate, the protein/fat ratio, and milk production in cows.

Callus   A cluster of undifferentiated plant cells that have the capacity to regenerate a whole plant in some species.

Cancer vaccines   Unlike other vaccines that provide preventative protection, cancer vaccines stimulate the lymphocytes of the immune system to attack existing tumor cells.

Capsid   Protein coat of a virus. Plants can be genetically engineered to be resistant to a virus by giving them the gene that encodes the capsid.

Carbohydrates   Biological molecules composed of simple sugars such as glucose. Also known as polysaccharides. Familiar examples include starch, cellulose, glycogen, and lactose.

Catalyst   A substance that speeds up a chemical reaction but is not itself changed during the reaction.

cDNA library   A collection of genetic clones that contains all of the cDNA derived from a source organism. The cDNA is "housed" in the library by splicing portions of the entire complement of cDNA into suitable vectors. (See **Complementary DNA**.)

Cell culture   A technique for growing cells under laboratory conditions.

Cell fusion   The formation of a hybrid cell produced by the fusing of two different cells.

Cell therapy   Treating a disease by providing healthy cells to replace dead or malfunctioning cells. Genetically engineered cells may also be used to deliver a constant supply of therapeutic compounds.

Chitin   A carbohydrate that is the principal component of fungal cell walls and insect exoskeletons.

Chromatin   The DNA-protein complex found in the nucleus.

Chromosomal aberration   An abnormality in chromosome number or structure.

Chromosomes   Components in a cell that contain genetic information. Each chromosome contains numerous genes. Chromosomes occur in pairs: one is obtained from the mother, and the other is obtained from the father. Chromosomes of different pairs are often visibly different from each other. (See also **DNA**.)

Chromosome walking   The step-by-step analysis of a long stretch of DNA by the sequential isolation of clones that carry overlapping sequences of this DNA. Used to locate unknown genes. Using a gene library and starting from a known sequence (usually the site of a restriction fragment length polymorphism), scientists isolate clones containing DNA that hybridizes to DNA probes taken from the ends of the known sequence. The ends of these clones are then used to screen the library for clones that hybridize to the ends of the first clone. This screening and isolation is repeated again and again until the unknown gene of interest is reached. If each clone covers a long stretch of DNA, the researcher can "walk" the chromosome quickly, because each "step" is like a "giant step." So researchers prefer to use clones or vectors that carry a large amount of foreign DNA. (See **Clone, Cloning**, and **Genetic library**.)

Clone   A cell, collection of cells, or collection of individuals containing genetic material identical to that of the parent cell and of each other. Clones are produced from a single parent cell and thus show little if any variation compared with that in similar organisms produced through sexual reproduction. The word "clone" also refers to the identical piece of DNA that a collection of cells (usually bacterial) contains.

Cloning   Isolating DNA sequences and incorporating them into plasmids or other vectors so that they can be inserted into a suitable organism (bacterium, yeast cell) for copying. Cloning also refers to the production of genetically identical cells from a single parent cell. These genetically identical cells are referred to as a clonal population.

Codon   A triplet of bases that specifies an amino acid.

Colony hybridization   A technique that uses hybridization to identify bacteria containing DNA that is complementary to a certain sequence.

Complementary DNA (cDNA)   A single-stranded DNA that is synthesized in vitro from an RNA template by reverse transcriptase.

Conjugation   The transfer of genetic material from one bacterium to another through physical contact between the cells. In *Escherichia coli*, the contact occurs through a special structure called a pilus.

Cosmid A plasmid that is packaged in a phage coat. Scientists use cosmids to transfer a relatively long stretch of DNA into host organisms.

Covalent A type of chemical bond that consists of two electrons shared by two atomic nuclei.

Crossing over A natural process that occurs during meiosis in which pieces of homologous chromosomes are exchanged.

Culture To grow living organisms in prepared medium.

Culture medium A nutrient system for growing bacteria or other cells in the laboratory.

Cytosine A nitrogen-containing base found in DNA and RNA.

Deletion A form of chromosomal aberration in which a portion of a chromosome is lost.

Denaturation The complete unfolding of a protein or the separation of the two complementary strands of a DNA double helix.

Deoxynucleotide A compound made up of the sugar deoxyribose, phosphate, and a nitrogen-containing base. Found in DNA.

Deoxyribose The five-carbon sugar found in DNA.

DNA (deoxyribonucleic acid) The chemical molecule that is the basic genetic material found in all cells. DNA is the carrier of genetic information from one generation to the next. Because DNA is a very long, thin molecule, it is packaged into units called chromosomes. DNA belongs to a class of biological molecules called nucleic acids.

DNA chip A small piece of glass or silicon that has arrays of DNA on its surface. Its usefulness as a gene identification tool is based on DNA hybridization.

DNA hybridization The formation of a double-stranded nucleic acid molecule from two separate but complementary single strands. The single strands can be two DNA strands or one RNA and one DNA strand. The term also applies to a molecular technique that uses one nucleic acid strand to locate another.

DNA library A collection of cloned DNA fragments that collectively represents the genome of an organism. A complementary DNA library collectively represents the messenger RNA species that were present in the cells when the library was made.

DNA ligase An enzyme that rejoins cut pieces of DNA.

DNA polymerase An enzyme that replicates DNA. DNA polymerases synthesize new DNA complementary to a template strand. Synthesis occurs in the $5'$-to-$3'$ direction only and requires a primer. (See **Primer.**)

DNA probe A relatively short single strand of DNA that is used to detect a specific sequence of nucleotides through hybridization.

DNA repair enzymes Proteins that recognize and repair abnormalities in DNA.

DNA sequence The order of nucleotide bases in the DNA molecule.

DNA vaccines Pieces of foreign DNA that are injected into an organism to trigger an immune response. The injected DNA is expressed by the host cell, producing an antigen that is recognized as foreign by both B and T lymphocytes. Unlike common vaccines, DNA vaccines contain the information for making the antigen and not the antigen itself or the whole organism.

Dominant allele An allele that is phenotypically expressed in the same way in individuals who are either homozygous or heterozygous for that allele.

Duplication A chromosomal aberration in which part of the chromosome is present in duplicate form in a cell.

Electronegativity The ability of an atom to attract electrons.

Electroporation A technique that uses an electrical current to create temporary pores in a cell membrane through which DNA can enter.

Embryo transplant A technique used in animal biotechnology. After in vitro fertilization, the zygote is cultured for a few days and then implanted into a female. The developing embryo is sometimes separated into individual cells at the four- to eight-cell stage, and each cell is implanted in a female.

Embryonic stem cells Cells of an early embryo that can give rise to any type of differentiated cell.

Endonuclease An enzyme that cleaves a nucleic acid at nonterminal phosphodiester bonds. (See **Exonuclease** for comparison; exonucleases cleave at terminal sites.) One class of endonucleases, the restriction endonucleases, recognizes specific sequences of bases along a DNA molecule and cleaves the molecule following recognition.

Endostatin An endogenous protein that blocks the proliferation of blood vessels by inhibiting the growth of endothelial cells.

Enhancers DNA-binding sites of certain transcription activator proteins that are important for maximal transcription of associated promoters.

Enzyme A protein that accelerates the rate of chemical reactions. Enzymes are catalysts that promote reactions repeatedly without being changed by the reactions.

Enzyme-linked immunosorbent assay (ELISA) A technique for detecting specific proteins by using antibodies linked to enzymes.

Erythropoietin A growth factor that stimulates the cells that give rise to red blood cells.

Escherichia coli A bacterium commonly found in the intestinal tracts of most vertebrates. *E. coli* is used extensively in recombinant DNA research because it has been genetically well characterized.

Eukaryote An organism whose genetic material is located within a nucleus. Yeast cells, fungi, protozoans, plants, and animals are eukaryotes.

Evolution Changes in the gene pool of a population over time. These changes in the frequencies of certain genotypes result primarily from natural selection. Other factors that may contribute to changes in the genetic composition of a population are genetic drift and migration.

Exons The regions of a gene that determine the amino acid sequence of a protein.

Exonuclease An enzyme that cleaves the terminal phosphodiester bond of a nucleic acid molecule, releasing a single nucleotide. Exonucleases must have access to the end of a molecule for activity; they will not cleave circular nucleic acid molecules.

Expression The physical manifestation (protein production) of the information contained in a gene.

Extremophiles Microorganisms that live at extreme levels of pH, temperature, pressure, and salinity.

Factor VIII One of many compounds involved in blood clotting in humans. If this protein is missing or defective, the resulting condition is hemophilia. Because factor VIII is a small protein, it can be produced in large quantities by bacteria genetically engineered for its production.

Fermentation In biochemistry, fermentation is the anaerobic breakdown of glucose. In biotechnology and biochemical engineering, fermentation is the process of growing microorganisms to produce various chemical or pharmaceutical compounds. Microbes are usually incubated under specific conditions in large tanks called fermenters. Fermentation is a specific type of bioprocessing.

Functional foods Foods containing compounds with beneficial health effects beyond those provided by the basic nutrients, vitamins and minerals. Also called nutraceuticals.

Functional genomics A field of research whose goal is to understand what each gene does, where it is located, and how it is regulated.

Gel electrophoresis A process for separating molecules by forcing them to migrate through a semisolid material (gel) under the influence of an electric field.

Gene A unit of hereditary information. A gene is a section of a DNA molecule that specifies the production of a particular protein.

Gene amplification The increase, within a cell, of the number of copies of a given gene.

Gene knockout The replacement of a normal gene with a mutated form of the gene by homologous recombination. Used to study gene function.

Gene mapping Determining the relative locations of genes on a chromosome.

Gene therapy The addition of genetic material to an individual so that a defect or disease can be corrected. To date, human gene therapy has involved changing the genetic makeup of somatic cells only. Genetic changes to germ cells are prohibited in humans and have been restricted to animals.

Genetic code The system of nucleotide triplets in genes that encode the amino acids in proteins. All living organisms on earth use the same genetic code.

Genetic engineering The technique of removing, modifying, or adding genes to a DNA molecule in order to change the information it contains. By changing this information, genetic engineering changes the type or amount of proteins an organism is capable of producing.

Genetic library A collection of DNA that, taken collectively, represents all of an organism's genome. The DNA molecules are "housed" in microorganisms as recombinant DNA molecules and are copied when the microorganism replicates.

Genome The total hereditary material of a cell.

Genotype The specific genetic makeup of an organism, as opposed to the actual characteristics of an organism. (See **Phenotype.**)

Glycoprotein A protein molecule that is bound to a simple sugar or carbohydrate.

Growth factors Naturally occurring proteins that stimulate growth and reproduction of specific cell types. For example, epidermal growth factor stimulates the production and differentiation of cells in the upper skin layer. Fibroblast growth factor stimulates growth of cells in connective tissue. Growth factors are being studied as possible therapeutic compounds to be used in the treatment of diseases or injuries. For example, the two growth factors just mentioned could be useful for treating burn victims.

Growth hormones Hormones that stimulate growth in plants and animals. The growth hormones in plants bear no chemical resemblance to the growth hormones in animals. In the vertebrates, growth hormone is a protein hormone secreted by the anterior pituitary. It stimulates protein production in its target organs. Also known as somatotropin.

Guanine A nitrogen-containing base found in DNA and RNA.

Hematopoietic A type of stem cell that can differentiate into all types of blood cells.

Homologous Two chromosomes are said to be homologous if they carry alleles for the same traits. In each cell containing homologous chromosomes, each member of a homologous pair is derived from a different parent. Nonhomologous chromosomes carry genes for different traits.

Hyaluronate A carbohydrate made of a chain of several thousand sugar molecules in a regular, repeating sequence.

Hybridization Production of offspring, or hybrids, from genetically dissimilar parents. In selective breeding, it usually refers to the offspring of two different species. (See also **DNA hybridization**.)

Hybridoma A type of hybrid cell produced by fusing a normal cell with a tumor cell. When lymphocytes (antibody-producing cells) are fused to the tumor cells, the resulting hybridomas produce antibodies and maintain rapid, sustained growth, producing large amounts of an antibody. Hybridomas are the source of monoclonal antibodies.

Hydrogen bonds Weak electrostatic attractions. Hydrogen bonds exist between paired bases in DNA and are important in determining protein structure.

Hydrophilic Favoring chemical associations with water molecules.

Hydrophobic Disfavoring chemical associations with water molecules.

Hypersensitive response (HR) One of the endogenous defense systems that plants use against pathogens. By causing localized plant cell death at the point of infection, the hypersensitive response denies the pathogen living host cells.

Immunoassay A technique for identifying substances that is based on the use of antibodies.

Immunotoxin A molecule that is toxic to the cell and is attached to an antibody.

Industrial sustainability Manufacturing processes and products that meet the current consumer demand for products without compromising the resources and energy supply of future generations.

Initiation codon The codon in messenger RNA that tells the ribosomes to start synthesizing a protein. Usually 5′ AUG 3′.

Initiation factor A protein necessary to begin translation. Initiation factors are not parts of the ribosomes and do not participate in translation once the process has begun.

Insulin A protein hormone that lowers blood glucose levels. The first commercial product derived from genetically engineered bacteria.

Interferon A protein produced naturally by the cells in our bodies. It increases the resistance of surrounding cells to attacks by viruses. One type of interferon, alpha interferon, is effective against certain types of cancer. Others may prove effective in treating autoimmune diseases.

Interleukin-2 A protein produced naturally by our bodies to stimulate our immune systems. There are at least six kinds of interleukins.

Introns Noncoding regions within a gene. They are transcribed into RNA but are removed by splicing before protein synthesis.

Inversion A chromosomal aberration in which a section of chromosome is reversed.

In vitro Performed in a test tube or other laboratory apparatus.

In vitro selection Selection at the cellular or callus stage of individuals possessing certain traits, such as herbicide resistance.

In vivo In the living organism.

Islet cells Pancreatic cells that are the source of insulin, glucagon, and somatostatin, three hormones involved in regulating glucose metabolism and absorption.

Keratins The family of structural proteins that make up hair, wool, feathers, claws, hooves, and so on.

Ligation The joining of the ends of two DNA molecules.

Linkage The tendency of pairs or groups of genes to be inherited together because they occur close together on the same chromosome.

Locus The position a gene occupies on a chromosome.

Lysate The mixture of cellular components obtained after cells have been broken open.

Lysis The breaking open of cells.

Macrolesion A genetic change in which a large amount of DNA is altered by changing the total amount of DNA or changing the relative position of genes on a chromosome. (See also **Chromosomal aberration, Duplication, Deletion, Inversion,** and **Translocation**.)

Macromutation See **Macrolesion**.

Marker Restriction fragment(s) seen only when a particular genetic disease is present.

Marker genes Genes that identify which plants, bacteria, or other organisms have been successfully transformed.

Melting temperature   The temperature required to denature a DNA or protein molecule.

Messenger RNA (mRNA)   The RNA molecules that carry genetic information from the chromosomes to the ribosomes.

Metabolic engineering   Changing cellular activities by manipulating the enzymatic, transport, and regulatory functions of the cell.

Microbial fermentation   The production of useful compounds such as vitamins, antibiotics, enzymes, and certain foods. (See also **Fermentation** and **Bioprocess.**)

Microinjection   Method of delivering DNA, primarily to animal cells, by using a microscopic needle to pierce the nucleus.

Microlesion   A genetic change, sometimes called a micromutation, that involves a small amount of DNA. (See also **Point mutation.**)

Micromutation   See **Microlesion.**

Molecular genetics   The study of the molecular structures and functions of genes.

Monoclonal antibody   Highly specific, purified antibody that is derived from only one clone of cells and recognizes only one antigen. (See also **Hybridoma.**)

Multigenic   A multigenic, or polygenic, trait is one whose expression is governed by many genes.

Mutagen   A substance that induces mutations.

Mutant   A cell or organism that manifests new characteristics because of a change in its genetic material.

Mutation   Any change in the base sequence of a DNA molecule.

Mycorrhiza   A symbiotic association between certain fungi and the roots of vascular plants.

Natural selection   The differential rate of reproduction of certain phenotypes in a population. If those phenotypes have a genetic basis, natural selection can lead to a change in gene frequencies in a population.

Noncoding DNA   DNA that does not encode any product (RNA or protein). The majority of DNA in plants and animals is noncoding.

Nonhomologous   Chromosomes are described as nonhomologous if they carry genes for different traits. Contrast with the definition for **Homologous.**

Northern blotting   A technique for identifying an RNA sequence by transferring it from a gel to a filter and hybridizing it to a DNA probe. Useful for measuring gene expression.

Nuclease   An enzyme that cleaves the phosphodiester bonds of a nucleic acid molecule. (See **Endonuclease** and **Exonuclease.**)

Nucleic acid   A biological molecule composed of a long chain of nucleotides. DNA is made of thousands of four different nucleotides repeated randomly.

Nucleoside   A nucleotidelike molecule containing only the sugar and a base.

Nucleotide   A compound made up of three components: a sugar (either ribose or deoxyribose), phosphate, and a nitrogen-containing base. Found as individual molecules (e.g., adenosine triphosphate, the "energy molecule") or as many nucleotides linked together in a chain (nucleic acid such as DNA).

Oligonucleotide   A polymer consisting of a small number of nucleotides. Oligonucleotides can be synthesized by automated machines and so are widely used as probes and primers.

Oncogene   A gene thought to be capable of producing cancer.

Oncology   The study of tumors.

Operon   A collection of adjacent genes that are transcribed together and whose products usually have related functions.

Origin of replication   A sequence of DNA bases that tell DNA polymerase and its helper proteins where to begin duplicating a DNA molecule.

Peptide bond   The chemical bond that links adjacent amino acids within proteins.

Phage   See **Bacteriophage.**

Phenotype   The observable characteristics of an organism as opposed to the set of genes it possesses (its genotype). The phenotype that an organism manifests is a result of both genetic and environmental factors. Therefore, organisms with the same genotype may display different phenotypes because of environmental factors. Conversely, organisms with the same phenotypes may have different genotypes.

Phosphodiester bonds   The chemical bonds that connect nucleotides in the backbones of DNA and RNA.

Phytoremediation   The use of plants to clean up pollution.

Plasmid   A small, self-replicating piece of DNA found outside the chromosome. Plasmids are the principal tools for inserting new genetic information into microorganisms or plants.

Point mutation   A change in the DNA sequence in a single gene. Most often, this term refers to a change in a single base or a single base pair in a gene.

Polyhydroxybutyrate (PHB)   A naturally occurring compound produced by bacteria. PHB may be used in the production of biodegradable plastic.

Polymerase chain reaction (PCR) A method of making millions of copies of a single DNA molecule by using a heat-stable DNA polymerase.

Porcine somatotropin (PST) The growth hormone found in pigs. (See **Growth hormones.**)

Primary structure The linear sequence of amino acids within a protein molecule.

Primer A single-stranded nucleic acid molecule (DNA or RNA) hybridized to a template strand in such a way that the primer's 3′ end is available to serve as the starting point for synthesis of a new DNA strand complementary to the template. Required for DNA synthesis by DNA polymerase enzymes.

Probe A single-stranded DNA or RNA molecule used to detect the presence of a complementary nucleic acid.

Prokaryotes Organisms whose genetic material is not enclosed by a nucleus. The most common examples are bacteria.

Promoter A special sequence of bases in DNA that is recognized by RNA polymerase enzymes. The promoter signals RNA polymerase to begin transcription of a gene.

Protein A complex biological molecule composed of a chain of units called amino acids. Proteins have many different functions: structure (collagen), movement (actin and myosin), catalysis (enzymes), transport (hemoglobin), regulation of cellular processes (insulin), and response to stimuli (receptor proteins on the surface of all cells). Protein function is dependent on the protein's three-dimensional structure (tertiary structure), which depends on the linear sequence of amino acids in the protein (secondary structure). The information for making proteins is stored in the sequence of nucleotides in the DNA molecule.

Proteinase An enzyme that cleaves the peptide bonds of protein backbones.

Proteomics A field of research devoted to discovering the structure and function of all proteins made by a specific type of cell.

Protoplast A plant or bacterial cell whose wall has been removed by artificial treatment.

Quaternary structure The arrangement of multiple protein subunits in a larger complex.

Recessive allele An allele whose expression is masked in the heterozygous state by a dominant allele.

Recombinant DNA (rDNA) DNA that is formed by combining DNA from two different sources.

Recombinant DNA (rDNA) technology The laboratory manipulation of DNA in which DNA or fragments of DNA from different sources are cut and recombined by using enzymes. This rDNA is then inserted into a living organism. rDNA technology is usually synonymous with genetic engineering.

Recombination The formation of new combinations of genes. Recombination occurs naturally in plants and animals during the production of sex cells (sperm, eggs, pollen) and their subsequent joining in fertilization. In microbes, genetic material is recombined naturally during conjugation, transformation, and transduction.

Regeneration The process of growing an entire plant from a single cell or group of cells.

Repressors Proteins that bind to DNA and block transcription.

Restriction endonuclease See **Restriction enzyme.**

Restriction enzyme An enzyme that recognizes a specific sequence of bases in a DNA molecule and cleaves the molecule at or near that sequence. The recognition sequence is called a restriction site. Different restriction enzymes recognize and cleave at different restriction sites. Also called restriction endonuclease.

Restriction fragment A short length of DNA that results from the cleavage of a large DNA molecule by a restriction enzyme.

Restriction fragment length polymorphism (RFLP; pronounced "riflip") A difference in restriction fragment lengths between very similar DNA molecules (such as homologous chromosomes from two different individuals). RFLPs are caused by relatively minor differences in the base sequences of the molecules. RFLP analysis is used to detect differences in DNA molecules that are, on a large scale, quite similar. Applications of RFLP analysis include DNA typing and prediction of genetic disease through DNA testing.

Restriction map A diagram of the sites on a DNA molecule cleaved by different restriction enzymes.

Reverse transcriptase The enzyme that uses an RNA molecule as a template for synthesizing a complementary DNA molecule.

Ribose The five-carbon sugar found in RNA.

Ribosomes The protein-RNA complexes that form the site of protein synthesis.

Ribozymes RNA molecules that catalyze reactions, often the breakdown of RNA molecules. Also called catalytic RNA.

RNA (ribonucleic acid) Like DNA, a type of nucleic acid. RNA differs from DNA in three ways: RNA nucleotides contain the sugar ribose instead of deoxyribose; RNA contains the base uracil instead of thymine; and RNA is primarily a single-stranded molecule rather than a double-stranded helix. The three major types are messenger RNA, transfer RNA, and ribosomal RNA. All are involved in the synthesis of proteins from the information contained in the DNA molecule.

RNA polymerase   The enzyme that synthesizes RNA using a DNA template.

Secondary structure   Local regions of alpha helixes, beta sheets, and unstructured loops within a protein molecule.

Sex-linked inheritance   A trait that is determined by a gene on a sex chromosome, most often the X chromosome. As a result, the trait shows a different pattern of inheritance in males and females. In humans, the ability to discriminate color is a sex-linked trait.

Somaclonal variant selection   A form of plant genetic manipulation that is analogous to selective breeding at the plantlet and not the reproductive stage.

Somatotropin   A synonym for growth hormone.

Southern blotting   A technique for identifying a specific DNA sequence by transferring single-stranded DNA from a gel to a filter and then hybridizing the DNA with a complementary nucleic acid probe.

Splicing   The process of removing introns from messenger RNA.

Stop codon   One of the three codons in messenger RNA that cause protein synthesis to stop.

Structural motif   Simple combination of a few secondary-structure elements frequently found in protein molecules.

Subcloning   Breaking a large cloned gene into smaller parts and making a new clone from each of the DNA pieces.

Systemic acquired resistance (SAR)   One of the endogenous protection mechanisms found in plants. In response to infection, the plant synthesizes a variety of defense-related proteins, such as the enzyme chitinase, which degrades fungal cell walls.

T cells   Lymphocytic cells of the immune system involved in cell-mediated immunity and interactions with B cells.

Terminator   Sequence of DNA bases that tells the RNA polymerase to stop synthesizing RNA.

Tertiary structure   The total three-dimensional structure of a protein.

Thymine   A nitrogen-containing base found in DNA.

Tissue culture   A procedure for growing or cloning cells or tissue by in vitro techniques.

Tissue engineering   The synthesis of organs and tissues under laboratory conditions. Dependent on cell culture techniques.

Tissue plasminogen activator   A naturally occurring protein that dissolves blood clots; currently being produced for commercial use by genetically engineered bacteria. Also known as tPA.

Tobacco mosaic virus (TMV)   A naturally occurring pathogen of tobacco. TMV consists only of RNA and protein. If the RNA is extracted from TMV and rubbed into a tobacco leaf, new viruses are produced. Therefore, RNA, and not DNA, is the genetic material of TMV. TMV was important in research that elucidated the relationship between DNA, RNA, and protein.

Transcription   The process of using a DNA template to make a complementary RNA molecule.

Transcriptional activator   A protein that helps RNA polymerase begin transcription at one or more promoters.

Transduction   The transfer of DNA from one bacterium to another via a bacteriophage.

Transfer RNA (tRNA)   The RNA molecules that match codons and amino acids at the ribosome.

Transformation   A change in the genetic structure of an organism as a result of the uptake and incorporation of foreign DNA.

Transgenic   A transgenic organism is one that has been altered to contain a gene from an organism that belongs to a different species.

Translation   The process of using a messenger RNA template to make a protein.

Translocation   A chromosomal aberration in which a segment of one chromosome breaks off and joins a nonhomologous chromosome.

Transposon   A mobile genetic element that can move from one location in a plasmid or chromosome to another location.

Tumor necrosis factor   An endogenous compound that slows cell growth and kills some tumor cells.

Uracil   A nitrogen-containing base found in RNA.

Vector   The agent used to carry new DNA into a cell. Viruses or plasmids are often used as vectors.

Virus   An infectious agent composed of a single type of nucleic acid (DNA or RNA) enclosed in a coat of protein. Viruses can multiply only within living cells.

Western blotting   A technique for identifying a protein by transferring it from a gel to a membrane and then probing it with a labeled antibody.

Index

Artificial insemination, 47
Aseptic technique, xiii-xiv
Asexual reproduction, recombination and, 67-69
Aspartame, 257
Astrocytoma, 268
Autoimmune disease, 19
Autorad, *see* Autoradiogram
Autoradiogram, 130, 183
Autoradiography, 177, 183
Avery, O. T., 59
Azidothymidine (AZT), 185

B

B lymphocytes, 4, 88
Bacillus licheniformis, 229-230
Bacillus subtilis, sporulation in, 88
Bacillus thuringiensis, 12-13, 39, *see also* Bt corn
    facts about, 286
    insect orders and species susceptible to, 286
Bacteria
    competent, 172
    DNA extraction from, 156-157
    as tools in biotechnology, 84
Bacterial sexuality, *see* Conjugation
Bacteriophage, 91, *see also* Transduction
    Hershey-Chase experiment, 59-60
Bacteriophage lambda, 91
    lambda repressor, 101, 105-107
    restriction analysis of DNA, 166-170
Bacteriophage T4, 88, 91, 219-220
    life cycle of, 219
    safe handling in laboratory, xi
Baculovirus, 23
*Bam*HI, 161, 164, 166-167, 172, 181
Barley malt, 229
bcl2 gene, 271
Bean, amylase activity in, 231, 234
Benign erythrocytosis, 110-111
Benign tumor, 268
Beta barrel, 99-100
Beta sheet, 97-99
Beta-alpha-beta motif, 99-100
Beta-turn motif, 99-100
Biocatalyst, 30-31
Biodegradable products, 31
Biodegradation, 7-9
Bioethics
    decision-making model, 301-305
    human cloning, 49
Bioethics case study
    gene therapy, 306-309
    genetic screening, 310-312
    honor code violation, 303-304
    withdrawal of life support, 304-305
Bioinformatics, 15, 227, 240-246, 265
Biolistics, 224
Biological control agents, 9, 12-13, 23, 37, 285-291
Biological molecules, 3-4
Biomass production, 7
Biomedical ethics, *see* Bioethics
Biomedical literature references, online, 245-246

Biopolymers, 16-17, 31
Bioprocessing technology, 6-9
Bioreactor, 6, 9
Bioremediation, 8, 28-30
Biosensor technology, 10-11, 31
Biotechnician
    general employability skills of, 318-319
    technical skills of, 318
Biotechnology, *see also* Technology
    agricultural, 22-28, 36-40
    benefits of, 33-34, 280-292
    careers in, 313-319
    collection of technologies, 4-5
    costs of, 34-35
    debating risks of, 297-300
        debate methodology, 297-298
        herbicide-tolerant crops, 299-300
    definitions in, 3-4, 297
    environmental, 28-31
    medical, 15-22, 41-43
    "new," 3
    public concerns about, 26
    risks of, 280-292
    societal issues in, *see* Societal issues
    technologies and their uses, 4-15
Biotechnology industry, 313
    preparing for careers in, 315-317
    types of jobs in, 313-315
BLAST, search to identify amylase source organisms,
        240-244
Blastocyst, 44-45
Blood cells, gene therapy using, 136-137
Blood type, 191, 251
Blotting, 122, 129
Botulism, 219
Bovine growth hormone, *see* Bovine somatotropin
Bovine somatotropin (BST), 26
BRCA1 gene, 134, 271
BRCA2 gene, 271
Breast cancer, 271
Brewing process, 229
B.S. degree, 316-317
BST, *see* Bovine somatotropin
Bt corn, 23
    facts about, 286
    monarch butterflies and, 285-291
        benefits to monarchs and others, 291
        Cornell study, 288
        risks to butterflies in nature, 289-291
    safety of bioengineered foods, 293-294
    U.S. regulatory process for approval of, 286-288
Bt toxin, 12-13
"Bubble boy," 17-18, 137, 306
Buffer, 162

C

Caenorhabditis elegans, model organism for Human Ge-
        nome Project, 263-264
Campylobacter infection, 211, 217-218
Cancer
    classification of, 268

diagnostics based on monoclonal antibodies, 5-6
immortality of cancer cells, 90
laboratory model of, 268-269
molecular genetics of, 268-272
mutations in cancer cells, 268
viruses and, 268, 270
Cancer therapy, 19-20, 268
Cancer vaccine, 20
Canola, genetically engineered, 294, 296
Carbohydrates, 4
Carcinogen, natural, 281-282
Carcinoma, 268
Careers in biotechnology, 313-319
Carrier of genetic disease, 42
Carrier protein, 148
Case of the bloody knife, 204-205
CCAAT sequence, 84
cDNA, 116
cDNA library, 124, 126
Cell culture technology, 6-7, 9-10
Cell division, 269-271
Cell fusion, 119
Cell growth, 269-270
Cell lysis, *see* Lysis
Cell membrane, 156
Cell specialization, 47
Cell therapy, 18-19
Cell wall, 156
Cellular cloning, 46
Central dogma, 61-62
Cephalosporin resistance, 216
Chain terminator, 182-184
    as antiviral drugs, 185-186
Chance, 71-72
Chaperone protein, 101-102
Chargaff, Erwin, 61
Chitin, 17
Chloramphenicol resistance, 216
Chloroplast DNA, 90
Cholera, 219
Chromatin, 89
Chromosomal protein, 90
Chromosome, 56, 89-90, 148, 153
    chromosomal nature of inheritance, 57-58
    comparing chromosomes with PCR, 195
    mapping of, 264
Chromosome defects, 64
Chymotrypsin, 107-109
Clonal population, 67
Clone, definition of, 46
Cloning, 124-125
    adult cells as source of genetic material, 47
    animal, *see* Animal cloning
    cells in culture, 50
    cellular, 46
    definition of, 46
    human, 49-50
    mammalian, 47
    molecular, 46
    of multicellular organism, 46-47

societal issues, 46-50
Clostridium botulinum, 28
Coat color
    in dogs, *see* Dog hair
    in mice, 260
CODIS, *see* Combined DNA Identification System
Codominance, 251
Codon, 80, 83, 92, 148
Coffee, decaffeinated, 282-283
Combined DNA Identification System (CODIS), 193
Common ancestor, 236
Competent bacteria, 172
Complementary base pairing, 75-76, 83, 121-122, 141, 152
Complementary DNA, *see* cDNA
Conjugation, 11, 68-69, 207
    transfer of antibiotic resistance in *E. coli*, 211-213
Conjugative plasmid, 211-213
Conservation biology, 130
Conserved region, of proteins, 236
Contig, 265
Corn
    Bt, *see* Bt corn
    genetically engineered, 13, 17, 22-23, 25, 135
Corn starch, conversion to high-fructose corn syrup, 229
Cotton, genetically engineered, 13, 22-23, 31
Covalent bond, 94-95
Crick, Frances, 61
Criminal case, *see* Forensic case
Crop plants, *see also specific crops*
    aluminum-resistant, 135-136
    antibiotic resistance genes in, 295
    defense chemicals made by, 281
    genetic engineering of, 22-25, 36-37, 223-225
    herbicide-tolerant, debating risks of, 299-300
    interspecific and intergeneric hybrids, 37-39
    nutritional value of, 25
    pesticide-resistant, 39-40
    production and protection of, 22-25
    safety of bioengineered foods, 39-40, 293-296
Crossbreeding, 37-39, 54
Crossing over, 11, 58
Crown gall disease, 223
Culture tube, xiii
Cyanobacterial toxin, 29
Cystic fibrosis, 17, 134, 177
    genetic screening for, 310-312
Cytomegalovirus infection, 152
Cytosine, 61, 75-76, 141

D

Darwin, Charles, 54, 71
ddC, *see* Dideoxycytosine
ddI, *see* Dideoxyinosine
Decaffeinated coffee, 282-283
Decision-making model, bioethics, 301-305
Deletion, 64-65
Denaturation
    of DNA, 121
    of proteins, 102, 233
Deoxynucleotide, 75-76
Deoxyribose, 60, 75-76, 144

Glycoproteins, 15
Government agencies, regulation of embryonic stem cell research, 45–46
Government-funded research, 45
GRAS, *see* "Substance generally recognized as safe"
Griffith, Frederick, 59
Growth hormone, 7, 26
    gene therapy for healthy short children, 308–309
    genetically engineered, 309
    protein for healthy short children, 309
Guanine, 61, 75–76, 141
Guidelines for Research Involving Recombinant DNA Molecules (NIH), xii

H

Hair color, 260
Hazard, toxicity versus, 283
HD, *see* Huntington's disease
Health insurance, 43, 285
Height, 259–260, *see also* Short stature
Helix-loop-helix motif, 99–100
Helix-turn-helix motif, 101, 105–107, 109, 112–113
Hematopoietic cells, 44
Hemoglobin, 109–110, 258–259
Herbicide, 281
    environmental benefits of, 300
    environmental impacts of, 299–300
Herbicide resistance, 293–294
Herbicide-tolerant crops, 23
    background on, 299
    debating risks of, 299–300
    effect on amount of herbicide used, 300
    movement of tolerance from crop to weeds, 300
Herpesvirus infection, 185–186
Hershey-Chase experiment, 59–60
*Hha*I, 161
High school diploma, 315–316
High-fructose corn syrup, 229
*Hin*dIII, 161, 164, 166–167, 172, 181
Home pregnancy kit, 5
Homeodomain, 102
Homeotic gene, 129
Homologous chromosomes, 11
Honor code violation, 303–304
Horizontal gene transfer, 38
Hospital mix-up of babies, 200–201
Hospital-acquired infection, 215, 217
HR, *see* Hypersensitive response
Human cloning
    acceptable situations, 49–50
    cells in culture, 50
    ethical issues involved in, 49
Human embryonic stem cell research, *see* Embryonic stem cell research
Human genome, 92
    size of, 153–155
Human Genome Project, 15, 260, 263–267, 310
    Ethical, Legal, and Social Implications, 266–267, 311
    genetic information explosion, 265
    genetic linkage map, 264
    goals and objectives of, 263, 265–267

    model organisms, 263–264
    physical maps, 264
    sequencing DNA, 264–265
    societal issues, 266–267
Human immunodeficiency virus, 185, *see also* AIDS
Human molecular genetics, 256–262
Human population size, 35
Human remains, identification of, 130, 198–199
Huntington's disease (HD), genetic screening for, 310–312
Hyaluronate, 17
Hybrid, interspecific and intergeneric plants, 37–39
Hybridization analysis, 129, 177–180, *see also* Polymerase chain reaction
    comparing whole genomes, 126–127
    DNA-based diagnostics, 131
    DNA-DNA, 121–122
    with restriction analysis, 179–180
    with RNA, 121
    Southern blot, *see* Southern blotting
Hydrogen bond, 75–76, 95, 97
Hydrogenated oils, 25
Hydrophilic molecule, 95
Hydrophobic molecule, 95
Hypersensitive response (HR), in plants, 23

I

Identification, DNA-based, *see* DNA typing
Immunoglobulin, 103
Immunosuppressive therapy, 19
In vitro fertilization, 47
Independent Assortment, Principle of, 55–57, 67
Individual identity, 129
Individual responsibility, 285
Individual rights, 285
Industrial sustainability, 30–31, 137
    biodegradable products, 31
    material and energy inputs, 30
    process application of biotechnology, 30–31
Inheritance
    chromosomal nature of, 57–58
    discrete particle model of, 55–57
    fluid-blending model of, 55
    sex-linked, 58
Inherited disease, *see* Genetic disease
Initiation codon, 84–85
Initiation factors, 85
Inner cell mass, 44–45
Inoculating loop/needle, xiv
Insect-resistant crops, 22–23
Insulin, 7, 17, 19
Integrase, 185
Interbreeding, 37–39
Interferon, 25
Intergeneric hybrid, 38
Interleukin-2, 16
Interspecific hybrid, 38
Intestinal microbes, antibiotic resistance in, 217
Intron, 62, 85, 117
Inversion, 64–65
Iodine test, for starch, 228–232
Islet cells, 19

as source of genetic variation, 65
types of, 64-65
Mutation rate, 214
myb gene, 271
myc gene, 269, 271
Mycobacterium tuberculosis, 131
Mycorrhiza, 25
Mycotoxin, 291

N

NAD, *see* Nicotinamide adenine dinucleotide
National Center for Biotechnology Information (NCBI), 240
National Institutes of Health (NIH), xii
Natural carcinogens, 281-282
Natural gene transfer, 214-218
Natural products
food safety, 280-283
as therapeutics, 16
Natural selection
for antibiotic resistance, 217-218
as honing device, 70-71
observable variation and, 70
survival of the fittest, 70
Natural toxin, 281
Natural transduction, 219
NCBI, *see* National Center for Biotechnology Information
Neel, James, 259
neu gene, 271
Nicholas (Czar of Russia), 199
Nicotinamide adenine dinucleotide (NAD), 102
NIH, *see* National Institutes of Health
Nitrogen fixation, 24-25
Nitrogen-containing bases, 60, 75-76, 144
Nonpolar covalent bond, 94-95
Nonreplicative transposition, 67
Northern blotting, 122
Nuclear magnetic resonance spectroscopy, of proteins, 93, 119-120
Nuclear transfer
from adult cells, 48
from embryonic cells, 48
in laboratory animals, 47
nonembryonic, problems with, 49
Nucleic acids, 3-4
Nucleotide, 4, 60, 146

O

Obesity gene, 256
Oil spill, 8-9
Oligonucleotide
antisense, 152
synthetic, 119
Oncogene, 19, 268-271
Online resources, 240-246
Operator, 86, 105-107
Operon, 29
Organ transplant, 19
Organic chemicals, manufacture of, 7-8
"Organically grown" food, 281
Origin of replication, 171

P

p53 gene, 270-271
Palindrome, DNA, 160
Pancreatic amylase, 229, 244-245
Papaya, genetically engineered, 135
Parasitic wasp, 23
Parthenogenesis, 68
Parts per billion, 281
Parts per million, 281
Paternity case, 191, 193, 202-203, 311-312
Pathogen, 131
Pauling, Linus, 61
PCR, *see* Polymerase chain reaction
Pedigree diagram, 256
Penicillin resistance, 214-216
Peptide bond, 93-94
Permanent hair wave, 104-105
Pesticide, registration with EPA, 287-288
Pesticide-resistant crops, 39-40
Petroleum-based production, 30
Phaeomelanin, 248-249, 252, 256
Phage, *see* Bacteriophage
Pharmaceuticals, *see* Therapeutics
PHB, *see* Polyhydroxybutyrate
Phenotype, 42, 56, 72, 93, 125, 250, 254
Phenotypic variation, 70
Phenylalanine, 257
Phenylalanine hydroxylase, 257
Phenylketone, 257
Phenylketonuria (PKU), 41, 257-258
Phosphate group, 75-76
Phosphodiester bond, 75-76
Physical map, Human Genome Project, 264
Phytoremediation, 8
Pilus, 211
Pipette, xiii
PKU, *see* Phenylketonuria
Plant(s), *see also* Crop plants
amylase activity in roots and leaves, 231
genetically engineered, 223-225, 293-296
Plant agriculture, 22-25
Plant biomass production, 30
Plant breeding, 293
Plant cell culture, 9
Plasmid, 20-21, 90, 153
with antibiotic resistance genes, 171-173, 208-213, 216-217
conjugative, 211-213
construction of pAMB and pKAN, 171
construction of pAMP/pKAN, 172
recombinant, 124-125, 171-173
Ti, *see* Ti plasmid
transformation with, *see* Transformation
as vector, 118, 124-125, 174-175, 208
Plastocyanin, 99-100
Point mutation, 65
Polar covalent bond, 94-95
Polarity
of amino acid side chains, 95
of water, 95, 97
Pollution, *see* Environmental issues

Restriction site, 160
Retinoblastoma, 270
Retinoblastoma gene (*rb*), 270–271
Reverse transcriptase, 90, 116–118, 124
RFLP, *see* Restriction fragment length polymorphism
Rheumatic fever, 215
Ribose, 80
Ribosomal proteins, 89
Ribosomal recognition element, 83–84, 93
Ribosomal RNA, 81, 89
Ribosome, 80, 83, 85, 118, 148
Rice, genetically engineered, 135
Risk
    control over versus no control, 280
    familiar versus unfamiliar, 280
    natural versus man-made, 280
    public perception of, 282
    voluntary versus involuntary, 280
Risk assessment, 276, 283–285
    emotional thinking and, 285
    toxicity versus hazards, 283
    unconscious, based on emotions, 280
Risk-benefit analysis, 276, 283–285
    assessing benefits, 284
    assessing risks, 283–284
    putting risks and benefits together, 284–285
RNA, 3
    catalytic activity of, 62
    compared to DNA, 81
    hybridization analysis, 121
    messenger, *see* Messenger RNA
    ribosomal, *see* Ribosomal RNA
    sequencing of, 124
    structure of, 80, 82
    transfer, *see* Transfer RNA
RNA polymerase, 82, 84–86, 88, 101, 116–117
RNA virus, 90

S

Saccharomyces cerevisiae, model organism for Human Genome Project, 263–264
Safety
    food, *see* Food safety
    laboratory, xi–xii
SAGE, *see* Serial analysis of gene expression
Sales and marketing division, of biotechnology company, 314
Saliva, testing for amylase activity, 230
Salivary amylase, 228–229, 234
    structure of, 244–245
Salmonella, 28
Sampling well, 162–163
SAR, *see* Systemic acquired resistance
Sarcoma, 268
Scarlet fever, 215
Science
    differences in science and technology, 278
    nature of, 278
    responsibility and, 278–279
    social context of, 278
    which came first, science or technology, 277

Scientific method, 278
Scientific models, 63
Scrapie, 110–111
SDS, *see* Sodium dodecyl sulfate
Secondary plant compounds, 281
Segregation, Principle of, 55–56, 67
Selective breeding, 11, 54, 135
    comparison with genetic engineering, 11–13
Sequenced tagged site (STS), 265
Sequencing gel, 182–183
Serial analysis of gene expression (SAGE), 266
Serine protease, 107, 109
Severe combined immunodeficiency disease, 136–137, 306
Sex-linked inheritance, 58
Sexual reproduction, recombination and, 66–68
Short stature
    growth hormone for healthy children, 309
    growth hormone gene therapy for healthy children, 308–309
Short tandem repeat (STR), 192–194
Sickle-cell anemia, 109–110, 258–259, 310
Sickle-cell trait, 258–259, 310
Signaling pathway, 269
Silent mutation, 92
sis gene, 271
Skin, artificial, 10
*Sma*I, 161
Smith, E. B., 58
Societal issues, 33–51
    in agricultural biotechnology, 36–40
    analysis of, 35–50
    in biotechnology, 275–276
    biotechnology's risks and benefits, 280–292
    cloning, 46–50
    human embryonic stem cell research, 43–46
    Human Genome Project, 266–267
    in medical biotechnology, 41–43
    science, technology, and society, 277–279
    strategy for analysis of, 46
Sodium dodecyl sulfate (SDS), 233
Soil organisms
    antibiotic-producing, 216
    testing for amylase activity, 230–231
Somaclonal variant selection, 27
Somatic-cell enhancement, 306
Somatic-cell gene therapy, 306
Southern blotting, 122
Southern hybridization analysis, 179–181, 202
    DNA typing by, 192
Soybean, genetically engineered, 17, 22–23, 294, 296
Sperm cell, 54
Splicing of RNA, 85, 117–118
Sporulation, in *B. subtilis*, 88
src gene, 269, 271
Stain, DNA, 162, 167–168
Staphylococcus aureus
    antibiotic resistance in, 214–215
    diseases associated with, 219
Starch, 228
    hydrolysis of, 228
    iodine test for, 228–232